T0296571

BOTANY

A SENIOR TEXT-BOOK FOR SCHOOLS

BOTANY

A SENIOR TEXT-BOOK FOR SCHOOLS

BY

D. THODAY, Sc.D. (Cantab.), F.R.S.

PROFESSOR OF BOTANY IN THE UNIVERSITY COLLEGE OF NORTH
WALES, BANGOR; FORMERLY HARRY BOLUS PROFESSOR OF
BOTANY IN THE UNIVERSITY OF CAPETOWN

CAMBRIDGE

AT THE UNIVERSITY PRESS

1950

CAMBRIDGE
UNIVERSITY PRESS

University Printing House, Cambridge CB2 8BS, United Kingdom

Cambridge University Press is part of the University of Cambridge.

It furthers the University's mission by disseminating knowledge in the pursuit of
education, learning and research at the highest international levels of excellence.

www.cambridge.org
Information on this title: www.cambridge.org/9781107586314

© Cambridge University Press 1950

First edition 1915
Second edition 1919
Reprinted 1920, 1921
Third edition 1923
Reprinted 1925
Fourth edition 1929
Fifth edition 1935
Reprinted with corrections 1938, 1950
First paperback edition 2015

A catalogue record for this publication is available from the British Library

ISBN 978-1-107-58631-4 Paperback

PREFACE

THIS book is intended primarily for use in connexion with the Senior Cambridge Local Examinations; but it is hoped that certain special features may make it of more general service, to teachers as well as to scholars in the upper forms of secondary schools.

No previous knowledge of Botany is assumed. The subject matter is divided into sections, each more or less self-contained, with cross-references, so that the more elementary parts of the book may be read in some other order if circumstances should make it desirable.

The treatment of physiology aims at giving each experimental fact and its interpretation a logical place in the whole. A necessary minimum of physics and chemistry is supplied. Special attention has been given to a clear exposition of certain matters (for example, the distinction between gaseous exchange and the processes of photosynthesis and respiration; geotropism; the balance between absorption and loss of water) which some years of experience as an examiner have convinced me are widely misconceived or imperfectly grasped. This has entailed a more fundamental consideration of some points than is usual in elementary books, a course which seemed educationally preferable to evading difficulties that many intelligent children feel. It is of course assumed throughout the book that the experiments will be performed by the students themselves, or at least demonstrated to them, and that specimens will be examined by each student individually.

The subject of classification is approached in close connexion with, and in illustration of, the problems of evolution. Special care has been taken (by emphasising the wide range of forms included in some of the Families) to combat the prevalent misconception, implicit if not explicit in much of the teaching of 'Natural Orders,' that certain 'typical characters' determine the limits of each Family. The special morphology of flowers and fruits is included incidentally.

Notes on common plants, other than those selected for the illustration of fundamental principles, and hints for extended work, as well as certain subsidiary or less elementary parts of the subject matter, are printed in smaller type.

I undertook to write this text-book after Mr A. Malins Smith had relinquished the task owing to the pressure of other duties. His draft MS, dealing with part of the subject matter of Sections I to III, was placed at my disposal and I have great pleasure in acknowledging my indebtedness to it. My scheme of arrangement is largely a modification of his; and in some parts, notably in Section I, I have been glad to avail myself of his simple and lucid descriptions.

I wish to record my gratitude to many friends who have put their special knowledge at my service and helped me with suggestions and criticisms. I am indebted to Prof. A. C. Seward, for suggestions made in the course of reading the proofs; to my colleague Mr R. S. Adamson, for revising Sections IV and V and part of Section III; and especially to my wife and workmate, who has at every point given ungrudgingly of her time and labour, and to whose criticism is due the removal of many blemishes which would else have disfigured these pages.

A certain number of the illustrations are from original line drawings. The sources of the others are acknowledged individually. Many are taken from Marshall Ward's *Trees*, Willis's *Flowering Plants* and other books already published by the Cambridge University Press. A few are reproduced from Strasburger's Text-book. A number of others are from Baillon's *Natural History of Plants*, by the courtesy of Messrs L. Reeve and Co.

My best thanks are due to Mr A. G. Tansley for three blocks from *Types of British Vegetation*; to Sir Francis Darwin, for allowing me to use the figures from his *Practical Plant Physiology*; to the Editors of *The Annals of Botany* for permission to reproduce Figure 202; to Mr Gurney H. Wilson, Editor of *The Orchid World*, for Figure 169, from Veitch's *Orchid Manual*; and to the Macmillan Company for Figure 104, from Osterhout's *Experiments with Plants*. I am also indebted for other illustrations to Prof. J. Shelley, Mr H. Hamshaw Thomas, Mr S. Mangham and my wife.

D. T.

BOTANICAL DEPARTMENT
UNIVERSITY OF MANCHESTER
June 1915

PREFACE TO THE SECOND EDITION

THE opportunity of a second edition has been taken to add a supplementary section on Cryptogams so as to cover the syllabus for the Cambridge Higher School Certificate and other similar examinations.

In this Section as in others it is assumed that the students will examine specimens for themselves. Stress is laid on features that can be seen with the naked eye or with the aid of a hand lens. Microscopic details of which the significance cannot be understood, or which are relatively unimportant, at this stage are omitted.

The illustrations with four exceptions are original or from books published by the Cambridge University Press. Figs. 219 and 227 I. are reproduced by arrangement with J. M. Dent and Sons, Ltd, Figs. 221 and 225 by arrangement with Messrs J. and A. Churchill. War conditions have necessitated the omission of some other illustrations which I hoped to be able to use, and circumstances have not allowed me to make substitutes; but the want of them should not be seriously felt if specimens are thoroughly examined.

I am indebted to Prof. W. H. Lang for valuable criticism and suggestions in the Chapters on Bryophyta and Pteridophyta; to Mr W. Robinson for help in the Chapter on Fungi; and once more to my wife who has assisted at every point, and on whom has fallen the task of seeing the new edition through the press in my absence.

D. T.

London.
2 *September* 1918.

PREFACE TO THE FIFTH EDITION

THIS edition has undergone a thorough revision. Most of the corrections are in relative details; but more extensive alterations will be found in parts dealing with the water-relations of plants, aimed at removing ambiguities and inadequacies without introducing a more advanced treatment of the subject. A new potometer is substituted, which is easier to set up in working order than Farmer's apparatus: I am indebted to the Editor of the *School Science Review* for permission to reproduce the figure of this potometer. A very simple form of Joseph Priestley's demonstration that plants renovate vitiated air has been added: it is taken from *The Food of Plants*, by A. P. Laurie (Macmillan, 1893; Expts. 27 and 28, pp. 44–46). It brings the actual performance of this fundamental experiment within easy reach of everyone. The treatment of evolution has been modified, but without adding any specific reference to Mendelism, which properly belongs to a more advanced stage.

Since this book was first written the system of school examinations has undergone considerable reorganisation and syllabuses have been overhauled. Nevertheless, after careful consideration, it seems best to leave the book covering the same ground as in the second edition. Some teachers may find less than they would wish in one respect or another; particularly perhaps in the treatment of chemical phenomena, and of the lower plants. The formal chemistry now required by many School Certificate syllabuses, however, could not properly be dealt with in a short introductory chapter; and the Supplementary Chapter on the lower plants provides as it stands a basis on which further detail can be added by the teacher if required. Considerable addition of detail to the book would involve a danger of loss of balance.

Many years of experience confirm me in the opinion that the general plan of the book is sound and meets a real need for a view of plant biology corresponding as closely as may be with the logical and historical construction of our knowledge. If for examination purposes it needs to be otherwise classified and tabulated, the process is best carried out by pupil and teacher in co-operation.

D. T.

DEPARTMENT OF BOTANY
   MEMORIAL BUILDINGS
      UNIVERSITY COLLEGE OF NORTH WALES
         BANGOR.
            November 1934.

CONTENTS

SECTION III
REPRODUCTION

SECTION IV
THE CLASSIFICATION OF PLANTS

SECTION V

PLANTS IN RELATION TO THEIR ENVIRONMENT

SUPPLEMENT

SEEDLESS PLANTS

LIST OF FIGURES

CHAPTER I

INTRODUCTORY

It is to plants that the country-side owes much of its character and charm. While hill and valley, river and plain, give to landscapes their general contour, the vegetation changes their aspect from place to place and gives them greater variety of form and colour. The plants which compose this vegetation differ greatly one from another. Great trees stand out as the most conspicuous objects, while the humble grasses cover wide expanses with their verdure and a variety of herbs and shrubs clothe our banks and form our hedges. Yet all these plants alike are green. They all possess green leaves and have roots underground. Closer examination will show other common features. We shall find, as our study proceeds, that a general similarity underlies their construction and the same principles are involved in the maintenance of their life.

Let us with this end in view compare carefully a few common plants.

We will begin our study with a common erect plant, A Sun-flower. the Sunflower. If we dig up carefully a young Sunflower plant, and shake the root free from soil, we shall be able to observe the following points:

The whole plant can be divided into two obviously different parts, that which was above the ground, the *shoot-system*, and that which was below the ground, the

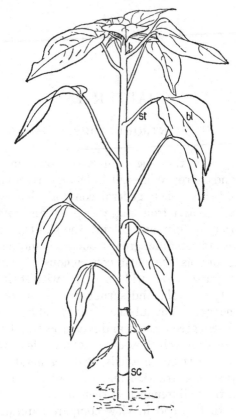

Fig. 1. Shoot-system of young Sunflower plant: *b*, axillary bud; *st*, leaf-stalk; *bl*, leaf-blade; *sc*, leaf-scar. The lower leaves or leaf-scars are in pairs: the other leaves are all arranged spirally.

root-system. The difference between these two parts which strikes us first is that of colour: the shoot-system is *green*, the root-system *whitish*.

There are great differences also in form. The *shoot-system* (Fig. 1) consists of an upright rounded stem along which are borne at intervals flat outgrowths, the leaves. The upper leaves are distributed in such a way that a spiral can be drawn, round the stem, through the points of attachment of successive leaves. Towards the top of the plant (Fig. 2) the leaves are smaller and closer together. The smallest leaves, numerous and

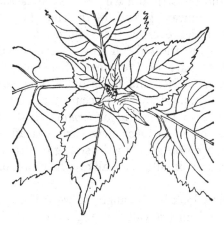

Fig. 2.   Apex of a young Sunflower plant seen from above, showing the terminal bud.

crowded, arch over the end of the stem and hide its tip. As we carefully lift the outer overlapping leaves we come to still smaller, softer and more delicate ones and we can see that the tip of the stem itself is very soft and tender. This crowd of tiny overlapping leaves enclosing the stem-tip is called a bud. We infer that its parts, being tender and small, are young; and by a few careful measurements at intervals of a week on a similar plant still in the ground we can readily prove them to be growing.

For this purpose a few of the outer leaves of the bud should be numbered with water-proof ink and their lengths carefully recorded. At the same time the lengths and breadths of the other older leaves may also be recorded, together with the distances between them on the stem. At the end of a week or a fortnight the outermost leaves of the bud will have expanded and turned back, their place being taken by younger leaves which they formerly enclosed; while the leaves just below will also have grown in length and breadth and will have been carried farther apart by the upward growth of the stem between them. Farther down, however, neither stem nor leaves will have grown: here are the old full-grown leaves on the oldest and toughest part of the stem. Some of the oldest leaves may have died and fallen off, leaving scars which mark the positions they once occupied.

Thus it is in the *apical* or *terminal* bud, at the tip or apex of the shoot, that new parts are added to the shoot. Such *apical growth* is in strong contrast with the general growth of all parts alike which we observe in animals.

If we now look in the angle between the base of any leaf and the stem (we call this angle the *axil* of the leaf) we see a small outgrowth (see Fig. 1, *b*) which on close examination we find to be composed of tiny closely overlapping delicate leaves like those of the terminal bud. This is in fact also a bud, and because it is at the side of the stem and not at the tip it is called a lateral bud. As it is in the axil of the leaf it is also called an axillary bud, and a careful search will prove that all the lateral buds are axillary. No lateral bud occurs except in the axil of a leaf. Often in vigorous old plants some of the buds grow out into branches, which, like the main stem, bear leaves with buds in their axils and end in the bud from which they have grown.

Let us now examine the *root-system*. Continuous with the main stem of the plant is the main or primary root. Like the stem, its tip is soft and delicate and, as we shall prove for ourselves later, is the part which is young and growing. The harder part farther up has ceased to grow in length. From the main root there come off, usually in a more or less horizontal position, side roots which, being branches of the primary root, are called secondary roots. These are of various sizes, the oldest and largest being at the top and the smallest and youngest nearer the tip of the primary root. From the secondary roots grow other roots in all directions, and these branch again and again; all these finer rootlets are called tertiary roots.

Except for their size and position there is nothing to distinguish these various roots in appearance one from another. They are all cylindrical; and in colour are whitish, at any rate towards the tip where the more delicate younger parts are situated.

In an older plant the root-system has grown. The main root has become thicker and tougher and is dark brown in colour. It has not grown far in a downward direction, but the root-system has developed on all sides by the further growth of some of the principal secondary roots and the tertiary roots arising from them, thus keeping in the upper layers of the soil.

A time comes in the life of the Sunflower plant when the apical bud of the stem no longer produces new leaves but opens out into the well-known golden 'flower,' the Sun-flower itself. This fades at last, leaving a mass of 'seeds,' which when sown produce new plants like the old one. We shall leave till later the study of the flower, merely noting here that the production of flowers and seeds is a very important point of agreement between most of our common plants.

Let us now take a grass plant and examine that. A
A grass   plant of the common wayside grass, *Poa*
plant.   *annua*, growing as a weed in the garden
will serve well; it can easily be taken up from the soft

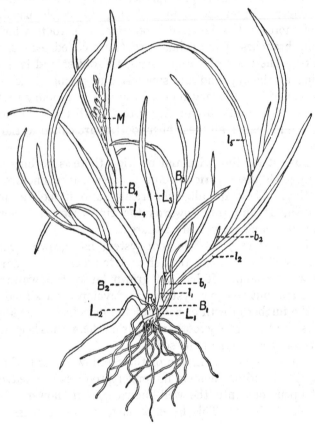

Fig. 3.   Grass plant (*Poa annua*). L_1, L_2, L_3, etc., successive leaves
of main stem, which ends in the inflorescence M; B_1, B_2, B_3, etc.,
the shoots in the axils of these leaves. l_1, l_2, l_3, leaves of axillary
shoot B_1; b_1, b_2, the shoots in their axils. R_3, the adventitious
roots springing from the node at which the leaf L_3 is attached.
(M. G. T.)

garden soil without damaging its roots. Here again we find a green shoot-system, and a bunch of whitish roots below it. This plant is not, however, tall and erect like the Sunflower, but low and tufted (Fig. 3) : let us see how it is built up.

The shoot-system consists of a number of shoots which come together at the base. Each of these shoots has a few bright green leaves, which are very narrow in comparison with their length and are called linear leaves. If we trace one of these leaves downwards, we find the blade attached, not to a stalk like that of the Sunflower, but to a sheath (Fig. 4, *sh*) which enwraps the sheath of another leaf, within which still other leaves may be visible. The innermost leaf is the smallest and most delicate; its sheath is hidden and its blade is folded. At the top of the sheath of each leaf, at the point where it joins the leaf-blade, is a small membranous or skin-like outgrowth, called a *ligule* (Fig. 4, *lig*), which fits closely against the leaf next within, and prevents rain that runs down the blade from getting into the sheath.

Low down on a shoot, we may find a leaf of which the sheath has been forced back by another smaller shoot arising within it (e.g. Fig. 3, l_1, with b_1). When this leaf is pulled down, we discover that the sheath is attached at the

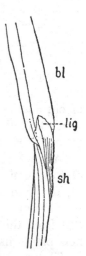

Fig. 4. Lower part of grass leaf: *sh*, sheath; *lig*, ligule; *bl*, blade.

bottom around a swollen place (a *node*) on a whitish stem. From this same node, on the same side as the leaf-blade—in the axil of the leaf—arises the small lateral shoot. The lowest leaf of this axillary shoot is a very thin transparent sheath without any blade.

A very short way above this node is another similar
node, and here arises the next leaf, with its blade on the
opposite side of the stem.   Pulling down this leaf we
shall find another younger axillary shoot (b_2), consisting
of very young and delicate leaves, nearly enclosed in the
lowest leaf, which as before is a thin transparent sheath.

The part of the stem which comes between this node
and the next (called an *internode*) is very short and tender.
Each successive internode is shorter and more delicate.
Thus we have here a terminal or apical bud, just as in
the Sunflower, and also axillary shoots and axillary buds
in the axils of the leaves.

This whole shoot is joined below to another similar
shoot.   Careful search at the point where they join may
reveal a dead or dying leaf with a torn sheath, surrounding them both (Fig. 3, L_1).   This is the leaf in the axil of
which our shoot started as a bud.   In fact, a comparison
of old and young grass plants will prove to us that the
whole tuft of shoots has been formed from a single shoot
by the production of lateral shoots from axillary buds.

As each shoot grows older its terminal bud, like that
of the Sunflower, stops producing leaves, and grows up
rapidly into a long stem ending in a plume of flowers
(like the main shoot in Fig. 3, M); this finally bears seed,
and from this seed, when scattered on the ground, arise
new plants like the old one.

The shoot-system of the grass plant is thus built up in a
similar way to the shoot-system of the Sunflower; but in
the grass the many lateral branches develop early from
the axils of the lower leaves, and as the leaves are close
together near the ground a tufted habit is produced.
The branches grow at first in a horizontal direction, so
that with each lateral shoot the plant spreads a little
farther over the surface of the soil.

If we now examine the lower nodes of the outer

shoots of our plant we find tiny roots beginning to
grow from them, or longer roots already well grown. A
close comparison of plants of different ages will prove
that the whole mass of roots is formed in this way
from the nodes of the stems: there is no large main
root continuous with the central shoot as in the Sunflower. But the difference is not so great as appears
at first sight, for nearly all plants have the power to
produce roots directly from the stem; gardeners take
advantage of this when they plant cuttings (i.e. cut
branches) in the ground to produce new plants. Moreover, we shall see later that even a grass plant begins life
with a main root but soon gives up the single root-system
in favour of these separate small root-systems produced
direct from its branches wherever required (*adventitious*
roots). By the formation of such roots each new shoot is
fixed firmly and closely to the ground. From each root
grow fine, hair-like, secondary roots, and from these still
finer tertiary roots; so the surface soil is penetrated as
far as the plant extends and, as the many roots soon
grow tough and fibrous, is converted into springy
turf.

As we have seen, the stem is short until the time of
flowering, and its tender tip is low down within the protecting sheaths of the outer leaves. In this position it
stands some chance of escaping the mouth of a grazing
animal. If, however, this terminal bud should be removed it is readily replaced by the growth of younger
lateral buds. Of such buds there are always many which
are not actively growing, and are therefore described as
dormant (i.e. sleeping) buds. If need or opportunity
arise they awake into activity and grow out into branches.
Dormant buds are also found in the Sunflower: even in
the oldest plant there are many buds which have not
grown out into branches and as the Sunflower does not

survive the winter these buds never grow out. In the grass, too, many buds remain always dormant.

The Dandelion is a plant which grows abundantly in
A Dande-   meadows amongst the grass, and in many
lion.   places besides. As this looks very different
from either Sunflower or grass, let us examine it (Fig. 5).

Fig. 5.   Dandelion, showing 'rosette' habit, and part of the long tap-root.   *S*, scars of fallen leaves.   *L, L*, leaves of an axillary shoot.

When we try to dig it up we find a stout tapering main root, which pierces the ground to such a depth that it would be very troublesome and difficult to remove the whole of it. We may be satisfied if we obtain six or eight inches of it. Compared with the root-system of the Sunflower, that of the Dandelion has few lateral roots, though these increase in number lower down. In contrast with the main root, all the secondary and tertiary roots are thin and fibrous.

Thus whereas the grass plant and the Sunflower are surface-rooted, the Dandelion is deep-rooted. Its long main root, straight and stout, piercing the ground so

deeply, is called a tap-root. The root-system of the
Dandelion is nevertheless built up in the same way as
that of the Sunflower, and its parts are similar in form
and method of growth to the roots of the Sunflower and
the grass plant.

The leaves of the Dandelion are grouped in a rosette
close to the ground. They spread out from a central
axis, continuous with the tap-root; but this stem is very
short, and the leaves are so close to one another that we
can scarcely distinguish internodes at all. The outer leaves
are larger than the inner ones, the leaves diminishing in
size towards the centre of the rosette. In the very centre
we find the small and tender young leaves arching over
the delicate tip of the stem. Thus, though the shoot of
the Dandelion is very short, it ends, like the tall shoot of
the Sunflower, in a bud. Moreover, the leaves are spirally
arranged, and in the axils of most of them are buds,
flower-buds, or flowers (Fig. 5, *LL*, etc.).

Around the bases of the oldest leaves are the rotting
remains of other leaves of the previous year; just below
these the stem bears the faintly marked scars left by still
older leaves which have died and fallen off. Farther
down the leaf-scars are less easy to detect, so that it is
difficult to tell where stem ends and root begins.

Let us as a last example consider a tree. Like the
A tree: Sunflower a tree is an erect plant, but
Horse- whereas the Sunflower dies in the autumn,
chestnut. the tree may live and grow for hundreds of
years. Its great main stem or trunk sends out strong
branches above, these again bear smaller branches and
finally twigs with leaves upon them.

Let us examine the twig of a Horse-chestnut. In
this particular tree there are two leaves at each node,
opposite to each other; each consists of a stout stalk

bearing seven distinct *leaflets*, and is therefore called a *compound* leaf to distinguish it from *simple* leaves like those of the Sunflower or Dandelion.   In the axil of each leaf is a bud.   The tip of the twig ends in a bud similar to the axillary buds but larger; or it may end in the well-known pyramid of flowers, or in a stalk bearing the few green fruits, according to the time of year.   Farther back on the twig, where now there are no leaves, are shield-shaped scars left by fallen old leaves, and just above many of them are tiny dormant buds, or lateral twigs.   Farther back still the branch is thicker and its surface rough so that the scars and buds are less easily detected.   Even the main trunk itself, now so large and furrowed, was once as thin and smooth as the twig we have been examining, and, like it, bore leaves.

There are many other features with which we must deal later when we study trees.   We have, however, seen enough to show us that in the Horse-chestnut tree, as in the other plants we have examined, the same mode of growth and construction is found; and all are made up of parts which, however different in many respects in the different plants, we have no hesitation in naming alike stem, leaf, and root.   Root and shoot grow always at the tip. The branch roots are similar to those from which they grow; and, likewise, the branch shoots, which grow always from buds in the axils of the leaves, are copies of the main shoot.   The leaves, though differing in many respects in different plants, have all a thin flat green expanded portion.   Differences in build, or *habit*, depend chiefly on whether the stem grows long or remains short, whether many or few axillary buds grow out into branches, and whether these are low or high on the plant.   The character of the root-system varies, too, according as the main root is long or short, persists, or early disappears, to be superseded by strong lateral roots (as in the Sunflower) or by adventitious roots (as in the grass) developed anew from the shoot-system.

Now from our own experience as living organisms we expect that each part of which a living plant is made up has its special work to do in the providing of food and all else that is necessary to the life of the plant—in other words, we expect that each member, stem, leaf, or root, is an *organ* with a *function*. We must now carefully inquire what these functions are.

SECTION I

FUNCTIONS OF PLANT ORGANS. FOOD OF PLANTS

CHAPTER II

CHEMISTRY OF NUTRITION

In searching for the functions of leaf, stem, and root, we may begin by inquiring how the plant obtains the food which is necessary for its growth. It is a matter of common knowledge that roots help in feeding the plant. A plant droops in dry soil, and when the soil around its roots is watered it recovers. Gardeners and farmers manure the soil to enable plants to grow better; and many people believe that the root is able alone to feed the plant entirely. We must, however, inquire into the matter more carefully, and find out exactly what food a plant requires, and how much of it is taken in by the root.

As we have seen, a plant in growing produces new shoots and roots like the old ones. If we can discover of what substances these are composed we shall know (1) what substances the plant must obtain in its food in order that it may build them up; we shall then be in a position to consider further (2) where the food comes from and (3) how it is absorbed.

The composition of plant-substance.

If we carefully heat a piece of a plant, such as a
Exp. 1   piece of twig, or a small shoot, in a test-tube,
we shall soon see drops of water condensed on the cool
upper part of the tube. If we drive off all the water from
the plant on a water-bath, so that it cannot be scorched
or burned, and compare its weight before and after drying,
we shall find that a very large part of the original weight
consisted of water—from 6 to 9 parts in 10, according as
the piece we have taken is woody, or green and juicy.

If we now heat it further in the tube, it will begin to
scorch and char, till it is wholly black; at the same time
gases and vapours are given off which can be set light to
and will burn. The wood of trees is treated in a similar
way to turn it into *charcoal*. Heated still more strongly
the charred mass gets red hot and gradually disappears
(as in a charcoal fire), leaving a small heap of *ash*. This
happens much more quickly if we hold it in the bare flame.
If instead we heat it in a tube of hard glass through which
a gentle current of air can be passed, and let the air, both
before and after entering the tube, bubble through bottles
of lime-water, we shall find that the lime-water in the
second bottle rapidly turns milky, while that in the first
shows hardly any sign of milkiness.

We gather from this experiment that, when plant
substance burns, a gas is produced which turns lime-
water milky. This is a well-known property of *carbonic
acid gas*, which is so called, indeed, because it is formed
from the black *carbon* of charcoal.

Carbonic acid gas is produced in the burning of many
substances. One is coal, which consists of the remains of
plants which have been packed together in the earth for
many millions of years. When coal is heated carefully
it gives off, like our dried piece of plant, vapours and gases
which will burn, before it glows and burns away to carbonic acid gas; part of these gases when purified form

the coal gas which we use to obtain light and heat. When coal gas burns it also produces carbonic acid gas, and so do the inflammable gases and vapours which came from our piece of plant. But none of these substances consist only of carbon, which is usually black and solid; they must therefore produce something besides carbonic acid gas when they burn. When a cold object, such as a flask of cold water, is held in the colourless flame of a Bunsen gas-burner, drops of water are observed to condense upon it. This water is the other substance produced when coal gas burns. It can also be shown, though less easily, that when a piece of plant is burned from which all free water has been carefully dried, water is produced as well as carbonic acid gas.

A plant, then, consists of (1) water, which easily evaporates, (2) gases and vapours, which produce carbonic acid gas and water when they burn, (3) carbon, which burns to form more carbonic acid gas, (4) ash, which does not burn away. It should therefore require in its food *water*, *carbon*, and the substances which form *ash*.

Clearly the ash must come from the soil, and there-fore must be absorbed by the roots. We know, too, that roots absorb water from the soil, because plants droop and eventually dry up in very dry soil, but recover if the soil is watered soon enough; more satisfactory proof of this we shall find later.

Sources of the food of plants.

We have still to find the source of the carbon. Accurate analysis has shown that about half the dry solid substance of plants is carbon. In fact carbon is the chief foundation of the bodies of plants, as also of animals; so that carbon is the most important part of their food.

Let us first examine the soil. Does soil contain carbon? If we take good dark-coloured soil and treat it as we did the piece of plant, burning it in a tube in a current of air, it turns a lighter and redder colour, while the air which has passed over it turns lime-water milky. Its dark colour was due to particles of *humus*, i.e. decayed and decaying vegetable and animal matter, and this when burned yielded carbonic acid gas. Does a plant, then, obtain its carbon too from the soil? To solve this problem we might grow plants from seed in burned soil; or more simply in clean sand, which contains no carbon, occasionally watering it with a solution of the substances which are present in the ash of plants. The best way of carrying out the experiment will be described later (p. 90). If the plants are well tended they will grow to a large size although they could have obtained no carbon by their roots. Yet in growing they have obtained from somewhere far more carbon than was present in the seeds. We are therefore obliged to turn to the *air* as a possible source of carbon.

We have seen that all our fires and furnaces and our gas flames are producing carbonic acid gas, which passes into the air. Candles, oil lamps and, in fact, most burning substances, also give off carbonic acid gas into the air. Not only so, but we and other animals are continually breathing out this same gas. If we blow through a tube into lime-water it rapidly turns milky.

That air does always contain a small proportion of carbonic acid gas can be shown by drawing air for a sufficient length of time through lime-water, or more simply by leaving a little lime-water in an open glass dish. Even in the open air a milky scum soon appears on the surface of the lime-water.

By careful analysis the proportion of carbonic acid gas in atmospheric air has been found to be about three parts in ten thousand. It may seem at first sight that so very small a proportion of this gas can hardly be of much use to plants as a source of their carbon; besides, a gas does not seem a likely food for anything. Yet we are obliged to consider this possibility; and so, as the shoot is in the air, we are led to inquire what is its function there.

Before we can understand the next step in our inquiry,
Combus- however, we must know more about carbonic
tion and acid gas and its production. Let us look
energy. more closely into the fact of burning or
combustion.

If a lighted candle be covered with an inverted jar, so that it is quite enclosed, it burns for a while and then the flame dies away and goes out: from this we conclude that the candle will only burn for a short time when shut up in a small volume of air.

Combustion therefore depends upon air. In order to
Exp. 2 *a* find out if any change takes place in the volume of the air we may burn a candle under a bell-jar standing in water. A bell-jar with an opening at the top is placed with its wide lower end in shallow water (Fig. 6); the upper opening is fitted with a cork, from the bottom of which a piece of candle is suspended by a wire stuck into the cork. The candle is lighted, and lowered into the bell-jar, which is then closed with the cork, and made air tight, if necessary, with plasticine or melted wax. When the candle has gone out and the jar cools down, the level of the water under the bell jar rises a little.

If the experiment is repeated, standing the bell-jar
Exp. 2 *b* in lime-water, or better still in a solution of caustic potash, which absorbs the carbonic acid gas rapidly, the level rises farther (Fig. 6). The carbonic acid gas must therefore have taken the place of a part of the air originally present which has disappeared. This part of the air in the absence of which the candle ceases

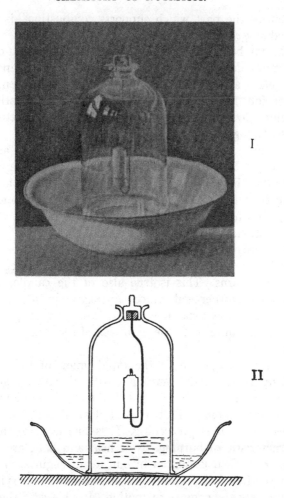

Fig. 6. Experiment 2 *b*. The apparatus is shown in external view in
I, and II is a diagrammatic sectional view of the same apparatus.
The bell-jar is closed by a stopper with a cork fixed into it below
to hold the wire and candle.　The level of the potash solution
has risen inside the jar owing to the removal of oxygen by the
candle in burning, and the absorption by the potash of the carbon
dioxide produced.　(I from a sepia drawing by Prof. J. Shelley.)

to burn—the part which supports combustion—is a gas
called *oxygen*.

It has been found by other experiments that a candle
goes out before all the oxygen has been exhausted, and that
altogether about one-fifth part of the air is oxygen. The
other four-fifths, which will not support combustion, is
chiefly *nitrogen*. Air consists mainly of a mixture of
these two gases.

When a candle burns in air, the oxygen of the air and
part of the candle disappear, and carbonic acid gas appears.
We infer therefore that the carbon of the candle and the
oxygen of the air are both present in the carbonic acid
gas; they have joined to form carbonic acid gas during
combustion. For this reason carbonic acid gas is also
called *carbon di-oxide*.

We have already seen that water is also formed when
coal gas burns. This is true also of the candle, which
contains another substance, *hydrogen*, in addition to
carbon; in the experiment a little of the oxygen that
disappears unites with the hydrogen of the candle forming
water.

Combustion, then, is the rapid union of the carbon
and hydrogen of the burning substance with oxygen, to
form carbon dioxide and water. The most obvious fact
about this union is that a great deal of heat and light
appear while it is proceeding. It is this feature which is
characteristic of burning. Now heat and light are forms
of *energy*. Coal is burned in our steam engines, and the
heat set free turns water into steam which in trying to
escape sets the engines in motion. In this way the heat
is expended in doing a vast amount of *work*, hauling
heavy trains, lifting huge weights, propelling steamers,
and so on. This is what is meant by saying that heat
is a form of energy: it can be used up in doing work.
Light, when absorbed by a black substance, warms it, in

other words, the light is changed into heat, and is therefore another form of energy.

Two important facts have, then, to be remembered about the production of carbon dioxide during combustion: (1) The carbon of the burning substance unites with oxygen from the air to form carbon dioxide. (2) During this process a large amount of *energy* is set free in the form of heat and light.

[NOTE: *Carbon dioxide and carbon monoxide.* When air passes through red hot charcoal, carbon *mon*-oxide is formed first. This burns with a blue flame (often seen above coke fires), combining with more oxygen to form carbon *di*-oxide. Carbon monoxide, the product of incomplete combustion of carbon, is a dangerous gas, because it is absorbed by the blood, forming a stable compound with haemoglobin (the pigment which carries oxygen throughout the body), preventing it from taking up oxygen and so putting the red corpuscles out of action.]

CHAPTER III

NUTRITION. THE LEAF

As a first step in our search for the work of the shoot,
Form and arrangement of leaves in relation to light and air. let us look at a shoot again to see if it affords us any hints, bearing in mind especially the question of the absorption of carbon from the air.

When we notice the broad green leaves of a Sunflower plant, were not the sight so familiar to us, we could not but be struck by the large surface which these leaves expose to the air compared with their volume. Though not universal this thin flat shape is very general in leaves.

We noticed in the first chapter how the leaves of the Sunflower are distributed around the stem. Now it is scarcely possible to imagine any other way of placing them in which they would shade each other so little; and they seem to spread themselves out horizontally to catch the *light*, for they would be equally well exposed to the air in almost any position.

If we turn to other plants we find their leaves also spread out to the light. When many leaves are nearly on the same level, as in the Daisy, Plantain, and other 'rosette' plants, in which the leaves form a tuft close to the ground, or in a young Ivy shoot clinging close to the trunk of a tree, the leaves fit in between one another so neatly that their groupings recall Roman mosaic pavements and are called *leaf-mosaics*. The Horse-chestnut tree shows the same thing very clearly: the lowest pair of leaves on a twig have long stalks which carry them far out from the twig; the second pair are not arranged immediately above the first, but in between them, also on long stalks; the next two pairs are above the first and second pairs, but are smaller and have shorter stalks so that their leaflets are roughly level with those of the lower leaves and shade only their stalks. The uppermost leaves are quite small, and often have only five leaflets

Fig. 7.   Prostrate shoots of Ivy, showing leaf-mosaic.   (After Kerner.)

instead of seven.   In each leaf, too, the separate leaflets practically touch each other, edge to edge, but do not overlap.

It is well known to all who have grown plants in a room that plants turn towards the light, and this fact suggests, like the position and arrangement of the leaves, that light is of great importance to plants. We may easily prove this to be so by attempting to grow plants in the dark.   If we sow Mustard seed, for instance, in

Exp. 3   boxes, and keep some in the dark and some in the light, we find that, whereas in the light the Mustard grows healthily, in the dark the plants are tall,

weak, and straggling, and instead of being deep green
are pale yellow. Eventually they droop and die.

We have then these facts: (1) leaves are so shaped as
to present a very large surface to air and light; (2) they
are so arranged that they shade each other as little as
possible from the light; and (3) light must be of great importance to plants, for they do not live long in the dark.

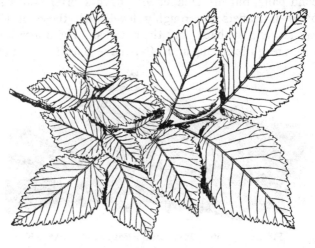

Fig. 8. Lateral shoot of Elm, showing leaf-mosaic. (After Kerner.)

The meaning of these facts was discovered in the
18th century. The first great step in the
discovery was made by an English man
of science, Priestley, who also discovered
oxygen and the part it plays in combustion. We can
easily repeat his important experiment in the following
way. We take the bell-jar which we used
for the experiment with the burning candle, and stand it
in rather deeper water than we used then. Now as before
we light a piece of candle, and lower it on the wire into
the bell-jar, closing the jar above with the cork which

Priestley's
experiment.

Exp. 4

holds the wire. When the candle has gone out we cut a
leafy twig, or other leafy shoot (e.g. Artichoke or Wallflower), and carefully pass it through the water up into
the air under the bell-jar, leaving it with the cut end of
the stem in the water. The experiment must be set up in
a well-lighted place, such as a greenhouse, or a south
window, or in the open air. After some hours at least in
the light, we carefully lift the cork and lower a lighted taper
into the jar: it does not go out. If we light the candle it
will continue to burn for a little while as it did at the
beginning.

To make sure that it is the plant which has brought
about this result we must set up another exactly similar
experiment, either by the side of the first or afterwards,
without the plant.   We ought to find that in this *control*
or *blank* experiment the taper still goes out, from which
we conclude that there is no oxygen to keep it burning
and no oxygen has got in through the water or through
the cork.

On the other hand *when the plant was introduced*
into the air in which the oxygen had been changed to
carbon dioxide by the burning candle, *oxygen reappeared ;*
and as no important change took place in the volume of
the air[1], the carbon dioxide must have disappeared.
Thus the plant has changed the carbon dioxide into
oxygen, and we suspect at once that it has kept the
carbon for itself.

If we repeat the experiment in the dark, we shall
find that the taper is extinguished even after several
days; therefore the plant can change carbon dioxide into
oxygen *only in the light*.   Here clearly is the explanation
of the form and arrangement of the leaves and of the
importance of light.

---

[1] Unless the temperature altered very much in the meantime.

The plant does, therefore, obtain its carbon from the air, and this is the work of the shoot.

Let us study this process further by means of a few more experiments, for it is of very great importance.

By using any green water-plant which spends its life submerged in the water, we can readily demonstrate the chief points in the process.

Oxygen given off.

The common Canadian Water-weed (*Elodea canadensis*) does well for this purpose. A good many shoots are put into plenty of water in a wide glass vessel; and placed in the sunlight.   Soon streams of bubbles are seen rising from the shoots through the water.   These bubbles continue to arise as long as the plant is in sunlight.   If the vessel is put in the shade, the bubbles soon become fewer and may even cease to appear.

Exp. 5

By a suitable arrangement of funnel and test-tube the bubbles may be collected.   The test-tube is filled with water and inverted, and its mouth slipped over the tube of the funnel under water.   If we test the gas that collects in the tube by lowering into it a splint of wood which is glowing but not in flame the splint is relighted and bursts into flame.   This test proves that the gas which comes off in bubbles from the green plant in the sunlight is much richer in oxygen than ordinary air, and the experiment demonstrates the first process going on in green leaves exposed to the light, namely, *the leaves give out oxygen.*

Set side by side with the first vessel another, containing a similar lot of water-plant shoots in water which has been recently boiled and cooled.

Carbon dioxide taken in.

The shoots in both vessels are green, both are in the light, yet only one gives off bubbles of oxygen: the shoots in the boiled water give off none.   How can we account for this?   We know that by boiling the water the air which was dissolved in it has been driven out, and with the air the carbon dioxide.   We can easily show that it

is for want of carbon dioxide that no bubbles of oxygen
Exp. 6   appear. Air is first bubbled through a solution
of potash to remove all the carbon dioxide from it, and

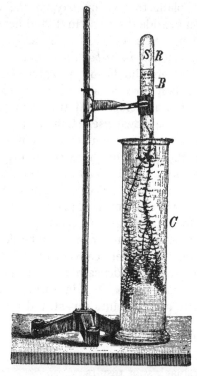

Fig. 9. Evolution of oxygen from assimilating plants. In the glass
   cylinder C, filled with water, are placed stems of *Elodea*: the ends
   of the stems are introduced into the test-tube R, which is also full
   of water. The gas-bubbles B, rising from the cut surfaces, collect
   at S. (After Strasburger.)

then through the boiled water containing the water-plant.
Although still in the light the plant gives out no oxygen.
If, however, we supply it with carbon dioxide as we

may do very simply by blowing our own breath through a tube into the water, the streams of oxygen bubbles begin to rise from the plant, provided it is all the time in sunlight. We have therefore shown that in order to enable the plant to give off oxygen the water must contain carbon dioxide dissolved in it, and hence conclude that a second process goes on in green leaves exposed to light, namely, *they take in carbon dioxide.*

Thus the green shoots of a water-plant take in carbon dioxide and give out oxygen. We **Starch** infer, as before, that they retain the carbon **formed.** for their own use. Can we prove this? If we could find carbon, or some substance containing carbon, appearing in the plant in presence of carbon dioxide and in the sunlight, we should have strong evidence for it. Now starch is a substance which occurs very widely in plants, especially in those parts of plants which are of great use to us as food, in wheat and other grain, in potatoes, etc. It is a substance which can easily be detected even in minute quantity by a very convenient and delicate test.

Exp. 7   To show how delicate it is, a small piece of washing starch may be dissolved in warm water in a test-tube. If a little of this solution is stirred into a tumbler full of water and a few drops of a solution of iodine added[1], the starch solution, although so dilute, turns a deep blue. If a drop of iodine solution is put directly on to the solid starch, a blue-black spot results.

If the water-plant shoots, or a leaf, were placed in iodine solution and such a colour were to appear, it would not be seen clearly because their green colour would mask it. Therefore, in order to test them for starch we first dissolve out the green colour and then apply the iodine.

---

[1] Iodine will not dissolve in water, but it will dissolve in alcohol or in a solution of potassium iodide. A 5 % solution of iodine in 20 % aqueous potassium iodide is a useful stock solution.

The procedure is as follows. A shoot or a leaf which has been in the sunlight is put into a beaker of boiling water for 2 to 15 minutes (the thicker and tougher the leaf the longer it should be boiled). This drives out the air, softens the tissues and renders the entry of the subsequent liquids easy. The leaf is now placed in a smaller beaker of methylated spirit, the gas under the beaker of water is turned out and the small beaker put into the hot water. In this way the spirit is warmed without danger of its catching fire, while the extraction of the green colour is hastened. After 20–30 minutes the leaf is colourless, the green colour having gone into the spirit. The leaf is now placed in iodine solution diluted to the colour of beer[1]. The whole leaf, or shoot, as the case may be, will then turn black, showing that starch is present in it.

Iodine test: method (margin)

We have still to show that the shoot of the water-plant formed starch when it was taking in carbon dioxide and giving out oxygen. Some shoots are put in the dark for a few days in a shallow open dish of water. A few of their leaves are then tested for starch. If no starch is found[2] a few of the shoots are put in the sunlight for a few hours, some in water freshly boiled and cooled, containing therefore no carbon dioxide, some in ordinary water, to which may with advantage be added extra carbon dioxide by pouring in a little soda water (in the making of which quantities of carbon dioxide are forced into the water under pressure). At the end of this time the shoots in each lot are tested for starch: those in the boiled water,

Exp. 8 (margin)

[1] The solution should be strong enough to enter the leaf rapidly but not so strong as to turn it brown. A few trials will soon indicate the best strength. For delicate leaves 0·25 % iodine is suitable.

[2] It generally takes about four days for the starch to disappear from a water-plant. From the leaves of some land-plants, like the Sunflower or Fuchsia, the starch often disappears in a single night.

which have therefore not been able to absorb any carbon
dioxide or give off any oxygen, contain no starch; but
those which have been supplied with carbon dioxide
contain plenty of starch.

We have thus demonstrated a third process which
Carbon   goes on in the green leaf in the sunlight,
assimila-   namely, *starch is formed*. This process is
tion or
photo-   not independent of the other two, but only
synthesis.   takes place when they also are going on.
All three together make up what is known as *carbon
assimilation* because by it the plant obtains its carbon.
It is also called *photosynthesis* because the starch is built
up, or synthesised, in the light. *The whole process of
photosynthesis includes*, then, (1) the taking in of carbon
dioxide, (2) the giving out of oxygen, (3) the building
up of starch from the carbon.

The method described above for detecting the pres-
                 ence of starch in leaves enables us not only
Conditions   to prove that photosynthesis takes place
of photo-
synthesis.   in green leaves but also to find out more
                 particularly than we have done so far what
are the conditions necessary for it to be maintained.

(1) *Light*. For instance we may prove very simply
that light is necessary in the following way:

A plant is prepared by leaving it in the dark over night or
Exp. 9 a   until its leaves are found when tested to be without starch[1]. In a piece of black paper, a little larger than a
leaf, holes are cut in a definite pattern; for instance, the word
STARCH is cut out. This stencil is placed on the upper side
of the leaf and a piece of black paper put on the under side to
shut out all the light from below, both pieces of paper being
fastened to the leaf at the edges by light clips. It is better to
put a piece of glass above the stencil so that the carbon dioxide

---

[1] Leaves of plants in the open air may be prepared by covering
them with black paper or tinfoil. A large twig of an oak or other
similar tree may be cut and put in water in the dark.

of the air cannot get more easily to those parts of the leaf not covered by black paper; otherwise there is no proof that the absence of starch in the covered parts is not due to lack of carbon dioxide. But the glass must be suitably supported if it is too heavy for the leaf. When all is ready the plant is exposed to sunlight. After a few hours' exposure the leaf is cut off and tested for starch. The parts exposed to the light are found to be black with starch, the other parts are yellowish, showing no starch. The pattern cut out of the stencil, in this case the word STARCH, appears in black on the leaf, while the rest of the leaf remains yellowish, proving that only where light fell upon the leaf has it been able to make any starch.

The experiment may be modified in an interesting way by **Exp. 9 b** covering the leaf with a photographic negative,— one in which there are well defined light and dark parts is best. On exposure to sunlight and subsequent testing for starch a print of the photograph is obtained in the leaf. Where the negative has allowed most light to pass through much starch has been formed and these parts of the leaf are deeply stained by the iodine. Under less transparent parts of the negative the leaf appears dusky, showing that some starch has been formed, but less than in more highly lighted parts of the leaf: from this we gather that the amount of starch formed depends upon the strength of the light which reaches the leaf. Parts of the leaf deeply shaded by the densest parts of the negative remain yellowish, since no starch has been formed there.

By means of the starch test we may also show that photosynthesis can take place in artificial light, for if a leaf devoid of starch be exposed to bright incandescent gas-light or bright electric light, or to any sufficiently bright artificial light, starch is formed as in sunlight.

(2) *Carbon dioxide.* We have already seen that a water-plant will not form starch or give off oxygen if the water in which it is placed contains no carbon dioxide. The following experiment shows that a land-plant similarly will not form starch in its leaves unless carbon dioxide is present in the surrounding air.

Two similar plants in pots should be prepared by keeping them **Exp. 10 a** in the dark until all the starch has disappeared from their leaves. One of them is then supplied with air containing no carbon dioxide, and the other with ordinary air. To do this the former is put on a glass plate under a bell-jar with an opening in the top, and along with the plant a small dish of potash solution which will soon absorb all the carbon dioxide from the air in the jar. To stop any carbon dioxide from entering

the bell-jar some melted wax is run round its bottom edge to fasten it to the glass plate, and in the opening at the top is fixed a tube packed with grains of soda-lime, a solid substance which absorbs carbon dioxide; through this tube air can enter but no carbon dioxide. The other plant, in order to equalise the conditions as much as possible for the two plants, is put under a similar bell-jar on a glass plate, melted wax run round as before, and in the top is put a tube filled with sawdust instead of soda-lime, so that carbon dioxide is not prevented from entering this jar. Now both bell-jars are placed in bright light for several hours, though not in strong sunshine as the plants, shut up under the glass bell-jars, might then get too hot. Finally the leaves of each plant are tested for starch; only the leaves from the plant that was supplied with carbon dioxide contain any. Thus a plant must have carbon dioxide in order to form starch.

Fig. 10.   Plant in the light without carbon dioxide: the tube T and the dish G contain soda-lime. (After Darwin and Acton.)

A simpler way of demonstrating this fact, requiring little
Exp. 10 b   apparatus, is to fit a wide-mouthed bottle with a split cork, and fasten a leaf of a plant which has been kept in the dark (without removing it from the plant) between the halves of the cork so that the tip of the leaf-blade is inside the bottle and the base of the blade is outside. In the bottle is put some lime-water or dilute potash, and the cork is made air-tight with vaseline; after a few hours in bright light (as before, the

experiment should not be performed in strong sunshine) the leaf
is tested for starch. No starch is found in the part of the leaf
which has been in the bottle, although it has been exposed to light
like the part outside, which is black with starch.

(3) *Green colour.* We have already observed that
the leaves of plants are characterised by their green
colour. We must now ask whether this colour has anything to do with their fitness for the work of photosynthesis. A variegated leaf, such as a leaf of the well-known
'Variegated Maple,' is partly green and partly white. The
white part is living and healthy just as the green part is,
the only difference being that in the former there is no
green colouring matter present. Such a leaf after it has
been a few hours in sunlight should answer the question
whether the green colour is necessary for the formation
Exp. 11   of starch. A careful sketch of the leaf
should be made, showing exactly the shape and size of
those parts which are green; then the leaf is tested for
starch. When the result of the test is compared with
the sketch it is found that starch is present only in those
parts which were green. Thus the green colour is essential
for photosynthesis; further, wherever the green colour is
found, whether in leaves or in stems, there photosynthesis
takes place in the light.

(4) *Pores in the epidermis.* If we tear a leaf across we
find that the green colour is confined to the interior tissues
of the leaf which are bounded above and below by a
colourless skin, called the *epidermis.* The carbon dioxide
must, therefore, get through the epidermis to the green
tissues; if a torn leaf is tested for starch, it is in the
internal tissues that starch is found, while the epidermis
shows little or none. How does the carbon dioxide get
through this skin?

In preparing leaves for the starch test we have seen

that by dipping them into boiling water air is driven
out of them in bubbles. Far more comes out than could
possibly have been held in solution by the small volume
of water in the leaf. In fact, as we shall see later for
ourselves, the internal tissues of the leaf are spongy, and
hold a great deal of air in tiny canals and spaces—*air-
spaces*; this air expands when heated by contact with
the boiling water, and escapes. It will probably have
been noticed that in most leaves the air escapes from one
side only, the under side—if not, it will be worth while to
gather leaves from various plants and dip them in boiling
water, carefully watching the result in each case.

If the underside of a smooth leaf is examined with a
good lens numerous white specks can often be seen: at
these points are pores, called *stomata*, in the lower epidermis. The position of the stomata is more easily seen in
the needle-leaves of pines and firs, where deposits of wax
occur around them, and they appear under the lens as
rows of conspicuous white dots. It is through these
pores that the air escapes from inside the leaf. We shall
examine them later under the microscope; for the present
let us try to show that it is through them that the carbon
dioxide of the air passes to the green tissue inside the
leaf.

The starch test enables us to do this. If it is through
the pores that the carbon dioxide is taken in we should
not expect any starch to be formed when the pores are
Exp. 12   blocked up. Two similar leaves are selected
from a plant, such as the Fuchsia, which has been in the
dark till the leaves contain no starch. The upper side of one
and the under side of the other are greased very carefully
with vaseline; the plant is then exposed to the sunlight
for a few hours. Now both leaves are labelled to distinguish them one from the other, and are tested for
starch along with a third leaf, which has not been greased.

No starch is found in the leaf with the under side greased
and the pores thereby blocked; on the other hand, the
second leaf contains as much starch as the third, so that
greasing the upper epidermis, in which there are no pores,
has made no difference to photosynthesis.

We may vary this experiment by greasing only one
half of each leaf, on one side of the midrib.   In this case
the leaf in which half of the under side is greased will
be half black, half yellow, when tested with iodine.   This
shows also that carbon dioxide does not pass from one
half through the midrib to the other half.

Let us in conclusion inquire what part the green colour

Chloro-   plays in photosynthesis.   Every time a leaf
phyll   is tested for starch a solution is made of the
absorbs   green colouring matter, *chlorophyll*, in the
light.   methylated spirit used to decolorise the leaf.
This solution is green (unless very strong, when it appears
deep red).   Thus light which has passed through chlorophyll
solution is green; and so is light which is reflected from the
chlorophyll in the leaf.   Now white light consists of many
parts, which when separated appear to us of many
different colours; a glass prism separates them, and so
do raindrops which, reflecting the sunshine, give us the
colours of the rainbow.   Glass and water appear transparent because they let all the colours through equally;
but stained glass and coloured liquids absorb, and so cut
off, some of the colours, while those colours which they
allow to pass and therefore reach the eye give what we
call the colour of the glass or liquid.   Similarly a solid
object is white if it reflects all the colours equally, black
if it absorbs them all and reflects none; while if it
reflects part and absorbs part, that part which it reflects
determines its colour.

A green leaf, therefore, and a solution of chlorophyll
absorb part of the light which falls on them, and reflect or

transmit green light.   We can find out just what parts of
white light are most absorbed by the chlorophyll by means
of a spectroscope.   A beam of white light, after it has
passed through the glass prism of the spectroscope, appears as a continuous series of colours running into one
another, forming what is called a *spectrum*; we roughly
distinguish in the series red, orange, yellow, green, blue,
and violet.   When the chlorophyll solution is put between
the light and the spectroscope, we find that part of the
red and orange disappears completely from the spectrum,
and much of the blue and violet; while the green is still
there as before.

Now why is the absorption of light necessary for photo-
Light-energy   synthesis?   Let us recall what the leaf
used in photo-   has to do.   It takes in carbon dioxide,
synthesis.   gives out an equal volume of oxygen and
makes starch.   Starch is found on careful analysis to
contain carbon, hydrogen and oxygen in the proportions represented by the formula $C_6H_{10}O_5$, and as water
contains hydrogen and oxygen in the proportion H_2O, we
can regard starch as made of carbon and water ($6C +
5H_2O = C_6H_{10}O_5$).   The plant gets water from the soil,
and the carbon it obtains by *splitting up carbon dioxide*,
letting go the oxygen.   We shall understand what a work
this separation really is if we recall what happens when
carbon and oxygen unite together.   When substances containing carbon—like coal, or charcoal, or candles—burn,
in other words, when they combine with oxygen, a great
deal of *heat and light* are given out; and both heat and light
are forms of *energy* (p. 20).   So energy in the form of light
and heat has been given off in large quantities when carbon
and oxygen combined to form carbon dioxide. This libera-
tion of energy shows the strength of their union, and
an equal amount of energy must be exerted and stored
up in them again in order to sever them.   This energy

the plant obtains from the sunlight by means of chlorophyll.

*In photosynthesis green plants use energy which comes
from the sun to manufacture food.* This is a fact of
stupendous importance ; for we and all other animals
depend for all our food on plants. If we eat the flesh of
other animals, they have fed on plants; while vegetables,
and the grain from which comes our bread, are parts of
plants and contain the food which plants have manufactured. Whereas animals must have ready-made starch
and other complete foods, the green plants are able to
manufacture food for themselves from the carbon dioxide
of the atmosphere and other simple substances which they
get from the soil.

The following very simple form of Priestley's fundamental
experiment does not require laboratory equipment and can be
Exp. 13 performed at home. Two tumblers are filled with
water, inverted over water in a bowl, and filled through a tube
with air from the lungs. A small leafy shoot is slipped under
water up into one tumbler and both are exposed to bright daylight
for some hours. A glass plate or piece of card is placed over the
mouth of each tumbler, when lifting it out to test the air as in
Exp. 4. Expired air, properly collected, does not support combustion (in breathing the air out, the first part should be allowed
to escape).

CHAPTER IV

CONVEYANCE AND STORAGE OF FOOD

In preparing plants for experiments on starch formation, we have already made use of the fact that starch, formed in the light, disappears from the leaves in darkness.

To find out whether this starch is used up within the leaf or passes out of the leaf into the stem, we try the effect of cutting leaves off and so preventing anything from passing through the stalk. A plant is exposed for Exp. 14 some hours in a bright light till its leaves have formed plenty of starch. A few leaves are tested for starch at once and kept afterwards in spirit for comparison. Other leaves are cut off and put in the dark, with their stalks in water, and the plant is put beside them. After about twelve hours of darkness the cut leaves and a few similar leaves which have remained on the plant are tested for starch and compared with the control leaves taken the night before. It is found that the leaves on the plant have lost all or most of their starch, while the cut leaves have lost far less. We infer from this that most of the starch which disappears from a leaf in the dark is transferred from the leaf into the stem.

Now starch is a solid, and as such cannot pass from one part of a plant to another; this we shall readily understand when we come to study the structure of plants under a microscope. Only when dissolved in water can substances travel through the plant. Starch, however, is insoluble; even in hot water it only forms a sort of jelly, which could not pass through the plant even if it could be formed there. The only possible explanation of its transference is that it has first been changed into a substance which *is* soluble in water.

In the animal body also there is the same necessity **Digestion** for changing insoluble solid foods into **and trans-** soluble substances. If we keep a piece of **location.** dry bread in the mouth for a few minutes it begins to taste sweet: this is due to the change of the *starch*, contained in the wheat flour of which the bread is made, into *sugar* by the action of the saliva. In the saliva is a ferment or *enzyme*, called diastase; though present only in very small quantity, this enzyme is able to cause a large amount of starch to change into sugar without itself being used up. It is the sugar so formed which can pass into the blood and be carried to all parts of the body, there to be used in various ways.

The disappearance of starch from a mixture of saliva **Exp. 14 a** and starch solution in a test tube can be followed by removing drops of the mixture at intervals on a clean glass rod and mixing each with a drop of iodine on a white plate. The process is hastened by gentle warmth. If the saliva is first boiled, no change occurs in the mixture: boiling destroys the enzyme.

A similar enzyme is formed in leaves. By the action of this diastase the starch is there also changed into sugar so that it can be carried or *translocated* to other parts of the plant, wherever it may be needed.

Some of the sugar which has been translocated, if it is not wanted for use immediately, may be changed back into starch and stored up round about the path by which it travels. We can readily convince ourselves of the existence of such temporary stores of starch by cutting across almost any fairly thick stem of a starch-forming

plant in summer and applying a drop of iodine solution
to the cut surface, which will turn dusky or black in parts.
We may even trace the path of the sugar right through
stem and root by means of the starch that is formed along
its track.

Sugar is also changed back again into starch in special
parts of the plants, and is there stored up for future use.
Potatoes are examples of such parts: if a potato is cut
open and a drop of iodine applied to the cut surface,
a black stain clearly indicates the presence of abundance
of starch. Grains of Wheat, Barley and other cereals are
packed with starch.

Sugar, like starch, is a carbohydrate, that is, it consists of carbon, hydrogen, and oxygen, the hydrogen and
oxygen being present in the same proportions as in water.
Sugar thus belongs to the same class of substances as
starch; but it is far less complex. This means that,
when starch is formed from sugar, the simpler sugar is
built up into the more complex starch; and when starch
is changed by diastase into sugar, the complex starch is
broken down into the less complex sugar.

Now we have seen that during photosynthesis starch
Photo- is built up from still simpler substances,
synthesis carbon dioxide and water; and it is found in
of sugar. chemistry that very complex substances are
usually formed in stages, simpler intermediate substances
being formed first, and afterwards built up into those
which are more complex. May it not be, then, that sugar
is such an intermediate substance in the formation of
starch, and that *sugar* is the *first product of photosynthesis*?

Certain other facts agree with this idea.

The leaves of most bulbous plants, like the Daffodil,
Snowdrop, or Lily, never form starch at all. In these,
clearly, some other substance must be manufactured in
photosynthesis. It has been shown, by more difficult

tests than the test for starch, that these leaves form *sugar* instead.

By the same tests it has been found that even leaves which form starch contain sugar too. It seems very likely therefore that, in these leaves also, sugar is first formed and that starch is formed from it afterwards.

Moreover, very different amounts of starch are formed by the leaves of different plants in the same time under the same circumstances: it has in fact been found that some plants change a large part, others less, of their sugar into starch.

Another fact about starch formation also calls for explanation. When the leaves of a plant that forms a great deal of starch, such as the Sunflower, are tested on a dull day, sometimes no starch is found at all. Now photosynthesis does not stop on a dull day, for it can be shown that the leaves are taking in carbon dioxide and giving out oxygen. Here again sugar is formed, but is not changed into starch, the explanation being that starch is only formed when sugar is rapidly accumulating.

When we consider all these facts, starch begins to appear of secondary importance. It is sugar which is first formed, sugar which is translocated. Starch is, as it were, merely sugar laid aside in a stationary form, in the leaf or elsewhere, until the time comes for it to be translocated or used on the spot.

It is probable that part of the sugar which is formed in the green tissue during photosynthesis passes at once into the veins on its way towards the stem. If it is only being formed slowly (in dull weather, for instance) it may be translocated as fast as it is formed. In bright sunny weather, on the other hand, it is formed more rapidly and cannot be translocated fast enough. Then, as it accumulates, the excess is turned into starch which at night is changed again into sugar and translocated.

Besides temporarily laying aside carbohydrate food **Storage of food-materials.** in their leaves and along the way from one part to another, plants which live through the winter lay in *special stores* of food. Trees which shed their leaves in autumn store food in abundance in all parts of their stem and roots during the summer: this is used the following spring during the early stages of growth, until the new leaves have expanded and are ready for work. Plants which die down at the approach of winter, but have shoots and roots that survive from year to year below the ground, store food in these underground parts; this food serves for the growth of new roots and shoots until the shoots come above ground and the newly expanded leaves are able to manufacture more food.

In a number of plants special enlarged receptacles **Storage organs.** are formed by the swelling of roots or underground shoots: in these, large quantities of food are stored. A potato, for example, is formed on the Potato plant by the swelling of the end of an underground branch: such a structure is called a *tuber*. When first the swelling begins the leaves appear as small scales with minute buds in their axils: later on only a long transverse ridge marks the former position of each leaf, and on one side of the ridge is the 'eye' or bud which in the following year may elongate into a green shoot and grow up into the air, feeding on the abundant starch and other food stored in the tuber.

In the Carrot and Turnip, food is stored in the enlarged tap-root. In the Dahlia lateral roots form tubers: such root-tubers may readily be distinguished from stem-tubers like those of the Potato by the absence of 'eyes.'

Other storage organs are the *corm* of the Crocus and the base of the stem of the bulbous Buttercup, which resemble each other in that the swollen portion is a short erect piece of the base of the stem. In the *bulbs* of the Tulip and Lily, on the other hand, most of the food is stored in thick swollen leaves, while the stem which bears them is small and flat.

Thus roots, stems, and leaves develop into special storage organs in different plants.

It must be remembered, however, that in the greater number of plants food is stored in roots and shoots but little modified or thickened. It is interesting, too, to notice that even special storage organs are often much smaller in nature. The wild Carrot, Parsnip, Radish, etc., have tap-roots but little swollen; the large size which these vegetables attain in our gardens is the result of prolonged cultivation and selection (p. 294).

All plants store food in their *seeds*, to support the young plants which grow from them. Until they have made their way up through the ground and spread out their leaves to the light, they are unable to make food for themselves and therefore depend entirely upon the store of food bequeathed to them by their parents.

These special stores of food give us an opportunity of finding out what kinds of food are needed by young plants or the growing parts of plants. It is clear that sugar or starch alone is not enough. Carbohydrates consist only of carbon, hydrogen, and oxygen; whereas a plant contains also substances which remain as ash when it is burned.

Kinds of food-substances.

Let us examine the substance contained in wheat grains which we can obtain ready powdered as *flour*. If we tie up some flour in a piece of muslin and knead it in water, we shall find that the water becomes milky while the flour inside the muslin becomes more and more sticky. Let us pour the milky water into a tumbler, and continue kneading the bag in running

Exp. 15

water. After a time the water which runs off will no
longer appear milky. By this time the water set aside
in the tumbler will be clear, with a white sediment at
the bottom. A drop of iodine will show that this sediment is starch. If we open the muslin bag and apply a
drop of iodine to the sticky mass inside, a yellowish brown
spot appears, but no blue or black colour if the washing
has been thoroughly done; no starch is left. This sticky
substance is called *gluten*, and is an example of a large
class of substances called *proteïns*. These are very complex substances which contain *nitrogen* and *sulphur*, in
addition to carbon, hydrogen, and oxygen. The living
matter of the plant itself consists of still more complex
proteïns which are built up in the plant from the less
complex food proteïns, some of them with the addition
of *phosphorus*; proteïns form therefore one extremely important part of the food stored by plants.

Proteïns, like starch, can be detected by certain tests.
Iodine, as we have seen, turns gluten yellow-brown: this
is one of the tests for proteïns. Another test is as follows:
a drop of nitric acid is first applied, producing a yellow
stain; then ammonia is added and the stain becomes a
much brighter golden or orange yellow. For a third test
a drop of 1 % copper sulphate solution may be first
applied, followed by 2 % potash; a beautiful mauve or
violet coloration develops.

We can use these tests to detect proteins stored in
seeds, tubers, etc. If for instance we cut across a broad
bean, and apply the tests to the cut surface, we shall find
that proteid food is abundant in this seed.

One other class of substances is important in the food-
Exp. 16   stores of plants. If we crush some hemp
seed in a mortar, add a little ether, then filter it off and
allow the ether to evaporate in a watch glass or small glass
dish, we shall find a greasy residue left in the dish. This is

an *oil*. Oils contain carbon and hydrogen, but only a small proportion of oxygen.

In the wheat grain there is very little oil; but in some seeds there is a great deal. Linseed, for example, the seed of the Flax plant, yields the well-known linseed oil, cotton seed also yields an oil, and castor oil is extracted from the seed of the Castor Oil plant. In such seeds oil takes the place of starch or sugar.

We have now found three classes of substances stored up as reserves of food in plants. These are: (1) carbohydrates, (2) proteïns, and (3) oils.

Of these the proteïns alone contain nitrogen and are spoken of as *nitrogenous* foods; carbohydrates and oils which contain no nitrogen are, on the other hand, called *carbonaceous* foods.

The different kinds of food occur, as we have already seen, in very different proportions in the stores which different plants lay up in reserve. Some nitrogenous food is always present, and some purely carbonaceous food; but while in the Broad Bean the greater part of the food stored is proteïn, in the Potato most of it is starch.

The carbonaceous food is often mainly starch; but sugar or oil may be stored instead of, or in addition to, starch. In the tuberous root of the Sugar Beet, and in the stem of the Sugar Cane, sugar is stored, and from these sources the sugar we use is extracted. Oils are found chiefly in seeds.

These stored reserves of food made by plants form the chief food of mankind. We and other animals need, like plants, both nitrogenous and carbonaceous foods. Only plants are able to manufacture these foods: plants and animals alike use them. What use they make of them we shall see in the next chapter.

CHAPTER V

TWOFOLD USE OF FOOD: GROWTH AND RESPIRATION

We have seen that starch, a solid insoluble substance,
Digestion. cannot be translocated from one part of the
Colloids plant to another for use until it has first
and crys- been changed into sugar which is soluble in
talloids. water. Most of the food substances which are
stored in reserve by plants are like starch in being insoluble
in water; or, if they dissolve at all, they form only a kind
of jelly and are therefore called *colloids* (Greek *kolla* =
glue). All such substances must be changed into simpler
soluble substances before they can be translocated and
used: these simpler substances, like sugar, generally
form crystals in the solid state, and are therefore called
crystalloids to distinguish them from colloids.

The same kind of change from colloids to crystalloids
is necessary in the animal body, and all the changes
of this kind which take place, together make up what
we call digestion. One digestive process, the change of
starch into sugar by the action of diastase in the saliva,
we have already mentioned. In the digestive organs the
food is mixed with other juices, of which the best known
is bile, secreted by the liver, and these juices also contain
enzymes which change proteïns into simpler nitrogenous
substances, and oils and fats into other soluble substances
which need not be mentioned here. Enzymes very similar
to those present in our digestive juices are formed by

plants, to change reserve proteïns and oils back again into soluble form. Owing to the digestion of the solid reserve materials stored in them, germinating seeds, sprouting potatoes, etc., become very soft.

Digestion, then, is the first process which reserve foods must undergo, whether in animals or in plants, before use can be made of them. The substances which both plants and animals use directly are the soluble crystalloids like sugar.

In seeking to understand the way in which plants make use of these soluble foods we shall find it helpful to look again to the animal or human body for suggestions. We may say at once that one use of food in both animals and plants is for the growth of the organism— in other words, as material from which to build up the plant or animal body. In plants it is required for this purpose chiefly at the growing points of roots, stems, or branches; that is, at the root tips and in the active expanding buds.

We and other higher animals require food also for *warmth*, and for strength or *energy*. Energy is expended in movement, and in all kinds of effort or work. How is this *energy* obtained from *food*? We have already learned that when we breathe we expire carbon dioxide. This gas is formed by the oxidation of part of the substances containing carbon in our blood. We have also learned that when carbonaceous substances undergo rapid oxidation, or combustion, they yield heat and light, which are both forms of energy. This energy was stored up, in carbonaceous foods, by plants during photosynthesis, being absorbed from the sunlight by the chlorophyll in their leaves. The oxidation of these substances takes place more slowly in respiration than when they burn with a flame, but it is none the less a slow combustion and the same amount of energy is gradually set free.

Part of this energy appears in the form of heat, and so our bodies are kept warm: part of it we use in doing work, much as a steam engine uses the energy liberated from burning fuel. Some, however, is used within the body in doing *chemical* work; for just as energy is required for the photosynthesis of sugar and starch, so also must energy be expended in the building up of the still more complex compounds of the body. During growth and all through life chemical work goes on.

Now plants do not keep themselves warm. They are in this respect like the lower, cold-blooded animals, the temperature of whose blood varies with that of the surrounding air, not like man and other higher animals whose blood remains at a certain fairly high temperature, however cold the air may be. Nor do plants move from place to place, or do any obvious mechanical work. They do, however, grow in size. The root thrusts its way into the soil, and must overcome great resistance when it grows in thickness. The stem must lift food and water to its growing points, the expanding buds: a tall tree raises large quantities for long distances, so that in time it must accomplish a very great amount of work. Moreover, as in animals, life and growth also mean chemical work. Do plants, then, liberate the energy which they require in the same way as animals, by slow combustion of carbonaceous food?

Do plants respire, absorbing oxygen and giving out **Do plants give out carbon dioxide?** carbon dioxide? Green shoots in the light do just the opposite of this; they give out oxygen and take in carbon dioxide vigorously in photosynthesis. We could not, therefore, tell if they use oxygen and produce carbon dioxide while photosynthesis is going on: but what happens in the dark, when they can no longer take carbon dioxide in? We can find this out by experiment.

Shoots or leaves of any plant may be used for this
Exp. 17 a   purpose. A handful of Grass will serve very
well. This should be put into a glass jar, not quite filling
it, as a small space must be left at the top so that a lighted
taper may be introduced afterwards. The jar is securely
closed with a stopper, cork, or glass plate well greased, so
that it is air-tight. It is then either covered carefully
with black cloth or paper, or put into a box or cupboard
in darkness. Next day the air should be tested. First
a clear drop of lime-water is introduced on a glass rod:
it rapidly becomes cloudy, showing that carbon dioxide
is present. A lighted taper plunged into the jar is extinguished. Oxygen has therefore been removed and
carbon dioxide produced by the green shoots in the dark.

The same experiment may more readily be performed
Exp. 17 b   with seeds that are beginning to germinate.
Germinating barley or beans, for instance, respire very
vigorously, and since they are not green it is unnecessary
to put them in the dark. Beans or barley are soaked overnight, and put next day into a bottle, closed and air-tight as
before. After several hours the air in the bottle is tested
with lime-water and with the taper. The result is the
same as in the case of the grass, whether the seeds have
been left in the dark or in the light.

We may repeat the experiment with roots, such as
carrots, or any other plant, or plant organ, which is alive
and growing; and, although some take longer than others
to exhaust the air in the same bottle, we shall always find
that eventually a taper is extinguished.

A better way of proving that carbon dioxide is
Exp. 18   produced is as follows. A series of jars or
flasks are fitted with stoppers and tubes, and connected
together as shown in Figure 11. Air is drawn through
the whole apparatus by means of an aspirator. The air
enters, by way of a tube T, filled with soda-lime, into the

first bottle, F, containing sticks of caustic potash; all carbon dioxide is thus completely removed. Bottle B contains lime-water to make sure that no carbon dioxide escapes the potash in F. The air next passes through bottle A, in which the respiring material is placed: when grass or other green shoots or leaves are used, this bottle and the tubes leading in and out of it are carefully darkened with cloth or paper. Finally the air passes through lime-water (or baryta water) in bottle P, where

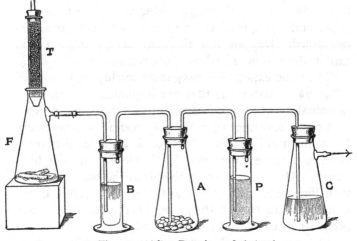

Fig. 11. (After Darwin and Acton.)

the carbon dioxide given out by the grass is absorbed and produces a milky precipitate.

The fifth bottle C shown in the diagram is only necessary if we wish to measure the actual quantity of carbon dioxide produced. This we can do by putting a known volume of baryta water of known strength into P, and likewise into C in case the baryta in P does not absorb all the carbon dioxide which is produced. If all is absorbed in P, the baryta water in C will remain clear. The baryta water in P becomes weaker as more and more precipitate is formed. By titrating this solution after a certain number of hours with standard hydrochloric acid we may tell how much it has been weakened, and thereby how much carbon dioxide has been produced by the plant.

Without, however, carrying out such exact measurements we can tell roughly from the amount of the precipitate that the amount of carbon dioxide produced in a given time varies with the conditions.

We shall find, for instance, that when bottle A is immersed in warm water, say at 35° C., much more precipitate appears in P than at the ordinary temperature of the room, 15–18° C.: while on the other hand, very little appears if A is surrounded by melting ice, and is therefore at 0° C. Temperature is the most important factor that affects the rate of respiration in plants.

It is clear that the production of carbon dioxide must involve a loss of carbonaceous substance during respiration. This may be demonstrated most readily with germinating seeds which respire very vigorously and so lose substance rapidly.

Substance lost in respiration.

Two similar lots of barley, for instance, are weighed out. One lot is dried in a steam oven at 100° C. for 24 hours, and when cool is weighed. The other lot is soaked in water, then spread on a plate and covered with a bell-jar or a second inverted plate to keep it moist. When it has been germinating for three or four days it is dried in the oven and weighed like the ungerminated grains. It is found that the dry weight is much less after the three or four days of vigorous respiration.

Exp. 19

If in respiration some of the carbonaceous food within the plant is slowly burned, not only carbon dioxide but also water should be produced, as in ordinary combustion. There is, however, a great deal of water always present

Water produced in respiration.

in the living respiring plant, and it would require very
accurate methods indeed to detect a small increase.

We can nevertheless obtain indirect evidence pointing to
the formation of water, by finding out how much oxygen is required in different cases for the formation of the same amount of
CO_2. We have seen that in some seeds the stored food is mainly
starch, while in others it is largely oil. Starch is a carbohydrate,
and so in order to burn it completely to carbon dioxide and water
only the carbon requires to be supplied with oxygen, the hydrogen
and oxygen already present being in the right proportion to form
water. In this case the volume of carbon dioxide produced by
combustion is equal to the volume of oxygen used up. Oils, on
the contrary, contain less oxygen and therefore, to burn them
completely to carbon dioxide and water, additional oxygen is
required to unite with hydrogen. Thus the volume of oxygen
absorbed is larger than the volume of carbon dioxide produced.

In order to find out whether the same is true in respiration,
Exp. 20   two similar large, wide test-tubes are fitted with
tightly fitting corks, through which pass glass tubes bent twice
and dipping below into coloured water (Fig. 12). Into one
test-tube is put a little germinating barley, and in the other a
similar quantity of germinating Hemp seed. The corks and tubes
are inserted firmly and the test-tubes are immersed together in a
jar of water so that the stoppers are covered. This keeps the
joints perfectly air-tight. It also makes it possible to observe
any changes of temperature that occur and ensures that they
affect both tubes equally. Changes of temperature alter the
volume of the air considerably and must be taken into account.
The best plan is to bring the water always to the same temperature, adding cold or warm water if necessary, before each
observation. The first effect of immersing the test-tubes is that
they are cooled down after contact with the warm hands: this
makes the coloured liquid rise in the vertical tubes. When it no
longer rises quickly the level is marked on each with a strip of
gummed paper (*a, a'* Fig. 12), and the temperature and time
recorded.

After from six to twelve hours the level of the water in the
tubes is again observed and the temperature noted. The barley
apparatus shows little change of level unless the temperature has

greatly altered; but in the case of the Hemp seed the water has
risen high in the tube (Fig. 12, left side) and may even have been
drawn over into the test-tube. Thus the oily seeds have absorbed
more oxygen than they have given out carbon dioxide, with the
result that the volume of air in which they have been respiring
has greatly diminished. On the other hand the starchy seeds
have only absorbed an amount of oxygen which is about equal

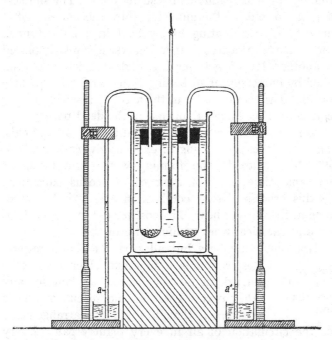

Fig. 12. Apparatus for Experiment 20.

in volume to the amount of carbon dioxide which they produce,
and so no change of volume has resulted. To make quite sure
that the barley has been respiring we may set up a third test-tube,
with barley, putting into it a small tube of potash to absorb the
carbon dioxide which is produced. This will show by the rise
of the water in the tube that much oxygen is absorbed, but that
its place is taken in the first test-tube by carbon dioxide.

We have remarked that plants do not, like warm-
blooded animals, keep up their temperature
Heat pro- above that of the air around them. Never-
duced in
respiration. theless some of the energy which is liberated
in respiration escapes in the form of heat.
In order to detect this it is necessary to prevent the heat
from escaping as readily as it usually does. The simplest
plan is to put a thermometer into the middle of a
quantity of germinating barley held in a large funnel.
For comparison[1] a second thermometer is similarly placed
in another lot of germinating barley which has been
killed by immersing it in boiling water, and subsequently
cooled. The thermometer in the actively respiring barley
registers 1–2° C. higher than that in the dead barley.

We have now seen that all plants, like all animals,
Con- respire, taking in oxygen, giving out carbon
clusion. dioxide and forming water, and losing at
the same time some of their carbonaceous substance.
By this process of slow combustion energy is liberated,
some in the form of heat, but most for use in the mechanical and chemical work of life and growth.

Not only all plants but all parts of plants respire.
Respira- Respiration is a function which cannot be
tion and carried on by any part of a plant for any
assimila- other part. There are no special organs of
tion. *respiration*, though there are arrangements
for the circulation of air in every plant organ. Every
living part of a plant respires for itself.

Parts which are not green are always giving out the
carbon dioxide which they produce and taking in the
oxygen which they require. Green parts, leaves and

[1] This control is necessary because a thermometer surrounded, for
instance, by a piece of wet rag registers a lower temperature than a
dry thermometer in the air, as the evaporating moisture robs the wet
thermometer of heat. Since the germinating seeds are moist the second
thermometer must also have its bulb in something moist.

shoots, do the same at night or whenever they are put in the dark. In the light, on the other hand, the oxygen which is produced in photosynthesis more than supplies their wants, so that they do not need to take in any from the air outside for respiration; while the carbon dioxide which they produce is all used up again in photosynthesis and none escapes.

Although, therefore, respiration is going on in a leaf during the day just as at night (or even faster owing to the higher temperature) yet the outward signs of respiration, the absorption of oxygen and evolution of carbon dioxide, do not appear, because in photosynthesis the reverse is taking place much more vigorously.

When the light is weak in the early morning, or when it wanes at evening, there is a short time when a leaf can only obtain just sufficient light to assimilate the carbon dioxide it produces by respiration: when this happens it cannot use any of the carbon dioxide in the air outside, and therefore takes in none; while the oxygen which is set free from the carbon dioxide just suffices for respiration, and none is either taken in or given out. In other words, there is at such times no exchange of gases between the air in the spongy tissue of the leaf and the air outside. Both respiration and photosynthesis are proceeding but they balance each other within the leaf, and no outward sign is given of the occurrence of either[1].

We have now seen that respiration and photosynthesis are two processes of the utmost importance in plants. In photosynthesis the green parts of plants take in carbon dioxide from the atmosphere and manufacture from it their food with the aid of the energy absorbed from sunlight by chlorophyll, storing this energy in the food and giving up again to the atmosphere the oxygen of the

[1] The same follows if the air supply contains no carbon dioxide.

carbon dioxide. In respiration all parts of a plant oxidise some of their food to free the energy contained in it for growth; the oxygen needed for this oxidation is absorbed by all parts of plants from the air around them (as we shall see later, air is even present in the soil around the roots). Green parts in the light are alone excepted, for they can supply themselves with more than sufficient oxygen by photosynthesis and do not absorb it from the surrounding air.

To enable us to keep our minds quite clear about these two processes they are contrasted together, point by point, in the following table:

Photosynthesis.	*Respiration.*
Green parts only.	All parts.
Only in light.	At all times.
Food (sugar and starch) manufactured.	Food oxidised and broken down.
Energy absorbed from sunlight by chlorophyll, and stored in sugar, etc.	Energy liberated for work of growth; some escaping as heat.
Carbon dioxide used.	Oxygen used.
Oxygen set free.	Carbon dioxide set free.
Water used.	Water produced.

Gaseous exchanges occurring.

Carbon dioxide enters from the atmosphere, oxygen passes into the atmosphere (except when the light is feeble, and the leaf cannot assimilate more CO_2 than is produced in its own respiration).	Oxygen enters from the atmosphere, carbon dioxide passes into the atmosphere (except in the case of green parts of plants in the light, when no exchange of gases with the atmosphere is needed for respiration, unless the light is very feeble, abundant oxygen being supplied by photosynthesis within the leaf itself, and all the carbon dioxide produced by respiration being used up again in photosynthesis).

CHAPTER VI

TRANSPIRATION

The need for water. It has been mentioned that for the manufacture of sugar and starch leaves need water, in addition to the carbon dioxide which comes from the air. They require, however, far more water than they use in this way. When a shoot is cut from a plant its leaves rapidly wither and dry up unless it is put in water. From this we infer that the leaves lose water rapidly, but that if the cut end of the stem is in water this loss is made good by the passage of water up the stem into the leaves.

Leaves give off water vapour. We must in this chapter study the behaviour of leaves in regard to water more closely; and in the first place we will obtain proof, by careful experiment, of the inference we have just drawn that leaves lose water in the form of vapour.

Exp. 21 This may be shown by putting a glass bell-jar or large bottle over a plant, after wiping the inside of it perfectly clean and dry. Moisture condenses in dew-like drops on the glass. But in order to be sure that the moisture comes from the leaves we must take precautions to prevent moisture passing from any other source into the air under the jar. There are several ways of doing this.

(1) If a potted plant be used, small enough for a bell-jar to cover both plant and pot, the pot and soil may be tied up in waterproof material. Gutta-percha sheeting is best for this purpose. It may be economised by putting the plant pot in a tin and covering the top of the tin with sheeting, tying the sheeting securely round the top of the tin and round the stem of the plant. If any doubt be possible about the air-tightness of the covering a control experiment should be set up, using a pot of soil with a piece of stick in place of the plant and tying on the cover exactly as before.

(2) If the jar used will only cover the plant and not

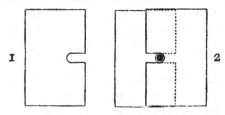

Fig. 13. Cardboard for Experiment 21 (2). 1, single piece; 2, two pieces overlapping, fitting round stem.

the pot, then the soil may be covered with two over-lapping pieces of thick cardboard, with places cut out of the sides which overlap, so that they fit round the stem (see Fig. 13). After stopping the hole round the stem and covering the joint between the two pieces of cardboard with plasticine or soft wax to make the cover quite air-tight, the glass jar is put over the plant, resting on the cardboard.

This plan can be adopted with plants growing in the ground. Only plants with erect stems leafless at the base are suitable, for otherwise a good joint cannot be made between cardboard and stem.

Several simple ways of demonstrating the escape of

moisture from leaves and shoots are possible with cut shoots, or single branches or leaves still on the plant.

(3) A tumbler of water may be covered with a piece of cardboard or tin having a hole in the middle, through which a cut shoot passes. The hole is blocked with wax, and a second tumbler inverted over the shoot.

(4) A branch or long-stalked leaf is selected, and covered while still attached to the plant with a dry bottle; the stalk or stem is securely fixed in the neck of the bottle by a split rubber stopper. The bottle must be suitably supported unless light enough not to injure the branch or leaf by its weight.

Whichever way the experiment is performed drops of water will soon be found clouding the inner surface of the glass jar, especially if the experiment is made in sunlight. Because of the precautions taken it may be inferred with certainty that this water cannot have come from any other source than the shoot.

To determine what part of the surface of the shoot plays the greatest part in this giving up of water vapour we may observe the effects produced by greasing different parts with vaseline and so making them waterproof.

First, two similar shoots should be selected and the **Exp. 22 a** stems put through holes in corks or rubber stoppers that fit into test-tubes of water, so that the cut ends of the stems dip well into the water. If each shoot so set up is weighed at once (care being taken that the test-tubes, etc., are quite dry on the outside) and again after say five hours, the weight lost by each will be a sufficiently accurate measurement of the amount of water vapour escaping during that time from the whole surface of the stem and leaves. If we now grease the stem of one and the leaves of the other, then weigh them again with the vaseline, and once more after five hours, we shall find that, while the shoot with greased leaves has lost very little in weight, the other has lost about as much as before[1].

By the same method it may be shown that most of the water vapour usually comes from the under surface of the leaves where, as we have seen in dealing with starch formation, the pores called stomata are present. When leaves having stomata only on their under side are well greased on that side so that the stomata are blocked, very little water is lost compared

Water vapour escapes chiefly through stomata.

Exp. 22 *b*   with the loss which takes place from similar leaves with only their upper side greased and the stomata therefore open. The leaves of some plants, like the Sunflower, have stomata on both sides and water vapour escapes from both surfaces; such leaves will, of course, give a different result.

These experiments show that water easily and quickly passes out through the stomata but slowly and with great difficulty through the skin of the leaf. Some leaves have thinner skins than others and water escapes through these thinner skins more quickly; but by far the greater part of the water vapour lost even by such leaves passes out through the stomata.

A much simpler way of showing how large a part the stomata play is to expose detached leaves to the air and watch their appearance, giving them meanwhile no water. Those which have been greased on the

Exp. 23

---

[1] If a balance weighing to the nearest milligram is available, the loss of weight of green herbaceous shoots during shorter intervals of one or two hours may be measured, especially if they are exposed meanwhile to sunshine; or single leaves with long stalks, like Lime leaves, may be used. With the large leaves of the Indiarubber plant, *Ficus elastica*, the experiment may easily be performed with much less sensitive balances and the leaves need not be supplied with water, but the cut end of the stalk must be sealed with wax or bound with rubber tubing. Leaves of evergreens, like the Cherry Laurel or *Aucuba*, will also serve, but must be left longer between the weighings.

side with stomata remain fresh for a long time, while those with the other side greased wither and dry up as quickly as leaves not greased at all.

A very rapid method is to apply to the surface of Exp. 24 a leaf pieces of filter paper soaked in a solution of cobalt chloride and thoroughly dried. When moist this substance is pink, but when dried it turns bright blue (cobalt blue). If blue cobalt paper is left in the air it absorbs moisture from it and turns pink, unless the air is very dry: contact of the paper with the air must be avoided in testing leaves. The paper is, therefore, freshly dried, and placed on the leaf between two pieces of glass which have been thoroughly wiped with a warm dry cloth. One leaf may be enclosed between two pieces of paper and two pieces of glass, held together by large spring paper-clips. The paper in contact with the stomatal surface turns pink in a few minutes, and the pink colour soon spreads beyond the edge of the leaf on both pieces of paper; but the part of the other piece which is in contact with the surface without stomata remains blue for much longer, especially if the surface of the leaf is smooth and even, without hairs, so that no space is left through which vapour could spread.

The stomata are, therefore, the chief paths by which water vapour passes out of the leaf. *This escape of water vapour through the stomata* is called *transpiration*.

We have next to show by exact experiment that a cut Cut shoot in water absorbs water to replace that shoots lost as vapour.
absorb   This may be demonstrated very simply water. by putting the stem through a hole in a rubber stopper which fits tightly into the neck of a Exp. 25 bottle of water. When the apparatus is made quite tight, soft wax being used if necessary, no

water can be lost by direct evaporation. The level of the water is marked at intervals on a strip of gummed paper stuck vertically on the bottle or read on a scale fastened securely to it.   The fall of level shows that the shoot absorbs water. This apparatus may be put under a bell-jar and absorption and escape of water both demonstrated together.

Another method is to take two similar jars of water, putting a shoot into one and stuffing the mouths of both with cotton wool. Here direct evaporation is retarded by the cotton wool, but not entirely prevented. The amount lost by direct evaporation is shown by the fall of level in the control jar. The fall of level in the other jar containing the shoot is much greater.

The water which is absorbed by a shoot passes in at **Path of water in the plant. Vascular bundles.** the cut end of the stem (this we can show by stopping up the cut end of a shoot with melted wax before putting it in water) and escapes as vapour from the stomata all over the under surface of the leaves.   It must, therefore, be conducted up the stem through the leaf-stalk to all parts of the leaf.

The path of the water can be traced by putting a shoot in water coloured by a dye, e.g. eosin, or red ink.

Exp. 26   A good plant for the purpose is the White Deadnettle. Shoots are cut and left over night in the coloured water[1]. Next morning the veins of the leaves are found to be coloured red, and if flowers are present red streaks are found on them also. In leaves, therefore, the veins conduct water, and similar veins are present in the flowers. When the stem is cut across some

[1] If cut under water and put where they will transpire vigorously they need not be left more than a few hours. If leaves of Tropaeolum (Nasturtium of the garden) are treated in this way their principal veins become stained in a few minutes.

distance above the level of the water, where the stain
cannot have soaked in from outside and coloured all the
tissues, it is found that the
red colour is confined to small
areas towards the outside of
the section (compare Fig. 14).
By carefully scraping away the
outer tissue from a node and
leaf-base of the stained shoot,
it can be seen that coloured
strands from the stem pass out
into the leaf. There they run
along the veins, dividing up
into finer and finer branches
so that they form an elaborate
network carrying water to all
parts of the leaf.

Thus the water is conducted
in the stem and leaves along
certain fixed channels. These
channels are small strands of

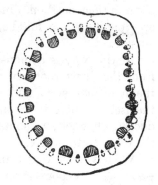

Fig. 14. Diagram of the appearance of the stem of a Sunflower when cut across,
after standing in coloured
water for some hours: only
the shaded patches, where
the bundles of water-con-
ducting tubes are situated,
are coloured.

hard tissue, consisting mainly of specially constructed
tubes. Each strand is a bundle of such tubes and is
called a *vascular bundle* by analogy with the vascular
system through which the blood of an animal flows. The
vascular bundles of a Plantain may easily be seen if the
leaf-stalk be broken across. They are so much tougher
than the rest of the leaf-stalk that they do not break
until the other tissues have been pulled widely apart, when
they are left hanging as thin strings of tissue.

The roots, too, have their conducting strands. This
can be shown if a plant is taken from the soil and placed
with only the lower part of its root-system dipping into
the coloured water. Wherever the roots are bathed in
the stain it gradually soaks in and colours all the tissues;

but, if the upper portion of the root is above the liquid, then as the liquid passes up it will mark out the conducting strands and on cutting the root across we can observe their position. In the root the conducting strands are found in the middle, not as in the stem near the outside.

In this way is demonstrated the important fact that from the finest roots up to the main root, then through the main root, stem, and leaf-stalks right up to the leaves themselves, there is a continuous system of special conducting channels, without a break, taking water up from the roots to the leaves.

We have seen that transpiration is the escape of water vapour through the stomata in the skin or epidermis of a leaf. The water vapour comes from within the leaf, where it is formed by evaporation of water from the spongy internal tissues into the air-spaces which are abundant there. Water standing in a saucer or dish evaporates at different rates under different conditions: it evaporates most quickly in the sun in dry windy weather, when the temperature is high and the air is dry and in rapid movement. We should, therefore, expect transpiration to be affected by these conditions.

Effect of conditions on transpiration.

In order to put this to the test of experiment we require to measure accurately the amount of water vapour which escapes in a given time under different conditions.

This may be done by measuring the loss of weight which accompanies transpiration. We have already used this method with a cut shoot in water to demonstrate the importance of stomata. A shoot set up as in Experiment 22 (page 59) may be weighed at intervals, being placed under a series of different conditions between the weighings. A plant growing in a pot (one with a

Exp. 27

large leaf surface should be selected) may also be used;
the pot and soil are covered with gutta-percha sheeting
as in Experiment 21 (1) (page 58).

We can, however, measure much more quickly and
readily the amount of water which a cut shoot takes up.
This is not exactly equivalent to measuring the actual
transpiration: a branch which is actively growing must
retain some of the water which it absorbs in order to
grow, for the young organs it is forming consist to a very
large extent of water, while a drooping branch absorbs
much more than it loses until it has completely recovered.
But after a branch has been in water for some hours it
usually absorbs about as fast as it transpires, and we
may use the rate at which a branch takes up water as
an indication of the rate at which it is losing water by
transpiration.

The methods we have already adopted for demon-
Exp. 28   strating that cut shoots take up water are
not delicate enough for our present purpose. We use instead an apparatus called a *potometer*, in which the water
absorbed is measured as it passes along a narrow tube.

A simple and convenient form of potometer is shown
in Fig. 15. It is constructed from a thistle funnel with
shortened stem, a bent tube, A, a T-piece, B, and a capillary tube, C, with a bore about a millimetre in diameter,
connected together with stout-walled rubber tubing and
mounted on a baseboard. The funnel and the shoot are
supported by tying them to the cross bar.

To use the apparatus, the end of the capillary tube is
first closed by attaching a short piece of rubber tubing and
a clip. Then the screw clip, S, is opened and water run
into the funnel till it fills the vertical arm of the T-piece
and its rubber connexion to the brim. The shoot[1] is now

---

[1] In summer, branches of many of our common trees or the shoots
of Sunflowers are suitable; in winter, branches of Portugal or Cherry

inserted, with care to avoid the trapping of air bubbles, and the rubber tube wired closely to it to make the joint air-tight. Finally the screw clip is closed, the end of the capillary unsealed and the whole apparatus filled with water from the thistle funnel by careful manipulation of

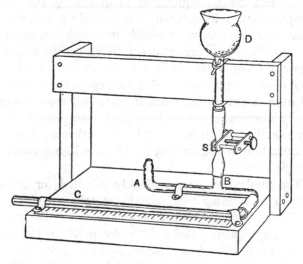

Fig. 15. Simple potometer. The shoot is attached at *A* and draws water along the capillary tube, *C*, over the graduated scale. (For description see text.)

the clip. The shoot now draws its water from the poto-meter and air follows the water as it recedes along the capillary tube. Under the tube is fixed a scale. When the air-water meniscus reaches the scale its position is read at regular intervals of time and the rate of its movement followed. If the intervals are so arranged that the meniscus only travels a centimetre or two between two readings, a

Laurel. It is best to cut them the day before they are required, leave them standing in water over night, and then, an hour or two before use, cut a small piece from the end, under water, so as to leave a clean fresh surface.

number of measurements of the rate may be made one after another. By cautiously opening the clip the meniscus can then be sent back again to the beginning of the scale and the readings continued. This can be repeated any number of times.

A potometer enables us to compare the rates at which water is absorbed by a cut branch under a series of different conditions such as the following, the average of several readings being taken in each case after the rate of movement has become steady: (a) in still air in a room near a closed window, not in direct sunlight; (b) after opening the window so that the plant is in a breeze—the meniscus moves much faster than before, showing that in moving air the leaves lose more moisture than in still air; (c) in still air in the shade, and then (d) in direct sunlight—the meniscus moves much faster in sunlight. Thus both wind and sunlight increase the rate at which water is taken up by the branch, and therefore the loss of water from the leaves. It is not surprising that when the apparatus is taken out into the open air and the branch is exposed (e) to sunlight and wind together, the meniscus moves faster still.

The leaves of many plants even in nature become **Stomata** visibly limp in hot dry weather, showing that **regulate** they have been losing water faster than the **trans-** roots could supply it. But plants cannot afford **piration.** an unlimited loss of water: under such conditions leaves close their stomata, and so protect themselves by reducing the loss of water to a minimum. After **Exp. 29** a leaf has been cut and allowed to become flaccid, a piece of cobalt paper applied to the under surface (as in Experiment 24, page 61) takes much longer to turn pink than a piece applied to a turgid leaf.

At night, when no photosynthesis is possible, many plants regularly close their stomata. In this way transpiration is reduced to a minimum and water is stored up in the tissues, replacing that which was lost during the day in excess of the amount provided by the roots.

The degree to which the stomata open depends moreover upon the intensity of the light; it is partly to this fact that the increased rate of transpiration observed in sunlight is due. By the cobalt paper method it may be shown

Exp. 30   that leaves taken out of the sunshine and put in deep shade gradually close their stomata. If a leaf be tested at intervals after the change of position, it is found that the blue paper takes each time longer to turn pink[1].

In this chapter we have seen that leaves, like other moist objects, are constantly losing water as vapour. A

Summary.   little is lost by direct evaporation through their skin, but most by transpiration through stomata, which occur in many plants only in the under skin of the leaves. Plants must therefore be supplied with water to make up for this great and continual loss, in addition to the amount they require for growth and assimilation. Cut shoots are able to absorb water for themselves through their cut ends. In a whole plant the supply of water must come from the soil by way of the root: we shall study the absorption of water by roots in the next chapter. The supply is not always fast enough to keep pace with transpiration; but the stomata are able to close, in order to diminish the loss of water, when the plant becomes in danger of losing too much.

---

[1] For these experiments small squares of paper are convenient, held to the leaf between ordinary microscope slides.

CHAPTER VII

THE WORK OF THE ROOT

Just as the shoot-systems of plants are remarkably
adapted to the work of photosynthesis, so
also their root-systems are constructed to
perform very efficiently the function of
taking up water from the soil. In the first
place, it would be difficult to imagine a better *arrangement of the roots* for this purpose. In the Sunflower and
similar plants a primary root grows vertically downwards,
and gives off secondary roots in a horizontal direction
on all sides. From these arise tertiary roots, running in
all directions between the secondary roots. The volume
of soil occupied by the roots is thus penetrated on all
sides by branches of the root-system. Some idea of the
large amount of soil which is thus brought within the
sphere of action of the roots may be obtained by measuring
the total length, as nearly as possible, of all the individual
roots together. The root-system of an Oat plant, spreading
through not more than a cubic yard or two of soil, has
been measured up to a total length of 150 yards by
putting the fibres end to end.

Further, roots come into still more intimate contact
with the soil by *root-hairs*. These are very delicate and
are often damaged in digging roots up and washing the
soil off them; but they are well shown by Mustard or

The large surface for absorption.

Radish seedlings grown on damp blotting paper (a dark coloured paper shows them up best).

In these plants the root-hairs are long as well as very numerous and form a white felt on the root extending for nearly half a centimetre outwards on all sides. In other

plants, such as the Bean, they are short and only just visible to the naked eye. If Mustard seedlings are grown in light soil they can be removed without tearing off the hairs. It can then be seen how the root is covered with particles of soil wherever it bears root-hairs, and that these particles cling so closely that it is difficult or impossible to wash them all off. This shows that the root-hairs come into very close contact with the soil, and that the particles stick to their some-what slimy surface.

If we examine older seedlings of Mustard we find that the soil clings only to the younger portions of the main root. The hairs which formerly covered the upper part have disappeared, and the less delicate secondary roots have taken their place. Soil clings to these roots, as to the main root, for a certain distance backwards from the tip, but the tip it-self, where growth is taking place, is free from hairs (Fig. 16). The meaning of their absence is not difficult to understand:

Fig. 16. Seedlings of Mustard: *A*, with soil adhering to the root-hairs; *B*, with root-hairs free from soil. (After Sachs.)

the hairs, being very delicate, would be torn off if they were produced by a part of the root which was still growing in length and pushing its way through the soil. Farther back, where the root has grown older, the root hairs have died; their work is finished. Thus as the root grows forward through the soil the zone of root-hairs also moves forward.

The root-hairs evidently increase very greatly the amount of soil with which the root comes into actual contact. Every one of the fine root-fibres of the Oat plant with 150 yards of root was clothed with numbers of such hairs, increasing its surface enormously. By means of these hairs the plant comes into the closest possible contact with the particles of soil.

We have now to find out what we can about the way **The pumping activity of roots.** in which this elaborate root-system works. Let us consider first the absorption of water. We know that a cut branch can absorb water for its own needs through its cut end. In the uninjured plant the water must enter by way of the root; but what part does the root play? Is it merely a passive channel or does it actively absorb the water and send it up into the stem? This we determine by the following experiment.

A healthy plant growing in a pot is selected, with a **Exp. 31** stem which for a short distance above the soil is smooth and free from leaves. The pot and stem are immersed in a pail of water and the stem cut through, under water, a few centimetres above the soil. While still under water a piece of rubber tube is slipped over the stump and a glass tube inserted in it. The pot is now taken out of the water, the rubber tube bound firmly with copper wire to make the joints water-tight, and the level of the water in the tube marked with gummed paper. It is sometimes found, in experiments such as this, that the level of the water falls in the tube for a short time. This is usually after the leaves have been transpiring very vigorously and have drawn heavily on the store of water in the stem. Water passes down into the stump, and even into the root itself, to replace that lost. But soon the water beings to rise in the tube, and may continue to rise for many hours or even days. Thus *the root pumps water up into the stem.*

It is well known to gardeners that the water which is pumped up by the root sometimes escapes from wounds or cut branches. This *bleeding* is most marked in spring when plants awake from their winter sleep and sap flows very abundantly to supply the opening buds and new shoots. Some plants, such as the Vine, may lose large quantities of water in this way. For this reason fruit trees are not as a rule pruned in spring, but in autumn when the flow of sap has diminished.

The force with which the root can pump can be
Exp. 32   measured by making it lift water or mercury.

(1) If a long straight narrow tube is attached as before, in a vertical position, water will be pumped up the tube by the root. Additional lengths of tube must be attached if the water rises to the top of the first length. It rises more and more slowly, however, as its height increases until it comes at last to a standstill. At this point the downward pressure of the water column just balances the pumping force of the root. By measuring the height of the column we therefore obtain a measure of the pumping force as equivalent to the pressure of so many centimetres of water.

This force is called the *root-pressure.* It varies widely among different plants, some having a very low, others a very high root-pressure. Fuchsia plants raise water often to a height of more than 250 centimetres, Vines to 1000 centimetres (10 metres) or more, and with such plants it becomes more convenient in measuring the root-pressure to use mercury, which is 13·6 times as heavy as water.

(2) To measure root-pressure by means of mercury the simplest way is to make use of a piece of ordinary glass tubing of not too wide a bore, bent in the form shown in Fig. 17. The long vertical arm of the U-tube must be higher than the mercury can be raised by the root-pressure (two feet is sufficient for most purposes) and the shorter arm about half that length. A funnel is attached to the end A by a piece of rubber tubing, and water poured in till it fills the whole tube. The other end C is then loosely inserted in a sleeve of rubber tubing attached to the stump of a plant and full of water. Sufficient mercury is now poured into the funnel, and the water it displaces allowed to escape from the rubber sleeve till it reaches the same level h in both arms of the U-tube. Finally the tube is pushed home into the rubber sleeve, the joints made secure with wire, and the funnel removed. As the root pumps water into the tube the mercury is depressed in B and raised in A. When no further change occurs the difference of level in centimetres gives the root-pressure in centimetres of mercury.

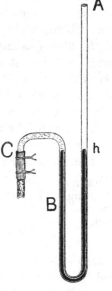

Fig. 17. Apparatus for measuring root-pressure. The root forces water through the stump into the bent glass tube ABC, depresses the mercury in B and raises it in A, its original level being h, the same in both sides of the tube.

It is often convenient to express the result in terms of the height to which the pressure of the atmosphere will raise mercury or water as the case may be. A barometer shows that the atmospheric pressure can raise a column of mercury on the average to a height of about 76 centimetres. The pressure of 76 centimetres of mercury is called *one atmosphere*. If the root-pressure raises mercury 38 cms. it is equal to half an atmosphere; if 16 cms., 0·21 atmosphere; and so on. On the other hand the pressure of the atmosphere will support a column of water over 1000 cms., i.e. about 10 metres high; so that a root, like those of Fuchsia plants, which raises water 250 centimetres has a root-pressure of a quarter of an atmosphere. A Vine whose root-system is capable of raising water to a height of more than 10 metres has a root-pressure of more than one atmosphere.

The roots which can pump against a great pressure are not always the roots which can pump fastest. Under natural conditions what most plants require is that their roots should supply water rapidly. In most cases when the leaves are transpiring vigorously the roots cannot absorb water fast enough to keep pace with transpiration, and the result is, as we have seen, that the water-content of the stem is reduced. Conditions in the soil that affect the rate at which roots can absorb are therefore of importance.

Conditions affecting absorption.

(1) *Temperature of the soil.* By applying the method of Experiment 31, one condition of the active absorption of water by roots may easily be demonstrated. If the pot is surrounded by ice practically no rise of water is observed. This fact is of great importance in countries like our own, for in winter, when the soil is cold, especially when it is frozen, absorption is very difficult. This is the chief reason why most trees drop their leaves in winter and most other plants die down (p. 107).

(2) *Available water in the soil.* The amount of water present in the soil varies very much in different places and at different times, and this fact influences the absorption of water by the roots.

If a plant growing in a pot is left unwatered it will in time begin to droop: the roots cannot obtain sufficient water from the soil. If some of the soil is taken from the pot and weighed, then dried in an oven and weighed again, it is found to have lost quite a considerable amount of water in the oven although it felt quite dry and powdery. Some of the water which is present in the soil is thus held too fast to be absorbed by the roots of plants. This part of the soil-water is distinguished as *non-available water* from the *available water* which plants are able to absorb.

The water held by soil that is powdery and apparently dry is present as thin films round each particle. In moist soil the water also fills the spaces between the particles and holds them together, so that well watered soil becomes 'sticky' and will often hold together in large lumps. Much of the water soon drains through, however, and after a time the soil readily separates into small crumbs. As more water gradually evaporates or is absorbed by plants, these crumbs separate into smaller crumbs and particles. Small particles retain water and hold together longer than larger ones: sandy soils, which do not contain much finely divided matter, allow water to drain through them more readily and hold far less than clay soils which consist in great part of extremely fine particles.

Soils may readily be compared in this respect by shaking up equal quantities with water and allowing the mixtures to settle in tall narrow glass jars. The larger stony particles rapidly sink to the bottom of the vessel followed by smaller and smaller particles, while the very finest remain suspended in the water for a long time, making it appear cloudy, and only settle very slowly. In this way the particles of different sizes are separated out into layers at the bottom of the jar and the composition of different soils may be compared.   In sandy soil there is very little fine sediment.

(3) *Supply of air in the soil.*   There is a third important condition affecting the absorption of water by roots, in addition to the temperature of the soil and the quantity of available water it contains.   If a pot of soil is immersed in water, bubbles of air arise from the soil. Clearly, therefore, when water penetrates soil it takes the place of *air.*   To find how much air is present in soil we may fill a jar with the soil and pour water into it from a graduated measure until the water just appears above the soil, i.e. until it has entirely displaced all the air. The volume of the water used is of course equal to the volume of the air that was present.   The total volume of the soil can be found by measuring the water required to fill the empty jar.

Thus the soil in which a root lives and performs its functions consists of (1) the solid particles, (2) water, and (3) air.   Water and air together fill the space between the solid particles.   If this space remains the same, the more water present, the less room there is for air.

Now air is necessary to roots in order that they may obtain oxygen for respiration.   It is, therefore, possible to give plants too much water because the water drives out the air and too little air may be left in the soil for the healthy life of the root.   Plants in pots, if continually over-watered so that the soil remains *water-logged*, become sickly and may sooner or later die.   Roots can, in fact, be drowned.

The need for keeping roots well supplied with air is well known to gardeners, who always keep the soil around plants well stirred and open so that air can penetrate and oxygen and carbon dioxide readily circulate between the roots and the atmosphere.

The absorption of water by roots from the soil depends,

then, on three conditions: (1) the amount of available water present in the soil, (2) the temperature of the soil, and (3) the aëration (or ventilation) of the soil. It follows from this third condition that only living healthy roots can absorb water efficiently.

The root-systems of plants with their numerous fine branches, as we have seen, explore the soil very intimately, and by means of the fine hairs that clothe the delicate ends of the root-fibres expose a greatly increased surface to the soil and to the water in it. These elaborate root-systems absorb water and pump it up to the stem. We must next inquire how these processes are to be explained.

CHAPTER VIII

HOW ROOT-HAIRS ABSORB WATER.   OSMOSIS

In order to understand in some degree *how* root-hairs
Structure absorb water it is necessary to examine
of root- closely the structure of these hairs.
hairs.   For this purpose a microscope is required.
A small piece of the root of a Mustard seedling, grown on
moist blotting paper, is laid in a drop of water on a
slip of glass, and a coverslip of very thin glass is laid over
it.   Care must be taken that sufficient water is present to
prevent the trapping of air-bubbles under the coverslip,
and that on the other hand there is not so much water
that it runs over the slide on to the stage of the microscope or allows the coverslip to float off the preparation.
If there be any excess of water it may be removed with
blotting paper applied at the edge of the coverslip.

If we look at the surface of the root itself we see a
chambered structure.   The root-hairs extend from the
surface of the root, appearing like long flexible tubes;
but they are closed at the end.   It is clear that absorption
is not suction, as through the tubular proboscis of a
house-fly, or of bees and butterflies, or through the mouth
of other animals, for such suction depends upon the
activity of an elaborate muscular apparatus.   No such
apparatus is to be discovered in the roots of plants, nor
is there any aperture at the end, or in any other part of
a root-hair, through which water might enter in this
fashion.

Towards the tip of the root the hairs are shorter, being young; those nearest the tip are only just beginning to grow out. Here it may be seen that each hair arises as an outgrowth from one of the chambers, or cells, which form the surface of the root. These cells are in fact growing out into contact with the soil.

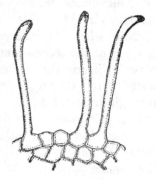

A very important property
Exp. 33 *a*   possessed by each of these cells can easily be demonstrated. If some five per cent. solution of common salt is put against one side of the coverslip, and drawn under it by applying a piece of blotting

Fig. 18.   A portion of a section through a young root, showing some of the superficial cells growing out into root-hairs. A thin layer of protoplasm (dotted) lines the cell-wall, and encloses the cell-sap.

paper to the other side, the root-hairs collapse. When on the other hand the salt solution is washed out again in the same way by fresh water, the hairs fill out and regain their former appearance.

This behaviour is common to all the living cells of
Osmosis.   which plants are chiefly built up. In order to study it more fully, it is convenient to use cells which can be obtained separate, and so more clearly observed under the microscope. The cells of a ripe Privet
Exp. 33 *b*   berry are loose and can easily be scraped out on the point of a needle. A few cells from just below the skin, where they are dark red in colour, are mounted in water, and salt solution drawn under the coverslip as before. Some of the cells may appear merely to collapse and become wrinkled like the root-hairs. Most of them, however, show something more: the red cell-sap appears to contract away from a colourless membrane, the *cell-wall*.

This shows that the cell-sap must be enclosed in a second inner membrane, which prevents it from mixing with the colourless liquid outside.

Now what does the collapse of cells, and the contraction of the second membrane containing the cell-sap mean?   Clearly the volume of the cell-sap is diminished: water must, therefore, have passed out from the sap, through the membrane, into the salt solution outside, without the colouring matter dissolved in it passing out too.

When the salt solution is washed out again with water the cell recovers its former size and appearance: the volume of the cell-sap increases again, because of the entry of water, so that the inner membrane is extended, and brought into contact with the cell-wall.   The cell-wall is finally distended again even where it had become wrinkled.

Water passes out of the cell-sap into a strong solution of salt, but into it from pure water, through the inner cell-membrane.

This curious phenomenon can be imitated on a large
Exp. 34   scale with animal membranes.   If a pig's bladder is filled with water, tied up, and put into salt solution, it collapses: water passes out of the bladder into the salt solution.   If on the other hand the bladder is filled with the salt solution and put into water, water enters the bladder and it swells, becoming *tense and hard*, just as a bicycle tyre becomes hard when inflated.   It may even be so much distended with water that it bursts. If a glass tube be tied into the mouth of the bladder, the salt solution rises in the tube as the water enters the bladder.

In the water, outside the bladder containing salt solution, only a little salt can be detected even after a long time.   We may repeat the experiment using sugar instead of salt, and again water passes into the solution through the bladder, but very little sugar passes in the reverse direction.

Animal bladder is thus a membrane which allows water to pass through it readily enough, but sugar and salt, and most other substances dissolved in water, only with great difficulty. The peculiar phenomena demonstrated in our experiments are due to this property. Were it not for the bladder (if, for instance, the solution were contained in a pot of porous earthenware), the sugar or salt would diffuse from the solution into the surrounding water, until finally a solution of uniform strength would result. This diffusion, however, is prevented by the bladder, but water passes instead into the bladder, tending to produce the same result, namely, to make the solution weaker.

Sugar or salt tends to diffuse not only from a solution into pure water but also from a strong solution into a weak solution. If the two solutions are separated by a membrane which hinders the diffusion of the sugar or salt, water again passes through instead. If a bladder is filled with a two per cent. salt solution and immersed in a five per cent. solution it collapses, though less quickly than when it contained pure water. Here salt tends to diffuse into the bladder, the bladder prevents this diffusion, and water passes through in the opposite direction instead. Whereas the dissolved substance would diffuse from the strong into the weak solution, *water* passes *from the weak into the strong solution.*

We have seen that the water entering the bladder distends and may even burst it. This means that the *water enters even against strong resistance,* for it needs a considerable force to burst so strong a membrane.

The passage of water through a membrane under these conditions is called *osmosis,* and the force which it is capable of developing (measured by the resistance which it can overcome) is called the *osmotic pressure*. We may call the membrane, which allows the water but not the substances dissolved in the water to pass through it, the *osmotic membrane*.

In the cell of the Privet berry the sap must be a solution, because, when it is surrounded by pure water, water enters: on the other hand it must be weaker than a five per cent. solution of salt. The osmotic membrane is the inner membrane, not the cell-wall: for the cell-wall allows the salt solution to pass through it when the inner membrane collapses. This delicate membrane prevents salt from diffusing into the cell-sap, and the substances dissolved in the cell-sap, including the colouring matter, from diffusing out.

Therefore, when a cell is in contact with water, water tends to enter it and press the osmotic membrane against the cell-wall. When there is plenty of water within reach of the cell, water continues to enter until the resistance of the stretched inflated cell-wall equals the osmotic pressure. It is thus the cell-wall which prevents the inner membrane from bursting. In this inflated condition a cell is said to be *turgid* (Fig. 19, 1).

So far we have not seen the inner membrane, but **The living proto- plasm.** have only inferred that it is present from the behaviour of the cell-sap. We can, however, see it in the inner cells of the Privet berry, in which the cell-sap is colourless. If a few of these cells are Exp. 35 mounted in water, and salt solution run in, a very thin, slightly granular layer is observed to contract away from the cell-wall. If a dilute solution of iodine is drawn in, the membrane stains faintly brown, because it consists of *proteïns* (see p. 44): it is called the *protoplasm*. At one part of this proteid membrane is a rounded, more deeply stained body, called the *nucleus*.

The protoplasm and nucleus are the living matter of

the cell. The behaviour of the protoplasmic membrane changes entirely when it is killed. If red cells from a Privet berry are mounted in water and *alcohol* run in, the protoplasm is killed, and the red colour escapes.

Exp. 36   This fact may be demonstrated still better with red beetroot which consists almost entirely of cells containing bright red sap.   When a small cube is cut from the inside of a beetroot, well washed in water to remove the sap of the damaged cells, and then held up,

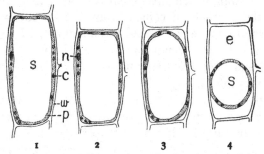

Fig. 19. Stages in the collapse of a cell when immersed in a strong solution: in (1), it is fully turgid, with the cell-walls distended and so pressing upon the protoplasm; in (2), the cell-wall is no longer stretched, and the cell is therefore flaccid but the protoplasm is still in contact with the cell-wall; in (3) plasmolysis has begun, and in (4) the protoplasm has rounded off and only remains in contact with a small part of the wall. *w*, cell-wall; *s*, cell-sap; *p*, protoplasm; *n*, nucleus; *c*, chloroplast; *e*, solution which has passed through the wall of the cell. (After De Vries, modified.)

the drop of water which drains from it is quite colourless. If it is then dipped for a few moments into boiling water and again held up the drop which collects is red and its colour quickly deepens. The dead protoplasm no longer prevents the substances dissolved in the cell-sap, including the colour, passing through it; and some of the sap which was distending the cell escapes. For the same reason beetroots which have had their outer skin broken lose their colour when cooked.

The living protoplasm is a delicate semi-fluid substance

Turgidity: the cell-wall. rather like a thin jelly. Such substances are, of course, incapable by themselves of forming strong bladders. The protoplasmic membrane, therefore, needs the support of the cell-wall in order to withstand great osmotic pressures. If the cell-wall were not present it would quickly burst. Thus although the protoplasm alone is the osmotic membrane, a cell becomes *turgid* because of the strong and elastic cell-wall, which offers great resistance to stretching. When the protoplasm contracts away from the wall, in a salt solution, the cell is said to be *plasmolysed* (Fig. 19, 3 and 4). In this condition the wall is no longer distended, and the cell is limp and flaccid, like a bicycle tyre that wants inflating.

Plants are chiefly composed, at least when young, of living cells, and are limp or turgid according as their cells are limp or turgid. If a long strip is cut from a potato and soaked in water it gets stiff, because all the cells get very turgid; but when it is immersed in salt solution so that the cells are plasmolysed it soon becomes quite limp. Similarly a potato which is left out in the air gradually becomes soft because water evaporates from it and too little is left in the cells to distend the cell-walls. Many leaves, too, on a hot sunny day become limp through too rapid loss of water.

Young green shoots and leaves, therefore, depend for their mechanical support largely on the osmotic pressure set up in their own cells.

What is the bearing of all this on the way in which

Osmosis and the absorption of water. root-hairs do their work? Root-hairs are outgrowing parts of the surface cells of the young root. They are in contact with the water which is present in the soil between and around the particles. Each hair will thus absorb water for itself, until the osmotic pressure cannot inflate the cell-wall any further.

But it is not enough to show how each root-hair can absorb water into itself. Somehow or other the water is pumped up the root into the stem. It must, therefore, pass from the root-hair through the inner cells of the root to the conducting tubes in the middle. If a piece of potato, consisting of numerous cells closely packed together, is put into salt solution the whole piece becomes limp through and through. Water passes from the outer cells directly into the salt solution; as a consequence of this the cell-sap becomes stronger and the walls less inflated, so that these cells are able to absorb water from those just within. These inner cells in their turn absorb water from cells farther in, and so water passes eventually from the innermost cells of the piece out into the salt solution. If the limp piece is now put into water the opposite process takes place: the outer cells absorb water, which inflates them and makes their cell-sap weaker; therefore water passes into the cells within, less squeezed by their walls and with stronger sap, so that eventually the whole piece becomes turgid again.

In the same way, as a root-hair absorbs water its wall is stretched more and squeezes it harder, also the cell-sap becomes weaker, and so the cells adjoining it are able in their turn to absorb water from it. Water would thus pass from cell to cell until all the root cells are fully turgid and can absorb no more.

We have still not explained how the root pumps water up into the stem, for the channels through which the water flows, in the vascular bundles, are dead tubes, without any living protoplasm in them. Even if we imagined them, in the root, to contain at first a strong solution which would be capable of attracting water from the neighbouring cells, the passage of the large amount of water, which we know does pass up through them, would very soon dilute the solution and wash it away.

Therefore, though osmosis helps us to understand how the root-hairs *absorb* water from the soil it does not explain root-pressure, the active part which the root takes in supplying water to the stem. The living cells of the root must be capable of giving water out again as well as of absorbing it. Cells are, in fact, known which can take in water at one end and exude it at the other; it is probable that cells are present in roots which can do this, and that these pump the water up the water tubes.

Just as the root-hairs absorb water from the soil by osmosis so also every other living cell of the plant absorbs water for itself, either directly from the conducting tubes of the vascular bundles, or from neighbouring cells. The cells of a leaf keep themselves turgid in the same way although the leaf loses water in transpiration. As water passes from the cells as vapour, so they become slightly less turgid and draw upon the supply of water in the water tubes. The more rapid is transpiration, the less turgid do the cells become and the more strongly do they draw water in. All the cells of a transpiring shoot absorb water in this fashion, and the result is that the shoot as a whole absorbs water through its cut end, as we have already demonstrated.

In conclusion, it is important to remember the following points:

Summary.   (*a*)   There are two factors which can be shown to take part in the raising of water from the root to the leaves in a plant: (1) the pumping force of the roots—root-pressure, (2) the suction force of the leaves which is set going when water is lost by evaporation.

(*b*)   Water *enters living cells*, whether root-hairs, the cells of leaves, or other cells of plants, *by osmosis*.

(c) This osmosis depends upon (1) the osmotic membrane—in plants the living protoplasm—which allows water but not the substances dissolved in the cell-sap to pass through it, (2) the presence, in solution in the cell-sap, of substances which cannot pass through the protoplasm, and (3) a supply of available water, whether it be in the soil, in the water tubes or in a neighbouring cell.

(d) Root-pressure and the active pumping of water up into the stem from the root are not explained by simple osmotic absorption but depend in part upon powers possessed by the *living* protoplasm which do not belong to dead osmotic membranes like pig's bladder.

CHAPTER IX

THE MINERAL FOOD OF PLANTS

Mineral substances absorbed in solution. Plants obtain from the soil not only water but the substances of their ash. Roots must therefore absorb some of the solid constituents of the soil, as well as water. How does this happen?

Roots are made up of cells, all of which, including the root-hairs, are enclosed by cell-walls. These walls will not allow solids but only liquids to pass through them. It is clear, therefore, that the mineral substances of the plant must enter, just as the manufactured foods can only be translocated, *in solution in water*.

Protoplasm permeable to some substances. The protoplasm of the root-hairs must, therefore, allow the necessary mineral substances to pass through it, although it prevents substances dissolved in the cell-sap from diffusing out. In other words the protoplasm is *permeable* to some substances and *impermeable* to others.

That this is so can be shown by experiment. If **Exp. 37** cells from the outer part of a ripe Privet berry are mounted in a fifteen per cent. solution of *glycerine*, they at first collapse. Later, however, they recover and gradually become turgid although the solution of glycerine still surrounds them. Cells plasmolysed in salt solution do not recover, however long they may be left. This shows that the glycerine slowly passes through the protoplasm into the cell-sap. Because it can only pass through very slowly, the strong solution at first draws water from the weaker cell-sap and plasmolyses the cell. As the glycerine slowly diffuses through the protoplasm, the concentration of the cell-sap is thereby increased and water passes back again into the cell. Finally, the concentration of glycerine within the cell becomes the same as outside and the cell becomes as turgid as when no glycerine was present.

Among other substances to which protoplasm must be permeable are the sugar and the simple nitrogenous food-substances which are translocated: these have to pass frequently through protoplasm on their way to and from the vascular bundles. This does not mean that always in every cell the protoplasm is permeable to these substances, for the living protoplasm is able to adjust itself to varying needs and it is possible that it may allow a substance to pass through at one time readily, at another time with great difficulty. Dead membranes, like pig's bladder, do not show this power of adjustment although they also allow some dissolved substances to pass through them fairly readily: pig's bladder even lets common salt through in time, though extremely slowly.

The protoplasm of the root-hairs, then, allows certain substances to enter in solution in water. The water in the soil contains very small quantities of mineral substances dissolved in it; it is thus a *very dilute solution* which is absorbed by the root-hairs and is carried up to the leaves. Large quantities of water are absorbed, only to be lost again as vapour; but the mineral substances are left behind and accumulate in the leaves (except for the portion of them that diffuses out into the surrounding cells of the stem, etc., on the way up). Thus transpiration, an in-

evitable consequence of the exposure of moist shoots to
the air and often fraught with danger, is not wholly useless
to the plant.

The substances absorbed must include in their com-

**Elements
present
in the
ash of
plants.**
position the chemical elements, other than
carbon, hydrogen and oxygen, which are
constituents of the protoplasm, namely
nitrogen, sulphur, and *phosphorus*; but the
composition of the ash of plants shows that
other elements also are absorbed. Among these are always
found the following: potassium, sodium, magnesium, calcium, iron, chlorine, and silicon.

It does not necessarily follow, because these elements
are always found in plants, that they are all essential
to its life. This point can only be determined for each
element by trying to grow plants without it. The
method which is used for this purpose is to grow plants
with their roots in dilute solutions; this method is called
water-culture.

Water-culture. It is best to use young healthy seedlings
Exp. 38   which have been allowed to germinate in clean,
moist sand till their roots are well grown. Ordinary soil sticks
to the roots so closely that it cannot be thoroughly washed
off: if left on, it would introduce into the solutions unknown
substances and so prevent any proper conclusion being reached.

If the seedling has a great quantity of reserve food in the
seed this will supply all the minerals required for several weeks
and the plants will flourish in any solution, even in distilled water.
Seedlings are therefore chosen with little reserve food (e.g.,
Wallflower or Willow Herb) or young cuttings of Fuchsia, or, if
seedlings like those of the Bean, Pea, and Maize are used the
reserve food in the seed or the grain is cut off after the shoot has
grown up and one or two leaves have expanded.

The plants are grown in bottles of about a litre in capacity
(ordinary glass jam jars answer well). The bottles must first
be very carefully washed. After ordinary scrubbing and washing
they are rinsed out with strong nitric acid to remove from the

surface of the glass traces of chemical substances, which contain some of the elements to be experimented upon. The nitric acid is removed by washing well with tap water and finally the tap water is washed out with pure distilled water.

Another precaution is necessary to ensure successful cultures. Failure is most often due to disease attacking the plants, which decay and die. In order to prevent this as far as possible, all germs of disease (the spores of bacteria and mould-fungi) lurking in the bottles or in the culture solutions must be killed—in other words, the bottles and solutions must be sterilised—and other germs floating in the air must afterwards be kept out. The bottles are sterilised by rinsing with a solution of corrosive sublimate (a deadly poison) and this is then removed by repeatedly washing with distilled water which has itself been sterilised recently by boiling.

An alternative method which answers most purposes is to coat the inside of the bottles with paraffin wax. After being washed, the bottles should be left to drain and when thoroughly dry placed in warm water reaching nearly to the brim, till well warmed through. Melted paraffin wax is then poured into each and the bottle is turned round and round on its side until completely coated; any excess of wax is poured out.

The bottles are afterwards left upside down until required for use, so that no dust or germs can settle in them.

The culture solutions are made up with pure chemicals and pure distilled water, and are sterilised by boiling for half-an-hour in flasks with their mouths loosely plugged with cotton wool. Any water which boils away is made up by adding boiled distilled water.

The seedlings are supported by corks which fit the necks of the bottles or pieces of sheet cork resting upon them. If the bottle has a wide neck, several vertical grooves may be cut into the edge of the cork and a seedling inserted in each, with only its roots dipping into the solution, and the lower part of its stem held in the groove by a small piece of clean cotton wool (or, better, asbestos wool, which can be dried and sterilised in the flame of a Bunsen gas-burner); or a single seedling can be fixed by the same means in a hole in a split cork.

In either case it is convenient to have a piece of glass tube passing through the cork nearly to the bottom of the culture solution, so that the roots may be regularly supplied with fresh air for respiration. A small plug of cotton wool in the upper end of this tube will prevent the entry of dust and germs. Air is slowly bubbled through the solution for a few minutes each day, by means of a rubber bulb, such as is used in spraying scent, etc. (or a bicycle pump), attached to the glass tube, without removing the plug of cotton wool. Great care must be taken not to splash the cotton wool or asbestos around the seedlings, for if this gets damp it enables germs to flourish, and the seedlings are 'damped off.'

The roots should be darkened by wrapping the bottle in opaque paper or black cloth.

The solutions should be changed every ten days or fortnight.

It is advisable to place by the side of plants growing in the solutions, one growing in distilled water, to make sure that this contains nothing in solution. If so, the plant will soon cease to grow.

The culture solutions. It has been found in water-culture experiments that either of the two following solutions enables plants to grow healthily and produce flowers and seeds as in ordinary soil; these solutions are therefore called *normal solutions*:

I.			II.		
Sachs' solution.			*Knop's solution.*		
Potassium nitrate ..	1	gram	Calcium nitrate ..	4	grams
Sodium chloride ..	0 5	,,	Potassium nitrate ..	1	gram
Calcium sulphate ..	0·5	,,	Magnesium sulphate	1	,,
Magnesium sulphate	0·5	,,	Potassium phosphate	1	,,
Calcium phosphate	0·5	,,	Distilled water, 50 cc., forming		
Distilled water ..	1 litre		a stock solution, 2 or 3 cc.		
A drop or two of a solution of			of which are diluted for		
iron chloride*.			use to one litre with dis-		
			tilled water, and a drop		
			or two of iron chloride		
* Or an iron nail.			solution* added.		

If we compare these two solutions we see that they both contain the same elements, though combined in different ways, with the exception of sodium and chlorine which are present, as sodium chloride, only in Sachs' solution. This means that neither sodium nor chlorine is really essential; for plants can grow equally well in Knop's solution which does not contain them. Similarly, neither solution contains any silicon.

These successful cultures also prove conclusively that

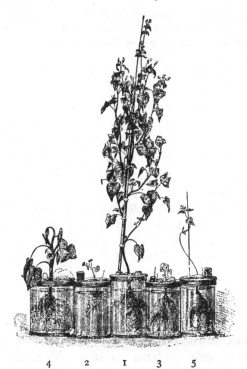

Fig. 20. Water-cultures of Buckwheat, after Nobbe. 1, in normal culture solution; 2, without potassium; 3, with sodium instead of potassium; 4, without calcium; 5, without nitrogen.

plants do not need to be supplied with carbon by their roots (see p. 17) for the normal culture solutions contain no carbonaceous substance.

On the other hand, if any other element be omitted the plants are small, feeble and unhealthy; sooner or

later they die without flowering.   Fig. 20 shows the kind
of results obtained in a careful experiment of this kind.
The plant in the centre was grown in a normal solution,
each of the others contained everything necessary with
the exception of one of the essential elements.

Potassium can be omitted from Knop's solution by substituting sodium nitrate and sodium phosphate for potassium
nitrate and phosphate; *nitrogen,* by leaving out calcium nitrate
and potassium nitrate and substituting calcium sulphate and
potassium chloride; *iron,* by not adding any drops of solution of
iron chloride.

The effect of omitting iron is very interesting.   The
new leaves that are produced are not green but
yellow.   In the absence of iron they are unable to make
chlorophyll.   If these yellow leaves are painted with a
dilute solution of iron chloride they soon become green
where the solution was applied.

As a result of water-culture experiments it has, then,
Essential   been found that plants cannot grow to
elements.   maturity, flower and set good seed, without
a sufficient supply of each of the following elements to
their roots: nitrogen, sulphur, phosphorus, potassium,
magnesium, calcium, iron.   Of these, the first three enter
into the composition of the protoplasm, and magnesium
is a constituent of chlorophyll.   In order to be able to
make chlorophyll, plants require iron as well, although
chlorophyll itself contains no iron; in a similar way
potassium and calcium may be necessary for other
important chemical processes.   Potassium, calcium and
iron are, in fact, necessary parts of the working machinery
of the living green plant, although we know little about
the part they play in the protoplasm.

Water-culture experiments also show that the form
in which these seven elements are supplied is very

important. Not all substances containing nitrogen are
suitable. Most plants can use best the nitrogen in nitrates,
such as calcium or potassium nitrate, but can also make
use of the nitrogen present in compounds of ammonia,
such as ammonium sulphate. Similarly, other elements
must be supplied in mineral salts like those used in making
up culture solutions. Of such salts there are only very
small quantities present in the soil-water. Some salts,
though abundant, are only very slightly soluble in water.
Other substances, particularly salts containing nitrogen,
are readily soluble, but the soil only contains a very
small amount of them at any one time: it is interesting
to inquire how the supply of these salts is kept up.

If the same kind of crop be grown year after year on
Manures. the same piece of land, the successive crops
become poorer and poorer, showing that the
soil is getting exhausted of the mineral salts suitable for
plants. One way in which this exhaustion is remedied
is by manuring. Farmyard or stable manure consists
largely of compounds containing nitrogen. Among these
are compounds of ammonia, which can be used to some
extent directly by plants; but most of the nitrogen is
present in more complex compounds, which along with
the ammonia are products of the decomposition of animal
matter, and these are of no immediate use to plants.
When mixed with soil, however, manure undergoes great
changes. The loss of its unpleasant odour is a sign of
these changes. They are brought about by minute
organisms called *bacteria*. Each bacterium consists of a
single very minute cell, only distinguishable when very
highly magnified; but there are billions of these bacteria
in a very small quantity of soil. They change gradually
the products of decomposition into *nitrates*, just the form of
nitrogen compound which green plants most readily use.

If an amount of nitrate equivalent to all the nitrogen

in the manure were used instead, the greater part of it would quickly be washed away before the plants could absorb it. The more complex nitrogen compounds that occur in the manure, on the other hand, being colloids (p. 46), are held in the soil, and nitrates are produced from them by the action of the bacteria little by little.

Mineral substances are, however, often applied directly to the soil in modern agriculture. The use of lime has long been known for 'sweetening' soil which has become 'sour' or acid for lack of it. Nitre (potassium nitrate) can be used with advantage under some circumstances. Bone manure, obtained by burning and crushing bones, is a valuable source of phosphorus; it contains phosphate of lime, which passes gradually into solution as it is absorbed by plants. Slags and other preparations which are rich in necessary minerals and yield them gradually in contact with water and air, are also much used.

Exhausted soil will recover its fertility in time if allowed to lie fallow (i.e. unsown). It is ploughed up to open it to the weather and so left. The carbon dioxide of the air has a chemical action upon certain constituents of the soil, freeing from it salts which are of use to plants, while alternate freezing and thawing during frosty weather break up the soil and still further expose it to this corrosive action. Other changes, however, also occur. In addition to bacteria which make the nitrogenous products of decay available to plants, there exist other bacteria which can absorb nitrogen directly from the air. By their activity the amount of nitrogen in the soil is increased and becomes available for green plants on the death of the bacteria or of the minute animals which devour them.

These changes are always proceeding even during cultivation, but in most soils are not rapid enough to keep heavy crops supplied. Rich soils, however, do occur—in parts of Germany, for instance—in which the nitrogen-fixing bacteria are so abundant that crops of Rye have been grown for many years in succession without exhausting the supplies of nitrogen.

Leguminous plants. Peas, Beans, Vetches, Clover, and other plants belonging to the Family Leguminosae have been found to increase instead of exhaust the supply of nitrogen in the soil.

When the plants are quite young, small swellings, called tubercles or nodules, grow on their roots (Fig. 21), caused by bacteria which, like the soil bacteria already mentioned, have the power when living in these nodules of fixing atmospheric nitrogen. The bacteria obtain other constituents of their food from the roots in which they live, and the plants are able to obtain nitrogen by digesting some of the bacteria. After a crop of leguminous plants the soil is richer instead of poorer in nitrogen, as the roots are left to rot in the soil.

Rotation of crops. This fertilising influence of leguminous plants is made use of in agriculture and gardening by growing them in the same soil with other crops in alternate years. Thus instead of growing Wheat in one field year after year, and Clover in another, Clover is grown one year and Wheat the next in the same field.

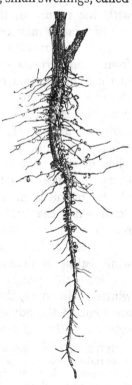

Fig. 21.   Root of Broad Bean, with nodules. (After Strasburger.)

The alternation of two crops would not, however, be the most economical way of using the soil. The supply of salts which weathering and other changes make available is not drawn upon by different plants equally or in the same way. Some plants absorb more of one element, others more of another. Some, like Barley, exhaust the surface layers of soil, others like Mangolds and other 'root crops' tap the deeper layers. Besides, some crops cover the ground closely, while others leave it more exposed to the weather. All these factors, along with the nature of the particular soil, affect the order in which crops may best be grown on any piece of land.

Any opportunity should be taken of ascertaining from local farmers or market gardeners the particular methods of rotation of crops in use in the district, and the special reasons for them. The following is a four-yearly cycle of crops commonly grown in rotation in the eastern counties: (1) Clover, (2) Wheat, (3) Mangolds or other root crop, (4) Barley, or Oats, (1) Clover, etc. Thus in this cycle each kind of crop is grown once in every four years on the same plot. The Clover leaves plenty of nitrogen for the Wheat. After the heavy drain which the latter makes on all the resources of the soil, the root crop leaves it exposed to the weather, draws on the deeper layers and provides food for sheep, which while eating it manure the soil. Barley and Oats are not sown till spring: the fallow, therefore, lasts all the winter. Moreover, these crops are surface-rooted. These are some of the advantageous features of this particular system. Other systems are very similar in principle.

NOTE: Modern work has revealed that, in addition to the essential elements already mentioned, there are others which are also essential, but only in traces, measured as parts per *million* of a culture solution. Such are boron, manganese, zinc and copper. In higher concentration they are poisonous.

CHAPTER X

BALANCE OF FUNCTIONS. ADAPTATION

We have now studied experimentally the chief functions of the leaves, stems, and roots of plants. Our experiments have shown that these organs are concerned mainly with nutrition; they are the *vegetative organs* of plants. They carry on the processes of photosynthesis or carbon assimilation, absorption, conduction or translocation, transpiration, and respiration. All these processes are essential to the life of a green plant, and it is necessary, moreover, that they should be properly balanced.

It is easy in some cases to see that plants cannot always take advantage of conditions specially favourable to one function and allow that function to proceed at the highest speed possible, without endangering their health through the effect on another function.

For instance, in order that photosynthesis may take place as rapidly as possible the stomata must be wide open; but then transpiration will of necessity also proceed at the highest rate possible under the atmospheric conditions existing at the time. If, however, those conditions are specially favourable to rapid evaporation, this highest possible rate of transpiration may be dangerously high, especially if the soil is dry. The plant has, therefore, to close its stomata sufficiently to reduce the rate of transpiration within safe limits, although it thereby diminishes the supply of carbon dioxide for photosynthesis.

In this example the need for limiting transpiration depends on the rate at which water can be absorbed at the time. The fact that the leaves of plants sometimes droop under natural conditions shows that they may lose water faster than they can obtain it, and under such circumstances they close their stomata in order to diminish the loss by transpiration. It is in this simple relation between absorption and transpiration that the need for balance is most easily understood.

The balance between absorption and loss of water.

If we recall the conditions which have most influence on each of these two processes and compare them, we shall understand better why there is need for a special mechanism for balancing them, as the two sets of conditions are very different.

Conditions which affect the rate of

Transpiration.	*Absorption.*
Temperature of air.	Temperature of soil.
Dryness of air.	Amount of available water in soil.
Movement of air.	
Sunshine (makes stomata open wide and warms leaves).	Character of soil-water (fresh, salt, or acid).
	Aëration of soil.

As shown in this table the conditions which have something in common are moisture and temperature, of air and soil respectively, but even these vary to a large extent independently. In winter, for example, the soil may be still frozen when a change in the weather has raised the temperature of the air, and when sunshine is adding still further to the warmth of the leaves. This is an extreme instance, but similar combinations of circumstances are liable frequently to occur, and all plants must be able to meet successfully the most adverse conditions to which they may be exposed in their particular habitats.

Not only, however, may circumstances sometimes combine to favour transpiration and hinder absorption, but at other times the conditions may be favourable to absorption but not to transpiration. Let us now consider each of these cases more fully, beginning with the latter.

A. Conditions more favourable to absorption than to transpiration.

This condition is less familiar than the reverse, but is nevertheless of common occurrence. During a warm and cloudy summer night, for example, after recent rain, the conditions are very favourable for the absorption of water. The soil is very moist and thus the supply of water is abundant, and it is also warm so that the roots absorb vigorously. On the other hand, the escape of water vapour is small because of the darkness and the moist atmosphere. Under such circumstances the root-pressure forces water up faster than it can escape in the form of vapour, and it emerges as liquid drops on the tips and edges of the leaves. Water which has been forced out in this way from the tips of grass-blades is often mistaken for dew: true dew is usually formed, on the contrary, not on warm, cloudy nights, but on clear, cool nights, when soil and plants lose their heat and moisture condenses in tiny droplets all over their surface.

The leaves of the Garden Nasturtiums (*Tropaeolum*) can readily be induced to exude water from the tips of each of their main veins by artificially reducing transpira-

Exp. 39   tion. This can be done by covering a plant with a darkened bell-jar: as a result of this (1) the air under the jar soon becomes saturated with moisture, (2) the light is prevented from reaching and warming the leaves and (3) in the darkness they close their stomata. Transpiration is, therefore, reduced to a minimum. The activity of the roots can be increased to a maximum by watering the soil with tepid water, and putting the plant, if it is in a pot, in a warm place.

Such exudation of water is very marked in the Tropics. The 'Rain Tree' (*Pithecolobium Saman*), which is very commonly planted in India and Ceylon, is so called because the water drips from its leaves so copiously as to resemble a shower of rain to anyone passing beneath. This occurs in the early morning, after the darkness of night has lessened transpiration and the roots have been actively absorbing in the warm soil.   A similar copious exudation of drops of water is met with also in many herbs, shrubs, and other trees under similar circumstances.

B.   Conditions more favourable to transpiration than to absorption.

This is very commonly the case, especially during the day.   In order the better to understand the consequences let us consider first a single example.

We have seen that the leaves of Sunflowers sometimes droop during a hot, sunny day.   In the evening, however, they soon recover again.

This recovery shows that the roots can still obtain water from the soil, and that the supply which they provide is sufficient to enable the leaves to become turgid again when the sun goes down and the temperature falls, and the conditions therefore become less favourable to evaporation.   During the day, on the other hand, this supply was insufficient.   Now we have seen that when leaves begin to wilt they close their stomata.   In this way they put a stop to any loss of water by transpiration through the stomata, or at least reduce this loss to a very little.   Nevertheless, the wilted leaves of the Sunflower must continue to lose water as fast as they absorb it from the stem, even after they have closed their stomata

—otherwise they would begin at once to recover their turgidity. How, then, does this continued loss of water take place?

In experimenting on the loss of water from leaves, we found that although leaves with their stomata blocked by vaseline lost far less than those with open stomata, yet they did lose a little. This they can only have lost directly through the skin or epidermis on the upper side of the leaf.

Now the leaf of a Sunflower, unlike the leaves used in that experiment, has stomata on both sides, and, as is usually the case with such leaves, the epidermis is not heavily waterproofed (it has not a very thick *cuticle*— see p. 115). Still more water will therefore evaporate from it than from the more heavily waterproofed upper epidermis of leaves of Privet, Laurel, and other evergreens, or even of Beech, Elm, and other trees with thin leaves (*cuticular transpiration*).

Thus leaves of Sunflowers remain wilted, even after they have closed their stomata, so long as hot sunshine and dry atmosphere combine to rob them through their epidermis of all the moisture they can absorb.

It is important to realise that, *even under such conditions*, if they are not to dry up and die, *leaves must be able to absorb as much water as they are losing*. In fact, a leaf continues to droop more and more until the rate at which it can absorb equals the rate at which it loses water.

If the leaves of a Sunflower are kept under observation on a hot summer's day it will be found that they begin to droop as the sunshine gets hotter and the air drier, and become more and more flaccid until they reach a certain degree of wilting. Up to this point they have been losing more water than they have absorbed, beyond it they absorb and lose equally and so remain unaltered in appearance until the conditions become less drying.

The maintenance of this condition of balance or equilibrium depends on two factors, in addition to the power possessed by the leaves of closing their stomata: the more flaccid a leaf becomes (1) the more firmly does it hold what water it has, and (2) the more powerfully does it draw upon the water in the stem.   In this way an automatic adjustment of the outgo and intake occurs, even under less extreme conditions, while the stomata are open. If the large leaves of the Sunflower are carefully measured[1] at intervals on a summer day it is found that they shrink a little when the sun shines and expand again when a cloud covers the sun.

After a long spell of dry weather, when the supply of water in the soil has run low, it may happen that the sun is so hot and the wind so strong and dry that a leaf is robbed of its water faster than it can absorb it, even though it has closed its stomata, and is quite wilted and therefore is absorbing water from the stem with the utmost force of which its cells are capable.   Under these extreme conditions the leaf gradually dries up.   It has failed to adjust the balance between the outgo and intake of water, and it dies.

Now it is important to observe that different parts of a plant are not equally but in various degrees able to withstand conditions such as these.   The stem of a Sunflower has a thicker cuticle than the leaves, and exposes a much smaller surface compared with its volume. When it has closed its stomata, it is therefore better able than the leaves to maintain the turgidity of its cells. The upper younger leaves are more robust than the lower, and always remain turgid longer, probably because they can absorb water with greater force.   The youngest

---

[1] This may be done by measuring exactly the distance between two marks made on the leaf with indelible ink, supporting the leaf below on a ruler covered with plush or velveteen to take the veins.

leaves of all, in the terminal bud, are packed tightly together; the outermost of these are, it is true, exposed to the sun, but they and the young leaves just beginning to expand are closely covered with a felting of hairs, which prevents the wind from removing moisture so rapidly. These hairs become separated as the leaves expand and afterwards can afford them little protection. Lastly, the axillary buds are protected by the bases of the sub-tending leaves, as well as by hairs. In a time of prolonged drought it is therefore the old leaves which dry up first, while the buds last longest, and produce new leaves when the drought has passed.

So far we have considered the case of a plant which
Special
means by
which
loss of
moisture
is re-
duced.
grows in garden soil under conditions which may be regarded as average and on the whole favourable to plant life. There are seasons and situations, however, in which plants are always liable to be exposed to drying winds or hot sun or to find absorption difficult, or to meet a combination of these conditions. Under such circumstances plants must either be able to absorb with special vigour, or be specially protected against loss of moisture, or have a store of water on which to draw.

Three ways in which loss of moisture may be reduced have already been illustrated by different parts of the Sunflower: (1) The epidermis may be heavily waterproofed (covered with a thick impervious cuticle). This is extremely important, for it diminishes the loss of water by evaporation after the stomata have closed, i.e. when the conditions are most severe. (2) The surface may be closely covered with hairs (Fig. 22), which prevent a dry wind from playing directly upon it. The water vapour must pass out between the hairs before it is snatched away. Thus in wind the direct loss of moisture from the epidermis and transpira-

tion through the stomata are both retarded. (3) The
surface exposed may be small. In this case too both
cuticular and stomatal transpiration are less; but photo-

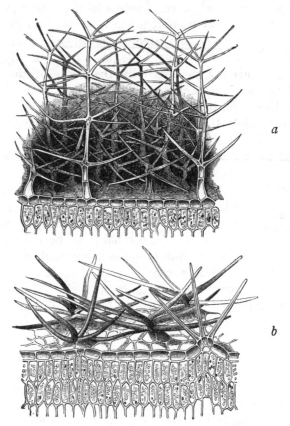

Fig. 22. Portions of leaves of (a) Mullein and (b) a Potentilla, magnified
to show the structure of two forms of woolly hairs. (After
Kerner.)

synthesis is also diminished. It is interesting to find,
therefore, that some plants with small leaves, like the
Gorse and Broom, have green stems.

Other features which reduce transpiration are fewer stomata, stomata confined to portions of the leaves only, stomata sunk in pits or grooves, sometimes provided with hairs.

The heaths—Purple Bell Heather, Ling, etc.—combine these features. The leaves are very small and partly rolled so as to form a groove on the under side (Fig. 23). In this

I   II

Fig. 23. I, small piece of a branch of Ling (*Calluna*) as seen under a lens, showing the small leaves, in four ranks, with grooves underneath. II, a single leaf, as seen from below, more highly magnified, showing the groove closely filled with hairs. The leaves are without stalks, and the two lobes (auricles) at the bottom project below the point of attachment and fit close to the stem.

groove the stomata are situated, further protected in many cases by hairs which fill the groove.

Shoot systems showing features such as these have been called *xeromorphic* (Gr. *xeros* = dry), since they are shown especially by plants living under dry conditions. It would however be a mistake to suppose that the transpiration of xeromorphic shoots is generally low. They may transpire very rapidly in their natural habitat, only reducing their transpiration under the most extreme conditions. The heaths, for example, although their leaves are small, have very many of them and with their stomata open transpire vigorously.

Reduction of transpiration is not the only way of adjusting the balance between transpiration and absorption. The root system may be extended, the flow of water within the plant made easier, or, most important of all, the plant may be able to resist the wilting effect of water loss, keep its stomata open and go on increasing its power of suction when other plants would shrivel. A store of water within the plant may tide over an emergency that does not last too long. So we often find alongside one another, in dry situations, plants showing various degrees and kinds of xeromorphy and others seemingly without any xeromorphy at all. Conversely also, we find xeromorphic features where we should not expect them.

We will now consider some examples of conditions under which loss of water tends to be excessive or absorption difficult.

In winter, when the ground is cold, the low temperature makes the roots less active and brings their pumping, like all their functions, nearly to a standstill. It is more difficult, too, for water to be absorbed even from wet soil if it is cold, and almost impossible if it is frozen. Accordingly the parts of plants that remain exposed above ground commonly have special protection from loss of water.

Effect of cold, in winter.

Trees have their stems covered by an impervious bark, and their growing points enclosed in special winter-buds. Most of our trees shed their leaves in autumn (*deciduous* trees: Beech, Oak, Elm, etc.), and so get rid of the greater part of their summer surface. The leaves of evergreen trees, which are not all shed in autumn but last through several winters, have tough waterproof skins, and are usually small (Holly, Evergreen Oak, Pine, etc.). Evergreen shrubs have similar though sometimes larger leaves; being lower, they are less exposed to drying winds. There are also many shrubs which are deciduous, shedding their less protected leaves in autumn.

The majority of *herbaceous* (i.e. not woody) plants die down at the approach of winter and so avoid the exposure of a delicate transpiring surface. Many of them have, as we have seen, underground shoots which last through the winter under the protection of the soil. Some plants, however, growing close to the ground may last through the winter although they show no obvious xeromorphic features. Green leaves of Grass, Dandelions, Daisies, and other similar plants may often be observed all through a winter season. Lying so near the ground under the shelter of the old dead grass, their growing points put forth fresh young leaves to replace those killed by frost and these grow during each mild spell and take advantage of any warmth and sunshine for photosynthesis. Near the soil the supply of carbon dioxide is increased by the respiration of roots and soil organisms. Moreover the air is usually moist and the wind has often little power; on the other hand the leaves are very near the source of their water supply, so that it has very little distance to travel and has not to be drawn up through any great length of stem. It must be observed, however, that rosette-plants like the Daisy and Meadow Plantain press their leaves close to the ground, and in this way transpiration is hindered.

In many tropical and sub-tropical countries, where the **Drought.** temperature of the soil is never low, rain falls mainly during one season of the year, the rainy season, and a dry season follows, during which the supply of water in the soil becomes less and less. In this season, too, the sun shines brightly and the air is hot and dry, so that transpiration tends to be excessive and absorption very difficult. In these countries grow deciduous trees which shed their leaves at the approach of the dry season, and most perennial herbaceous plants die down, as they do before winter in temperate climes like our own.

On the other hand, in warm countries where there is no dry season or winter, the shrubs and tress do not shed their leaves at any particular season, but usually a few at a time, and so are evergreen, though not necessarily xeromorphic like our own evergreens.

There are many kinds of situations in which an adequate supply of water is difficult to obtain not only during one season of the year but all the year round. It is clearly impracticable for plants growing in such situations to rely mainly upon closing their stomata and so reducing the loss of water by stopping transpiration, for they would have to keep them closed for the greater part of the time and would be unable to assimilate. In order that photosynthesis may go on, they must keep their stomata open, unless under exceptionally severe conditions, and at the same time must maintain a sufficient supply of water to their leaves.

On very sandy soils rain rapidly runs through, so that after a shower the water supply lasts but a short time. Rain runs quickly, too, from rocks and walls, and any that does not run off soon evaporates. Plants growing in such habitats have therefore to depend on a very *precarious* supply of water, abundant enough while it lasts but soon gone. In addition to plants with xeromorphic features of the kinds already mentioned are succulent plants. Stone Crop, for instance, found commonly on old walls, has very thick, small, fleshy leaves, which expose a very small surface compared with their volume. Such plants gradually lose their water during intervals of drought, and, when rain falls, quickly absorb as much as they can hold and store it in their tissues to last them through the next period of scarcity.

In other places there may be plenty of water but it contains in solution substances which hinder absorption. The sea-shore, for instance, is salted with spray, and the water in the sand may contain much salt. Root-hairs cannot absorb from such a solution, by osmosis, as readily as from the very weak solution contained in ordinary soils, and the salt probably interferes not merely osmotically but also in other ways with absorption.

We shall meet with many more examples in studying the distribution of plants (Section V). This is always very greatly influenced by the need for keeping a balance between absorption and the escape of water vapour, and between the promotion of photosynthesis and the control of transpiration; for plants grow successfully in those situations to which their powers of adjustment are most closely adapted.

SECTION II

FORM AND STRUCTURE

CHAPTER XI

INTERNAL STRUCTURE OR ANATOMY

(i) Cells and tissues.

We have so far studied the work that is done by different organs of plants and how they are fitted for it by their *external* form and construction. We will now examine their *internal* structure, to find how this is related to their various functions.

For this purpose a microscope is necessary, and as nearly all plant organs are too thick to be seen under the microscope they must be cut into very thin slices, or *sections*. Sections are made with a sharp razor, which is flooded with water so that the sections shall not stick and be crumpled or torn. An interesting object to practise upon is a Potato tuber: this is a greatly swollen stem in which much food is stored in reserve. A small piece should be cut from the potato with a knife, and held between the thumb and first finger of the left hand, while the razor is held horizontally in the right hand. First the top of the piece is removed, so as to make a horizontal smooth surface, and then very thin shavings are cut, not always right across, but beginning or ending on the surface instead of at the edge: in this way the sections, though not always of an even thickness, have thin parts where the structure is more readily seen.

The sections are washed from the razor into a watch glass full of water.   One or two of the thinnest are picked out and mounted in a small drop of water on a glass slide, and covered by a clean coverslip[1].

When the thinnest part of the section is examined under the microscope, it is seen to consist of many *cells*, each packed with whitish egg-shaped grains.   A section mounted in iodine appears black to the naked eye, and under the microscope the grains are seen to be stained dark blue: they are, therefore, grains of starch (p. 29).

At the very edge of the section the starch grains may have fallen out, so that the outlines of the cells can be more easily seen.   They are all very much alike in shape and size, and have thin walls.   In whatever direction sections are cut the cells appear of the same irregularly rounded shape, so that each cell must be of nearly the same width in every direction.   They are not, however, as round as the free cells of the Privet berry, for they are packed together; yet they do not fit everywhere quite closely to each other, but have small spaces between them.   In thicker parts of the section black lines can be seen between the cells.   This is air which fills the narrow spaces.   If an air bubble has been enclosed under the coverslip its edge will be found to appear dark, for it reflects back the light and does not allow it to pass up into the microscope: the dark lines in the section are the edges of narrow air bubbles in the narrow spaces which run between the rounded corners of the cells.   If a section be put into alcohol the air is seen escaping in tiny bubbles: if it be then plunged into water, held under till the water has soaked in again, and once more mounted in water, many of the black lines are found to have disappeared.

The starch grains are not the only contents of the

---

[1] For directions for mounting see p. 78.

cells. Around the starch grains on the outside of each cell is an almost transparent granular substance. In iodine this is stained yellowish brown, being proteid. It is the living protoplasm, the most important part of the cell. The cell-wall and the starch grains are but dead products manufactured by this protoplasm. Here and there cells may be found in which the *nucleus* is not hidden by the starch grains: this is a very important part of the protoplasm, a small rounded body staining more deeply than the rest, which we have seen already in the cell of the Privet berry (cf. Fig. 19, p. 83).

The inner part of the Potato tuber is thus composed of similar cells, each with thin walls, protoplasm, and starch grains, with cell-sap filling the space between the grains. Such a mass of similar cells is called a *tissue*, and a tissue of thin-walled living cells like these is called *parenchymatous tissue* or *parenchyma*. All the cells of such tissue need to respire, and therefore must be supplied with oxygen. The air-spaces between the cells allow air to bathe part of the surface of every cell: oxygen can thus diffuse up to the walls of each cell and the carbon dioxide produced can escape.

Besides carrying on this function of respiration, which is essential to every living cell, the cells of the parenchymatous tissue of the Potato tuber are serving the special function of storing food-material for future use.

If a section be made through the outside skin or 'peel' of the potato the cells are seen to be brown in colour and very regularly oblong in shape, fitting closely together. They differ from the cells of the parenchyma also in having no contents but air. When first formed they contained living protoplasm, but as they became mature this disappeared leaving behind only cell-walls, of special composition, tough and impervious to water, containing only air. This is another kind of tissue, called *cork*, composed of *dead* cells, which serves as a protection to the tuber against injury and loss of moisture. If a tuber be cut in half, new corky skins are formed over the exposed surface.

We have thus found in the Potato tuber two very different kinds of tissues, performing different functions.

(ii) The leaf: assimilating tissue.

Let us now begin our examination of ordinary plant organs with the most important, the leaf.

In order to cut sections across a leaf, a slit is cut in a piece of dry pith and a strip of the leaf, not wider than the pith, is put into the slit, so that by cutting across the pith the strip of leaf-blade is also cut, accurately at right-angles to its surface. With a sharp razor, flooded with water, the surface is first smoothed, and then thin shavings are cut from pith and leaf together. When transferred to water, the sections of the leaf are easily picked out from among the sections of pith.

A good leaf with which to begin is the thick leaf of London Pride (*Saxifraga umbrosa*). In a section of this leaf mounted in water a pattern in green and black is seen under the microscope. We have already found by experiment (p. 34) that the green tissue of the leaf is spongy, containing air, and here in the section the air accounts for the dark lines and patches amongst the green.

Examining the section more closely we notice the upper and lower *epidermis* of colourless cells, closely fitted together, with straight walls between them. The outer, surface walls are thick: in a very thin section mounted in iodine the outermost layer of these walls is seen stained bright yellow—this is the *cuticle* protecting the leaf from loss of moisture through the epidermal cells.

Below the upper epidermis are two or three rows of
green cells, oblong in shape, arranged like rows of stakes
at right-angles to the epidermis. This tissue is, therefore,
called the *palisade tissue*. The cells do not fit closely
together; but the air-spaces between the cells are rather
narrow, and the air still in them accounts for the straight
black lines parallel with the green cells.

Between the palisade tissue and the lower epidermis
much more air is present and large spaces are clearly
seen between the cells: this tissue is called the *spongy
tissue* of the leaf. In order to see the structure of this
tissue, and to study the section as a whole more closely,
it is necessary to remove the air. A simple way of doing
this is to dip the section for a few seconds into alcohol or
methylated spirit, as we did the section of potato. It is
then put into water and carefully kept from floating to
the surface, until the water has once more soaked into it
and washed out the alcohol. The section is then mounted
again in water.

It can now be clearly seen that, whereas the cells of
the palisade tissue are roughly oblong and fairly regularly
arranged, the cells in the spongy tissue are very irregular,
each projecting on several sides into a short arm, which
joins a similar arm of another cell, or a cell of the palisade
tissue. Such cells are roughly star-shaped, or 'stellate.'
As we have seen, the air-spaces included between the
arms are large.

All the cells of the spongy and palisade tissues contain
little round bright green grains[1]. Apart from these
grains the cells are colourless. We know that photosynthesis only goes on in green parts of plants. It must,
therefore, be these green grains or *chloroplasts* which are
essential to photosynthesis. It is they which contain

[1] If the section has remained long in alcohol some or all of the green
chlorophyll will have been dissolved out of the grains.

the chlorophyll which we found could be dissolved out in alcohol. If a section be cut from a leaf, which has just been in warm sunshine and has therefore been carrying on photosynthesis vigorously, and mounted in iodine, the whole section turns dark, and attached to each chloroplast can be seen one or more minute black granules: these are starch grains formed during photosynthesis. The green chloroplasts are, in fact, the organs of photosynthesis. It is they which change the carbon dioxide which reaches them, making sugar and starch, and setting free oxygen.

In a section stained with iodine it may be possible to make out the nucleus stained yellowish brown in some of the cells. If a fresh section be mounted in a five per cent. solution of salt, the cells are plasmolysed like those of the Privet berry: water is drawn out of the cell-sap by the strong solution outside, and the protoplasm contracts away from the cell-wall, carrying with it the chloroplasts. In water the plasmolysed cells recover again. Except for their shape, colourless sap, and the large number of chloroplasts present in the layer of protoplasm, the cells of the leaf are indeed very like those of the Privet berry. Each cell keeps turgid, and helps to hold the leaf outstretched, because the protoplasm does not allow the substances in solution in the cell-sap to escape, and water therefore enters the cell and inflates the elastic cell-wall.

Leaving the veins, which will be seen here and there in the section, often cut obliquely, let us examine the appearance of the different tissues in sections parallel to the surface of the leaf. By tearing the leaf across, a strip of the lower epidermis may easily be removed. Part of this should be mounted with the outer surface, part with the inner surface uppermost. The former shows that the cells of the lower epidermis are irregular

in outline, but fit closely together. In parts (between the
veins where the cells are smaller) many small oval stomata
may be found (Fig. 24): under the high power of the
microscope they are seen to consist of two crescent-shaped
cells, with a pore between them which may be open or
closed: its sides are thickened and appear like two lips.
Many of the pores appear black owing to the air which
they retain, showing that they are air-passages.

The piece of epidermis mounted the other way up
shows the cells of the spongy tissue that lie just within

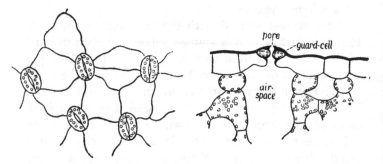

Fig. 24. Surface view of piece of lower epidermis of London Pride,
    showing five stomata.
Fig. 25. Section through a stoma, showing the guard-cells and pore,
    and the large air-space immediately below the pore.

the epidermis. These appear stellate (Fig. 26, I), much
as in the cross-section. By focusing down below them,
the stomata will be found always to lie under spaces
between the cells.

If we now cut a few shavings from the upper surface
of the leaf, wrapping it round the finger for the purpose,
and mount some one way up and some the other, as
before, we shall find that the cells of the upper epidermis
are in this leaf more regular in form, and that no stomata
are present; while the palisade cells[1] below the epidermis,

---
[1] For the palisade cells a portion of the section should be examined

now seen cut at right-angles to their length, appear round, with small air-spaces between them (Fig. 26, II). Sometimes the chloroplasts can be seen clustered next these air-spaces, from which their supply of carbon dioxide comes. From the appearance of the palisade cells in this section compared with their appearance in the first section we can tell that they are cylindrical cells which do not fit together closely but so that narrow air-channels are left between them.

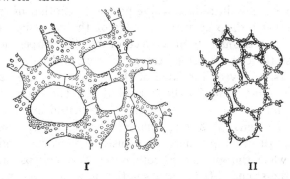

I   II

Fig. 26. Portions of the assimilating tissues of a leaf of London Pride, as seen in sections parallel to the surface. I, spongy parenchyma, of stellate cells with large air-spaces; II, palisade parenchyma, showing the narrow air-spaces between the round (cylindrical) cells.

The stomata of the London Pride are small, and in order to understand the structure of these important organs it is better to examine them in other plants, like the Iris, Fuchsia, Hellebore or Ivy-leaved Toad-flax, where they are larger. If a leaf of one of these plants be torn across, a piece of epidermis may be removed in the same way as from the leaf of London Pride, and mounted with its outer surface upwards in water. Dotted about among the epidermal cells are many small pores, each

which is a little removed from its thinnest edge and shows a thin layer of green tissue in addition to colourless epidermis.

guarded by two cells of special shape and size. These *guard-cells*, unlike the other cells of the epidermis, contain chloroplasts. Their outer edge is curved, so that the whole stoma has an oval outline. If the leaf has been freshly plucked on a bright day, and the strip of epidermis mounted in water, most of the stomata will probably be open; in this condition the inner edge of each guard-cell is also curved. The wide pore between the guard-cells often appears black, owing to air which is entangled in it. If we now introduce a five per cent. solution of salt by placing a drop against one edge of the coverslip and drawing it through with a piece of blotting paper held against the opposite edge, the guard-cells will be seen to close, especially at the edge of the strip where they are reached most easily by the salt solution. The effect of a salt solution, as we have seen, is to withdraw water from the cells (if strong enough causing plasmolysis), leaving them limp where previously they were turgid. Thus when the guard-cells lose water they collapse and close the stoma. If the salt solution is washed out with water again the guard-cells can be seen to swell and curve away from each other, opening the pore again.

The leaves of most plants have a structure which is similar in important features to that of the London Pride. They have an epidermis of colourless closely fitting cells, with an outer cuticle which protects the leaf from drying up. The cuticle of the upper epidermis, which is exposed to the sun, is usually the thicker. Stomata occur in the lower epidermis, allowing gases to diffuse to and from the air-spaces within the leaf. Some leaves (Sunflower, Broad Bean, etc.) have stomata also in their upper epidermis, so that they are supplied with carbon dioxide from both sides. Such leaves usually assimilate carbon dioxide very actively and belong to plants which grow rapidly and vigorously. They also, of course, transpire faster than leaves with stomata only on the under side.

Most leaves, too, have both palisade and spongy tissues containing chloroplasts. The palisade cells are often longer and narrower than in the London Pride. The spongy tissue very commonly consists of stellate cells, though these are not always found. In some cases (e.g. the common Cherry Laurel) the cells are more regular and are arranged to form the walls of air-chambers. Whatever the exact arrangement may be, the large air-spaces in the spongy tissue make it possible for carbon dioxide to diffuse rapidly from the stomata to the cells

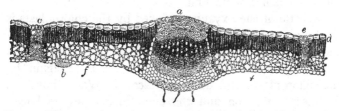

Fig. 27. Diagram of a section across part of a leaf of the Birch, passing through the midrib (*a*) and two smaller veins (*c*, *e*). It shows the palisade and spongy assimilating tissues, and the mechanical tissue in the veins above and below the vascular bundles. *f*, *f*, stomata in the lower epidermis, and *b*, a gland; the upper epidermis is thick-walled. (After Hartig.)

which are assimilating it. Water vapour can also escape readily from the cells to the stomata and out into the atmosphere.

The cells of the palisade tissue are elongated in the direction in which most light falls upon them: this tissue therefore allows the light to pass into it readily, having few cross-walls by which light is reflected and scattered, so that it penetrates deep into the leaf and is absorbed by the numerous chloroplasts which crowd the long walls of the palisade cells. The numerous though narrow air-spaces between the cells allow carbon dioxide to diffuse into the near neighbourhood of all the chloroplasts. The oxygen which is liberated passes out by the same route which the carbon dioxide followed when entering it, diffusing first into the air-spaces between the palisade cells, then into the larger air-spaces of the spongy tissue, through these to the usually still larger air-spaces beneath the stomata, and through the stomata into the atmosphere.

Photosynthesis is also carried on by the cells of the spongy tissue, but we may infer that, as they contain less chlorophyll and receive less light, they usually manufacture far less food than the palisade tissue, although the carbon dioxide reaches them first.

A little of the oxygen produced by the chloroplasts is used up by the protoplasm of the cells in respiration, the rest diffuses out into the air-spaces. The carbon dioxide produced in respiration is used up again by the chloroplasts and does not escape. When, however, darkness comes on and the chloroplasts can no longer carry on their work, then the carbon dioxide produced in respiration escapes into the air-spaces and out into the air through the stomata, now nearly closed; while oxygen passes in through the stomata from the atmosphere to take the place of that now absorbed by all the cells from the air-spaces.

Water evaporates from the wet cell-walls into the air-spaces and its vapour passes out, by the path followed by oxygen in the light and carbon dioxide in the dark, through the stomata into the air. More water is absorbed by the cells to take the place of that which evaporates, those cells next the veins passing it on to the rest. In this way the mineral salts absorbed by the roots are carried to the green cells of the leaf where the food of the plant is being manufactured.

The cells of the different layers of palisade tissue

(when there are more than one layer) are joined end to end; these are connected to the arms of stellate cells below, so that the products of photosynthesis readily pass from cell to cell towards the conducting tissue in the veins.

(iii) Stem structure—conducting, mechanical and storage tissues as illustrated by a typical Dicotyledon.

The experiment in which a cut stem is placed in red ink has shown us those portions of the stem which act as conducting channels for the transpiration stream (see p. 63). These channels are coloured by the red ink as it

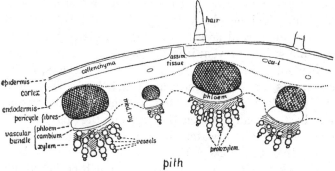

Fig. 28. Diagram of part of a transverse section from a Sunflower stem, showing the distribution of the different tissues.

passes up into the leaves. If such an experiment be carried out with the Sunflower and a section cut transversely across the stem, it will be seen that the conducting strands are in a circle near the outside of the section (Fig. 14). If the section is sufficiently thin to be examined under the microscope the cells of these red strands can be seen. Most of them are thick-walled and contain nothing but water. The most conspicuous of them are round and are arranged in rows, the smallest cells in a row being nearest the centre of the stem. The most characteristic

feature about them, however, is not seen until a longitudinal section of the stem is cut[1]. They are then seen
to be very long, as well as wider than the cells next them;
and though the section will probably not have passed
right down the middle of a tube through its whole length
it may be seen that cross-walls only occur at rare intervals.
They are, in fact, long cylindrical tubes. In a stem which
has been only a short time in red ink a cross-section
shows that it is these tubes, the *vessels*, which are coloured
first, and therefore that it is through them that water
passes most rapidly up the stem.

In the longitudinal section the walls of the vessels show
characteristic markings. Those nearer the pith have a
spiral line, looking like a spring (Fig. 30, III): this is a
strengthening band laid down on the inside of the otherwise thin wall to prevent it from collapsing. Larger vessels
farther out are marked with numerous small dots, really
thin places, or *pits*, in the wall, which is elsewhere uniformly thick (Fig. 30, II). These pits in the pitted vessel,
and the thin wall between the turns of the spiral in the
spiral vessel, allow water to pass through much more
readily than the thickened parts of the wall.

If a section from a stem which has not been in red ink
is mounted in iodine, the thick walls of the vessels appear
bright orange or yellow. A shaving of a match stick is
similarly coloured by iodine. In a twig, red ink travels up
the wood. The water-conducting tissue is, therefore, called
wood or *xylem* (Gr. *xylon* = wood).

Even under a lens (Fig. 14, p. 63) it is evident that on

---

[1] In order to cut suitable longitudinal sections a piece of stem is
split down near the middle into two slightly unequal parts. Sections
are now cut from the larger part. In this way the section will pass
through, or very near to, the centre of the stem. Some of these sections
will pass through a bundle; the bundles can be recognised in the tissues
with the naked eye and sections should be cut with especial care when
a bundle is reached.

the outside of each strand of wood is another distinct
group of tissue. In a section mounted in iodine the outer

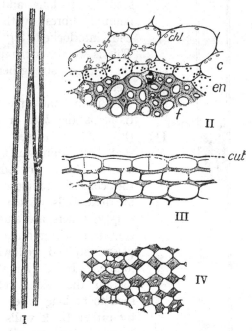

Fig. 29. Stem of Sunflower. I, fibres in longitudinal section; the
overlapping ends of two fibres and the middle part of a third are
shown. II, fibres of the pericycle, *f*, and adjacent parts of the
endodermis, *en*, and cortex, *c*, in transverse section. Some of the
fibres are cut across near their points. The cells of the endodermis
contain starch grains and show corky spots on the (radial) walls
separating them. The large cells of the cortex contain chloroplasts,
chl, and between them are air-spaces; *n*, nucleus. III. Portions
of epidermis and collenchyma from a well-grown stem: *cut*, cuticle.
Some of the cells of the epidermis have recently divided by walls
(still thin) at right-angles to the surface to keep pace with the
growth of the stem in diameter. The collenchyma has also been
stretched (compare IV): it consists of cells with thickened but not
woody walls, with thin places left for the ready passage of water
and food from cell to cell; small air-spaces are present here and
there. IV, collenchyma from a young stem.

part of this group (Fig. 28) appears bright orange, and is
seen under the microscope to consist of small cells all of

which have very thick woody walls (Fig. 29, II, *f*). In a

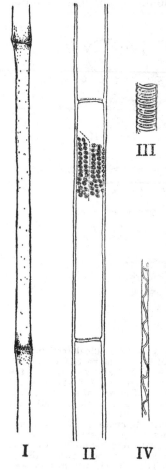

longitudinal section they are seen to be long and finely pointed fibres with their ends interlocked (Fig. 29, I). This is woody *mechanical tissue* which strengthens the stem.

Within the mechanical tissue is the *bast* or *phloëm* (Fig. 28), composed of cells that have thin walls, which are not stained by iodine for they consist of cellulose. Many of these cells appear nearly empty, others are small and contain much granular protoplasm, stained yellowish by iodine. In a longitudinal section the wider cells are seen to be long tubes crossed by rather thick walls at intervals (Fig. 30, I). The cross-walls are, however, perforated like sieves, and so they are called *sieve-plates*, and the tubes are called *sieve-tubes*: an indication of this sieve-like structure can be

I   II   IV

Fig. 30.   From longitudinal sections of the stem of a Sunflower. I, part of a sieve-tube; starch grains are clustered near the sieve-plates.   II, part of a pitted vessel, cut in half lengthways, showing the remains of two cross-walls; the pits are shown only in a small area of the wall.   III, a small part of a spiral vessel.   IV, part of a spiral vessel that was formed very early and has been greatly stretched during the growth in length of the stem: the spiral band has been pulled out and the whole vessel has collapsed.

seen in longitudinal and transverse sections of the stem
of the Vegetable Marrow, where the sieve-tubes are very
large, but to demonstrate clearly that there really are
holes (pores) through the wall requires very thin sections
and special methods of treating them. The sieve-tubes
are not really empty, but are filled with a transparent
solution of sugars and nitrogenous substances, which are
prevented from escaping by a thin layer of protoplasm
lining the walls of the tubes. The protoplasm and other
albuminous substances are often more abundant near the
sieve-plates: in longitudinal sections through the phloëm,
mounted in iodine, the dusky brown stain which these
substances take marks very clearly the neighbourhood of
the sieve-plate.

These sieve-tubes conduct the foods manufactured by
the leaves to all parts of the plant. The pores in the
sieve-plates allow sugars and other crystalloid substances
to diffuse more rapidly through from one cell of the
tube into the next: colloids also, like proteïns, which
are stopped by cell-walls can probably pass through the
sieve-plates.

Between the wood and the phloëm is found a narrow
band of tissue different from either. In transverse
section the cells are small and rectangular, are closely
packed without intercellular spaces, have thin cell-walls,
and are full of living protoplasm. In longitudinal
section it is seen that they are very much lengthened
vertically like the cells of wood and bast on either side.
This tissue is called the *cambium.* Cells like these in
several respects compose the growing point of the stem:
there all the cells have very thin walls and are full of
protoplasm; they are also closely packed without air-
spaces, their walls being flat and their corners not rounded
off. The cambium, like the tissue of the growing point, is
capable of growing very actively, as we shall see presently.

The whole strand of conducting tissue, consisting of wood and phloëm, with the cambium between them, is called a *vascular bundle*. Its cells are marked out from the surrounding parenchyma by their great length in a vertical direction. Associated in this stem with the vascular bundle is the strand of fibres protecting the delicate phloëm as well as helping to support the stem.

Besides the vascular bundles the following tissues are seen in this herbaceous stem (Figs. 28 and 29):

(1) The *epidermis*, a single layer of cells of regular shape, on the outside of the stem. As in the epidermis of leaves the outer wall is thickened and its outer, surface layer, stained yellow by iodine, forms an impervious cuticle that protects the stem from drying up. At certain points there arise from the epidermis outgrowths consisting of one tapering row of three to five cells. These are the hairs that clothe the surface of the plant. Here and there stomata occur; but they are few, and are most easily found in surface view in a piece of epidermis stripped from the stem.

(2) Within the epidermis is the *cortex*. In transverse section the cells on the outside are small and closely packed, and have their walls thickened at the corners; farther in the cells become larger, more rounded, with thinner walls and have larger air-spaces between them. In longitudinal section the outer narrower cells are seen to be also longer: with their thickened walls they form a mechanical tissue which helps to support the stem, being especially useful when it is young. This mechanical tissue is not woody: it is called *collenchyma*. Below the stomata it is interrupted by looser tissue with large air-spaces. The cells farther in become shorter, so that the inner cells of the cortex are little if any longer than they are broad; here the tissue is *parenchyma*, like that forming the Potato tuber. There is no sharp boundary between the parenchyma and collenchyma; they merge into one another[1]. The cells of both contain chloroplasts, which give to the herbaceous stem its green colour. They are, therefore,

[1] Here and there resin-canals occur in the cortex: these are long tabular spaces, appearing circular in transverse section and bounded by delicate cells which secrete oil and resin into them.

capable of photosynthesis, and absorb the carbon dioxide which
reaches them through air-spaces, either from respiring cells within,
or from the outer atmosphere through the stomata. Mineral salts
are brought to the assimilating cells with water which is absorbed
from the vascular bundles, just as in leaves.

The cortex is bounded on the inside by a layer of cells which
are marked out from the rest, in sections mounted in iodine, by
the presence of blackened starch grains, although the rest of the
cortex contains no starch. This layer is called the *starch sheath* or
endodermis.

(3) The tissues within the endodermis together form the
central cylinder, in which the vascular bundles are found. The
outer zone of tissue, between the vascular bundles and the endodermis, is distinguished as the *pericycle*; it is in this zone that the
strands of woody fibres are formed, outside each bundle, and they
are known therefore as *pericycle fibres*. In the stems of some plants
a continuous band of pericycle fibres is found.

In the centre of the stem is found parenchymatous tissue called
the *pith*. The bands of similar tissue between the vascular bundles
are known as the *medullary rays* (from *medulla* = pith). The cells
both of these and of the pith are used in many plants as storage
cells, like the cells of the Potato tuber, and are often found crowded
with starch (though not in the Sunflower). The medullary rays
connect the living tissue of the pith with that in the cortex and
both water and manufactured food-substances may pass inwards
and outwards along them.

(iv) The arrangement of the tissues in a root.

Most roots are so thin that it is difficult to prepare
sections in which their structure can be clearly seen. The
root of the Broad Bean seedling is, however, fairly stout.
When this is cut across and looked at with a lens it presents
a very different appearance from a stem, for instead of a
ring of bundles near the outside only a central core of
tissue can be distinguished from the rest.

In a thin, accurately transverse section, this core is
found to contain the conducting tissue, and to correspond
to the central cylinder of the stem. Outside is the cortex

of parenchymatous tissue, bounded by a layer of cells, some of which have grown out into root-hairs (Fig. 18): this outermost layer is therefore called the *piliferous layer*. On the inside of the cortex is the endodermis, much more clearly marked and regular than in stems although the cells contain no starch.

In the outer part of the central cylinder are generally four or five wedge-shaped strands at equal distances from one another, with the point of each wedge, comprising the narrowest cells, directed outwards, not towards the centre

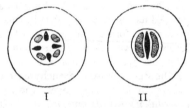

I   II

Fig. 31. Diagrams to illustrate the arrangement of the tissues in young roots of Dicotyledons. In I there are four groups of xylem (black) alternating with four groups of phloëm (shaded). II represents a case, frequently met with, of a root with only two opposite groups of xylem which meet in the middle and form a band, on either side of which are the two groups of phloëm.

as in the stem. These are strands of vessels, which can be recognised by their emptiness and the thickness of their walls, stained yellow in iodine.

Alternating with these groups of wood vessels are an equal number of groups of small thin-walled whitish cells, quite distinct from the larger cells which surround them on all sides and have air-spaces between them. These are the strands of phloëm (in the Bean root often with fibres outside them). Thus in a root the wood and phloëm are not arranged as in the stem, but in separate strands of wood and phloëm, which alternate round the central cylinder, with parenchyma between. Moreover, there is at this young stage no cambium.

Between the conducting tissue and the endodermis is a thin pericycle. Inside is a small pith of parenchymatous tissue like that of the cortex; but in many roots the xylem strands meet in the middle and form a solid core of xylem (compare Fig. 31, II).

(v) Relation between distribution of tissues and conditions of life in root and stem.

Comparing the root with the stem it is seen (1) that most of the tissues of the root are like those of the stem: wood and phloëm contain vessels and sieve-tubes just as in the stem; while throughout the tissues, except just around the conducting strands, air-spaces are found between the cells which allow the oxygen that is necessary for respiration to reach the near neighbourhood of each cell, by diffusion from the air in the soil, and the carbon dioxide produced to escape rapidly outwards. But (2) in the root of the Bean the mechanical tissues—collenchyma and woody pericycle fibres—so prominent in the stem of the Sunflower, are absent or scanty. There is no cuticle, and, of course, there are no chloroplasts in the cortex. It is also seen that the relative proportions and arrangement of the tissues are very different.

Now if other roots and green herbaceous stems be examined it is found that, although they differ in details from those already examined, yet these points of contrast are very general. Not only are the conducting tissues in roots always near the middle, and arranged in alternate strands of wood and phloëm, but roots are generally thin and have usually very little mechanical tissue other than the thick-walled cells of the xylem, which are chiefly water-conducting. In stems, on the other hand, special mechanical tissue is nearly always present, at the least collenchyma and generally woody fibres. Erect stems are generally stout and their mechanical tissue is always near the outside.

If we consider the conditions of life of stems and roots we shall see how these points of contrast in structural plan may be explained. Erect stems have to bear the weight of leaves and branches, which tend to bend them to one side or another; and they must be strong enough not to break down or bend too much when the wind blows. They must, therefore, be especially able to resist bending. Roots on the other hand have their own weight supported by the soil all around them, while their business is to keep the plant firmly fixed in the soil. When the wind sways the plant to one side it bends away from the roots on the opposite side so that there is a tendency to pull the plant out of the ground. Thus a strong pull is exerted on the roots and they must therefore be adapted to withstand pulling.

In order to understand how these different require-

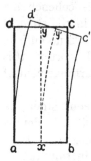

Fig. 32.

ments are met, let us make a very simple experiment. A sheet of paper, like a thin root, bends very readily though it resists stretching. If, however, we roll it up into a tube, it becomes at once rigid and will not bend without crumpling. A study of the diagram (Fig. 32) will make clear the reason for this. *abcd* represents a section through a cylinder of wood or metal. When this is bent, the end $c'd'$ still keeps at right angles to the axis of the block at that end and takes up an inclined position; while as the cylinder is not pulled, only bent, its average length must remain unaltered. It is clear from the figure that ad' is longer than ad, bc' shorter than bc, so that, in bending, the side ad has been stretched, and the side bc compressed. Between these extremes is a region *xy* which is neither stretched nor compressed, while on either side of it the amount of stretching or compression increases towards the outermost parts, *ad* and *cb*, where the change of length is greatest. It is these outermost parts, which have to undergo the greatest change of length, which therefore resist bending most, and in fact it is found that such a cylinder as this is nearly as rigid if the centre is removed so that it becomes a hollow tube.

Now when the sheet of paper was flat, it bent very readily, because, being very thin, it had to be neither compressed nor extended appreciably. As soon, however, as we rolled it up, it became rigid because it could only bend if one side were stretched and the other shortened.

It follows that in making a supporting pillar out of material such as iron, especially to resist bending, the most effective and economical plan is to make it hollow, for in this way a pillar of greater diameter and therefore of greater rigidity is obtained than if the same amount of material were made into a solid pillar. This principle is followed in the making of iron supports for piers, for instance, where they have to withstand the lateral pressure of flowing water, or of wind, which tends to bend them.

The mechanical construction of stems and roots is in remarkable accordance with these principles. We have seen that in the Sunflower stem all the mechanical tissues, including the woody vessels, are placed near the outside, that is, just where they will offer most resistance to bending. Many green stems become hollow very early, and nevertheless are still very strong (e.g. Hog's Fennel, Chervil and other Umbelliferae).

In roots the woody tissues are near the centre, where they will be least compressed or stretched if the root bends, and thus will hinder bending very little, but will be on the other hand quite as useful in enabling the root to resist the direct pulls to which it is exposed. It is, of course, a great advantage to the root to be as thin as possible, so that it can push its way more easily between the particles of soil.

Erect stems are thicker than roots, so that even if no special mechanical tissue has been developed stems are less easily bent. All thin-walled cells are capable of

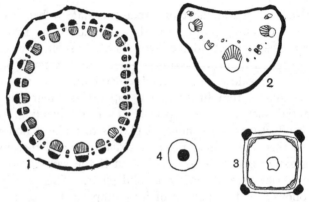

Fig. 33. Diagrams illustrating the distribution of mechanical tissues, in cross-section: (1) Stem of Sunflower. (2) Leaf-stalk of Sunflower. (3) Stem of Deadnettle, hollow in the middle. (4) Root of Iris. All are magnified about five times. The mechanical tissues, including collenchyma, are in black; and the xylem is shaded, except in the root, which is too small for the separate strands of vascular tissue to be shown.

standing a certain amount of strain when they are *turgid* and their walls are distended with sap owing to the entry of water through their semi-permeable protoplasm. Besides, these turgid cells are packed inside the strong cuticle which forms a tightly stretched skin over them: in this condition the cuticle itself resists further stretching and so, being on the outside like the mechanical tissue, tends to keep the stem straight; a rubber tube tied to a tap and distended with water becomes rigid in a similar way. Very young stems (e.g. those of young cabbage plants) depend entirely on their turgidity for strength, and when they lose too much water (as when cabbage plants are transplanted) they collapse.

Why the wood and phloëm are arranged differently in the root and in the stem is too difficult a question to discuss here, but one point may be noted. The root when young absorbs water from the soil. This water enters the root-hairs and passes through the cortex to the wood in which it is carried upwards. In the root the wood is near the outside of the central cylinder very near the cortex. The passage of water from cortex to wood is therefore more direct than it would be if the wood were separated from the cortex by the phloëm as it is in the stem.

(vi) The conducting and mechanical tissues of leaves.

We have seen that the veins of a leaf conduct water, for they become red when the stem is cut and put into red ink for some time. If a section is cut across the midrib of a Sunflower leaf we find a large bundle, with smaller ones on either side of it which will pass out to lateral veins; in each bundle are seen under the microscope an upper part, the wood, containing empty vessels with thick woody walls, and a thin-walled part below it, the phloëm, of thin-walled cells. Under the epidermis, above and below the bundle, collenchyma is found, just where it will be stretched or compressed most if the leaf is bent, and therefore just where its resistance will help most to keep the leaf outstretched.

Leaves of the Sunflower depend, however, very much on their turgidity for strength, for on a hot sunny day they may often be found hanging limp like wet rags against the stem, even the leaf-stalks themselves not

having enough mechanical tissue to support them when
flaccid.   Many other leaves have much more mechanical
tissue and some have woody mechanical tissue as well.
It is generally found above and below the bundle, and
in the strongest leaves it also surrounds the bundles of
the larger veins, and so forms a vertical plate or strip of
mechanical tissue which is very efficient; such a vertical
strip could readily be bent sideways but strongly resists
bending upwards or downwards, as may be illustrated
with a strip of card.

> Some plants, such as the Onion and Rushes, have leaves that
> are erect and cylindrical; these have their mechanical tissue
> arranged, as in stems, near the outside.   They are, in fact,
> usually hollow.   The sheathing base of a grass leaf is an example
> of a thin structure which is rigid because it is rolled into a
> tube.

The veins of a leaf thus serve (1) to supply the leaf
with water, and to carry away the products of photosynthesis, (2) to support the leaf-blade outstretched.   For
both these functions they are very effectively distributed.
In broad leaves large branch veins run out like ribs in all
directions; from these, smaller veins branch off, and from
these again finer and finer branches, so that there is no
area of the leaf, be it ever so small, in which some of the
finest branches at least are not present.   This is well seen,
for instance, in a leaf of the Lime or the Beech when
it is held up to the light: the translucent veins then
appear as a very fine network on a green ground.   It is
possible to obtain skeletons of many leaves if they are
soaked in water for some weeks until the softer tissues
rot.   The epidermis can then be torn and carefully
lifted off, and the other tissues washed away with a soft
brush, leaving only the veins.   A skeleton of a Birch
leaf is shown in Fig. 34.

(vii) Secondary growth in the stem.

When an old Sunflower stem is cut across, the conducting tissue is not found in detached vascular bundles, as in a younger stage, but forming a complete ring. In a transverse section under the microscope it is seen that the inner part of this ring consists of wood, with large

Fig. 34. Leaf of Birch showing pinnate venation.
(After Ettingshausen.)

vessels scattered among many smaller thick-walled cells, and that the outer part is thin-walled phloëm. Between these zones of wood and bast is a thin complete ring of cambium. Cambium, wood and phloëm must therefore have been formed between the vascular bundles, in the medullary rays.

If the section is carefully examined (Fig. 35), the wood of the original vascular bundles, the *primary* wood (x_1), is seen at intervals on the inside of the zone of wood; and, on the outside of the phloëm, opposite to each of these groups of wood, are the corresponding groups of primary phloëm (p_1), each with its protecting strand of fibres (f). The two portions of the original bundle have evidently been separated by the formation of new or *secondary* wood and phloëm, between which cambium is still found. The cambium has, in fact, formed new wood inside (x_2), between it and the old

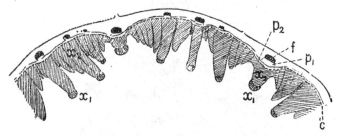

Fig. 35. Part of a transverse section of an old stem of a Sunflower: x_1, primary xylem, p_1, primary phloëm, of the original vascular bundles, and f, pericycle fibres; x_2, p_2, secondary xylem and phloëm, formed by the cambium, c. Some parts of the cambium have formed more wood than others.

wood, and new phloëm outside (p_2), between it and the old phloëm.

The cambium consists really of only one layer of delicate cells, which are rectangular in transverse section. These divide repeatedly, cutting off cells on their inner side which form the vessels and other constituents of the new xylem, and similarly on their outer side cells which form the sieve-tubes and other parts of the new phloëm. These cells look at first so like the cells of the cambium that it is difficult to tell which is the true cambium, especially as they may go on dividing further; all these similar cells together form a zone of delicate growing tissue which may be called the cambial zone. On the inside of this zone the cells grow in width and finally become thicker walled and assume their mature form and structure, while on the outside corresponding changes occur.

The new cambium which has appeared across the medullary rays, connecting the cambium of the vascular bundles into a complete ring, has also formed wood and phloëm. It arose by the division of parenchymatous cells of the medullary ray, which formed walls parallel to the circumference of the stem, and so cut off thin cells, rectangular in transverse section, which are similar to the cambium cells of the vascular bundle.

Fig. 36. A cambium cell, *c*, with two cells which have been cut off by it, on the one side, *x*, to form xylem, on the other, *p*, to form phloëm. The cells are represented with the upper part removed to show their appearance in transverse section.

In the Sunflower stem, secondary wood is formed in the bundles first; then new cambium appears on either side of some of the bundles so that the added layers of wood get broader and broader and the bundles appear to widen. Meantime secondary growth often begins half-way between two bundles and a little group of wood and phloëm appears, forming a little secondary vascular bundle, and from this as a centre the growth spreads sideways again. Eventually the cambium ring is completed and continuous zones of wood and phloëm are then formed right round the stem.

It must be remembered that, although we speak of a *ring* of cambium because we are referring to transverse sections, the cambium really forms a hollow *cylinder*. In longitudinal section it is found between the secondary xylem and phloëm, on each side of the pith, continuous through the length of the section. The secondary wood and phloëm are formed in *cylindrical* layers on the inner and outer faces of the cambium cylinder.

The further growth of a stem brought about by the division of the cells of the cambium cylinder is called *secondary growth* in thickness or *secondary thickening* to distinguish it from the *primary* growth of a young stem. In *primary growth* the growth is not localised in one region, the living cells generally are growing and dividing; while in *secondary growth* the growth is chiefly localised in the cambial zone. In the primary growth of each new portion of stem formed at the growing point the *primary tissues* are laid down, including the vascular bundles, pith, cortex, epidermis, etc. The new tissues formed later on by the cambium cylinder are called *secondary tissues*.

By this process of secondary growth more conducting tissue is provided to meet the increasing needs of the plant as it develops new leaves and branches, or flowers and fruit. At the same time, as the secondary wood consists largely of strong thick-walled cells, the amount of mechanical tissue is increased and the stem strengthened to bear the growing weight of the new parts that develop above.

In many large plants, especially in trees, new secondary tissues are added year by year. In the secondary wood, zones formed in successive years can be distinguished from one another: they are called *annual rings*. These will be seen when we come to examine trees.

The new zones of wood and phloëm increase considerably the thickness of the stem. Growth in thickness, of course, also means a corresponding increase in circumference. The epidermis and cortex are, therefore, very much stretched. For a time they yield to this stretching, the cells growing in width and dividing by walls at right-angles to the circumference of the stem (radial walls): in Fig. 29, III, some of the cells of the epidermis have just divided, and the new cell-walls are still very thin. Such growth does not, however, go on indefinitely: in trees and shrubs, in which secondary growth continues for many years, the epidermis eventually splits. Under the cracks so formed the tissues would be exposed were no other means of protection forthcoming. In these plants a second cambium ring arises, usually in the cortex just below the epidermis, which forms a tissue on its outer side of very regular oblong cells which lose their protoplasm and become filled with air. This tissue is *cork*, similar to that seen on the outside of a potato (p. 115). It replaces the epidermis as a protective layer. The walls of the cells are altered so that they will not allow water to pass through them, becoming somewhat similar in composition and properties to cuticle. Hence everything outside this waterproof layer, being cut off from the water supply, dies: the dead layers form the bark of old trees.

(viii)   Secondary growth of roots.

In their primary tissues roots have no cambium. Nevertheless old roots are found to be much thicker than they are when young. If a cross-section of an old root of a Sunflower be examined, zones of wood and phloëm are found, very like those in the stem, separated by a cambium ring. In the centre is the small pith, and the strands of primary wood can often be found around it, still with the points of the wedges directed characteristically outwards (cf. Fig. 37, 3).

Alternating with the wedges of primary wood are great blocks of secondary wood which dip down between them. The primary phloëm is thus no longer in its old position, but has been displaced outwards. The masses of secondary wood are separated from one another opposite the small wedges of primary wood by medullary rays which appear in the section as paths of

parenchymatous tissue down which the eye travels to find the
wedges of primary xylem.

This change of structure has been arrived at, as in the stem,
by growth from a cambium. A layer of actively dividing cells
develops just inside the phloëm strands and spreads between the

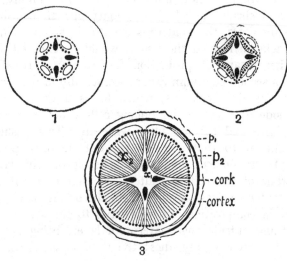

Fig. 37.   Diagrams of stages in the secondary growth of a root as seen
in transverse sections. Cambium is represented by heavy dots,
primary xylem black, secondary xylem shaded, phloëm unshaded:
the broken line represents the endodermis bounding the central
cylinder.

In stage 1, cambium has arisen on the inner side of each
primary phloëm strand. This cambium has in stage 2 already
formed secondary xylem and phloëm, but the cambium ring has
only just been completed, by cell-division outside the primary
xylem. In stage 3 much more secondary xylem (x_2) and secondary
phloëm (p_2) have been formed, so that the primary phloëm (p_1)
has been pushed out and stretched: cork has been formed in the
pericycle, and outside it are the remains of the cortex.

phloëm and xylem till it is completed just outside the wood
strands; it thus forms in transverse section a wavy ring (Fig. 37, 2).
Its growth is at first most active inside the phloëm strands, so that
the cambium there is pushed outwards by the wood which it
forms, and, sooner or later, the once wavy ring is smoothed out

and becomes round (Fig. 37, 3). Opposite the primary wood the cambium only forms broad medullary rays of parenchymatous cells, which maintain the connexion between the pericycle of the root and its inner parenchymatous tissues. The rest of the cambium, round the inner side of the phloëm strands, forms the masses of secondary wood on the inside and secondary phloëm on the outside. By this secondary growth the strands of primary phloëm, which alternate with the strands of primary wood and are at first quite near to them, are pushed outwards, becoming widely separated from them just as in the stem; but they remain, of course, opposite the position they originally occupied, that is they alternate with the primary wood strands.

The cortex very soon becomes crushed by the growth in thickness of the central cylinder, and cork is formed in the pericycle by a cork cambium, like that which arises in stems usually in the cortex. Thus old roots lose the whole of their cortex, for when it is shut off by the impervious layer of cork from all supplies of food it dies and decays.

(ix) The stem and roots of Monocotyledons.

The Sunflower stem illustrates the kind of structure and arrangement of tissues found in the stems of a very large class of plants called Dicotyledons. There is another large class, called the Monocotyledons, in which the stems show certain differences from the dicotyledonous type. When a transverse section of the stem of a Maize plant is examined, it is seen that the vascular bundles are not arranged in a ring, but are scattered, those near the centre being larger, those toward the outside smaller and more numerous. In some similar stems, like that of the Lily or Solomon's Seal, it is possible to see that there is a little cortex on the outside, bounded within by endodermis, and that the vascular bundles are scattered in the general tissue (similar to the pith of Dicotyledons) of the central cylinder; but in the Maize stem an endodermis cannot be detected.

Just within the epidermis is found a zone of strong woody mechanical tissue; Monocotyledons depend very greatly on such tissue as this for their rigidity, as they do not add to their woody vascular tissue by secondary growth.

Each bundle consists of phloëm, on the outside, and xylem, on the inside, as in a dicotyledonous stem. The xylem in the Maize is very characteristic. In a transverse section of the bundle the xylem is Y-shaped: the base of

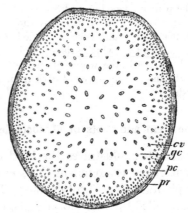

Fig. 38. Diagram of a section across the stem of a Maize plant, showing the mechanical tissue on the outside and the scattered vascular bundles. (After Strasburger.)

the Y is formed by a few annular vessels, the innermost of which has broken down and left a gap, and there are two very large pitted vessels, one on either side of the bundle, which form the arms of the Y. These are not all general features of Monocotyledons; but their xylem is often more or less V or Y-shaped and sometimes completely encloses the phloëm: the bundles are in the latter case called *concentric* bundles.

The phloëm is composed of sieve-tubes very regularly arranged to form a pattern like a draught-board with small cells filled with dense protoplasm plainly seen at the corners of the squares.

No cambium can be found between the wood and phloëm, nor does cambium ever develop there. The bundles are, therefore, called *closed* bundles because they are not capable of further growth like those of dicotyledonous stems. Moreover, they are often partly or completely surrounded by a sheath of thick-walled cells. There is no secondary growth in Maize stems, and in Monocotyledons as a whole growth in thickness by the activity of a cambium is very rarely found: the increase in diameter of the stem is due to the general growth of the parenchymatous tissues. The stems of Palms, as an extreme example, reach a very large diameter near the growing point, by primary growth: the stem apex and likewise the leaves are very large.

In the erect stems of many Grasses, and the assimilating stems of Rushes, which are hollow or contain a large pith, the vascular bundles are crowded into a narrow belt near the outside, sometimes forming a single ring. Such a stem is, however, easily distinguished from one of the dicotyledonous type, for the bundles are always closed, containing no cambium, and are often surrounded completely by woody mechanical tissue.

In the *roots* of Monocotyledons the same arrangement of alternating strands of wood and phloëm is found as in those of Dicotyledons, with the difference that there are usually a larger number of strands present.

The roots, like the stems, do not add to their mechanical tissue by secondary growth, and, in order to supply their need for strengthening tissue in the central portion of the root (compare p. 133), a great part of the parenchymatous tissues of the central cylinder becomes thick-walled and woody. Very soon after the root-hairs have finished their work, the layer of cells just below the hair-bearing (piliferous) layer develops somewhat thicker walls similar in composition to cork and forms a protective covering called, by analogy with the epidermis of the stem, the *exodermis*; moreover the endodermis often becomes thick-walled.

(x) The structure of growing points and the development of the tissues.

We have remarked in speaking of the cambium that growing points consist of cells closely packed and full of protoplasm. This may be seen without much difficulty by cutting a longitudinal section through the tip of a shoot, say of a young Sunflower or an Artichoke. At the very tip, all the cells are about as long as broad. Here is the growing point where the cells are actively dividing to form the rudiments of new leaves, nodes, and internodes.

Immediately below, where rapid growth in length is taking place, strands of elongated cells can already be distinguished, which are continuous with the vascular bundles lower down and would become vascular bundles later on: these cells are like those of the cambium in the mature bundle, and the strands are called at this stage procambium strands. In each such strand the outer cells become the phloëm, the inner cells form the wood. Between the wood and the phloëm is a layer of cells of the procambium strand which remain little altered and full of protoplasm; this becomes the cambium. In Monocotyledons all the cells of the procambium strands form either wood or phloëm.

In the pith and cortex the development of ordinary parenchymatous tissue can be traced. While the cells in the procambium strand as they grow longer divide but seldom transversely, the cells of the pith and inner cortex divide repeatedly and so appear at every stage about as long as broad. As the cells grow larger, cavities full of sap appear in the protoplasm and eventually join to form one large cavity bounded by the protoplasm, which is then spread as a thin layer over the wall of the much enlarged cell. The cells also become rounded off, separating from each other at the corners to allow air to pass between them.

By following a procambium strand down from the apex, successive stages in the formation of vessels and other conducting cells of the mature stem may be found. Certain rows of cells grow in width, and later on the walls are strengthened, by the activity of the living protoplasm, with added layers, which are at first of cellulose, but afterwards become woody and then stain yellow with iodine. The cross-walls are dissolved away and leave only ridges running round the vessel at intervals, marking the positions they once occupied. At long intervals thickened cross-walls remain, usually oblique, separating one long vessel from another. Finally the protoplasm, which has exhausted itself in this process, disappears, leaving the complete vessel a tube containing water. A *vessel* is thus not a single cell, but is formed from a number of cells in a row; it is, therefore, better in speaking of the parts which go to make up a vascular bundle to speak of *elements* rather than cells, a term which includes vessels and dead cells without protoplasm, as well as living cells.

In most stems separate cells, long and usually more or less wedge-shaped at the ends, become thick-walled and lose their protoplasm, like the vessels do; these water-conducting elements formed from single cells are called *tracheids*.

In a longitudinal section it may be clearly seen that the walls of vessels and tracheids are not thickened over the whole of their surface, but that thin places are left. In the largest vessels and most tracheids these thin places appear as numerous scattered dots called pits; such vessels are described as *pitted vessels*.

The small vessels nearest the pith are strengthened by fine spiral bands, the rest of the wall remaining thin; these are called *spiral vessels*. These spiral vessels can be traced farthest upwards, in a longitudinal section through the tip of the stem: it is they, therefore, that are formed first, and they are distinguished as *protoxylem* from the rest of the primary xylem. The stem continues to form spiral vessels while it is still growing in length: the spiral band can be pulled out and so does not interfere with this growth. Sometimes (as in the stem of the Maize) *annular* vessels occur in the protoxylem, strengthened by separate rings: these vessels also can be stretched. The more continuously thickened walls of pitted vessels cannot be stretched and these are not formed until growth in length has ceased.

Apices of root and shoot compared. In the root the development of the tissues proceeds in a similar way;

but, in the procambium strands that form the wedges of wood, the earliest spiral vessels are found on the outside, not as in the stem on the inside, of the strand. In other words the protoxylem in the root is on the outside, in the stem on the inside of the primary wood. (The term protoxylem is not to be confused with primary xylem:

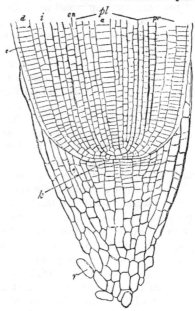

Fig. 39. Longitudinal section through the middle of the tip of a root of Barley, showing the growing point (k), the root-cap below it, with cells that have been rubbed loose (r), and the tissues of the root-tip behind the growing point: pl, young central cylinder; i, intercellular spaces forming between the cells of the cortex. (After Strasburger.)

the protoxylem is that *part* of the primary xylem or primary wood which is formed first.) Where the root and stem join there is no sharp line between them, and a more or less gradual change occurs from the structure characteristic of one to that of the other.

There are other important respects in which the growing regions of roots and stems differ. The growing point of a stem is situated at its very tip; but that of a root is covered by a more or less conical or thimble-shaped cap of cells, called the *root-cap*. The outer cells of this are loosened and rubbed off by friction with the soil as the root-tip grows and pushes its way through. While the root-cap is thus continually being worn away, it is renewed from within, for the growing point adds to the root-cap on the one side as well as to the length of the root on the other.

The growing point of the stem, being in the air, is not in need of the same kind of protection. The chief risk to which it is exposed is loss of moisture and desiccation: from this it is protected by the young leaves which overlap it. Shoots which grow underground have various special means of protection (see p. 174 and Ch. xv).

Another important difference between shoots and roots is in the formation of lateral organs.

The lateral organs of the shoot, the leaves and branches, arise quite near the apex as outgrowths on the outside of the stem. The rudiments of the leaves appear first as tiny bulges, which increase in size and develop the form characteristic of the mature leaf, while the tip of the stem grows on beyond them and forms more leaf-rudiments. At any one time during active growth leaf-rudiments are therefore found in all stages of development, the youngest nearest the apex. In the axils of the leaf-rudiments other bulges appear, on which in turn leaf-rudiments sooner or later arise; these are the axillary growing points which form axillary buds and, later, branches.

In the root, no lateral organs are formed close to the growing point. It is easy to understand why this should be, for if lateral roots grew out from a part of the root

which is still pushing its way through the soil they would
be broken off or badly injured.   Therefore even the root-
hairs are not formed except on parts of the roots some
little distance from the tip where growth in length has
ceased.   The lateral roots are formed still farther back.
Moreover, they do not arise on the outside, like leaf-
rudiments, but deep down in the tissues of the root, in
the pericycle.   Cell-division begins in cells of the pericycle.

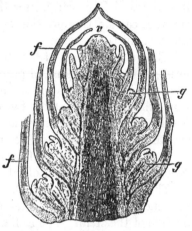

Fig. 40.   Apex of a shoot of
Mare's-tail(*Hippuris*) dissected from the bud and
much magnified: at the
top is an apical dome
of tissue and below this
are successively older
and larger bulges which
grow into leaves. (After
Sachs.)

Fig. 41.   Longitudinal section through
the apex of a shoot, showing the
apex (*v*), bare of appendages; leaf-
rudiments (*f*); growing leaves overlapping the apex; and developing
axillary growing points and buds (*g*).
(After Strasburger.)

A growing point thus arises which forms a root-cap and
gradually pushes and dissolves its way through the cortex
of the main root till it reaches the soil outside.   Where
a lateral root emerges a vertical split can usually be seen
in the cortex (for instance, in the root of the Bean).
This mode of origin from within is called *endogenous* in
contrast with the *exogenous* origin of leaves and branches.

In the seedlings of the Broad Bean it is readily seen
that the lateral roots are arranged in vertical rows. This
is because they arise in this as in many other plants
opposite the xylem strands which run straight along the
root; there are therefore as many rows as there are
strands of xylem.

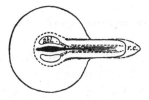

Fig. 41a. Diagram of a cross section of a root passing through a lateral
root: the latter arose in the pericycle opposite the protoxylem and
has perforated the cortex of the parent root. (Compare Figs. 31, II
and 37, 1.) Broken line, endodermis; xylem shown black; *phl.*, phloem;
r.c., rootcap.

CHAPTER XII

SPECIAL FORMS AND FUNCTIONS

In the course of our study we have met with leaves, stems and roots of very varied forms. Some of them differ very markedly from what may be regarded as the ordinary type of structure, the peculiarities being usually associated with the performance of some special function. We may recall in illustration the swollen tap-root of the Carrot and the tubers of the Potato, which have as their special function the storage of reserve food-materials.

'Modified' organs.

As a rule, these peculiar organs are at first indistinguishable in form from other shoots or roots, but as they grow they begin, some soon, some later, to develop differently.

The young tap-root of a Carrot plant is very like other main roots, and bears lateral rootlets; only as it grows older does it increase in thickness.

The branch which will produce the Potato-tuber arises like other branches in the axil of a leaf, but as the leaf is usually one that is borne on a part of the stem below ground it has not expanded or become green. This branch grows along more or less horizontally and remains below the ground; the leaf-rudiments which arise at its growing point therefore never expand into foliage-leaves but develop into small scales closely pressed against the branch. Later, the end of the branch swells:

the bases of the little scale-leaves take part in its growth and become wider, their axils grow into broad hollows, and the buds are also modified in form, the little leaf-rudiments becoming separated from one another so that they no longer closely overlap the growing point.

If a well-grown Potato plant be dug up it is easy to find underground branches in various stages of development, some still thin and of uniform thickness, others just beginning to form tubers, others again with well grown tubers. There is therefore no difficulty in observing for ourselves all the features of structure and development just mentioned.

Fig. 42, I is a drawing of a small young tuber, which bore leaf-rudiments in which the leaf-base and the leaf-blade were distinguishable. Clearly, the long ridges on

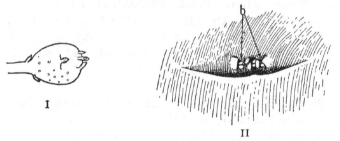

I

II

Fig. 42. I, young Potato-tuber (slightly enlarged) showing leaf-rudiments with blade and leaf-base distinguishable: the small spots are lenticels (p. 390). II, eye of Potato-tuber magnified, showing the ridge formed by the leaf-base and seven leaf-rudiments of the axillary bud, which have been separated by the growth of the tuber; the two outermost of these also have axillary buds, *b*.

an old tuber, which mark the positions once occupied by leaf-rudiments, represent only the leaf-bases. If blade-rudiments developed at all, they have been rubbed off against the soil during the growth of the tuber.

Fig. 42, II shows what the 'eye' of a Potato-tuber is like when it has been carefully washed and all soil

removed with a soft brush. The leaf-rudiments of the
bud in the axil of the scale-leaf are plainly seen. In
this specimen there were also still other buds (*b*) in
the axils of the two oldest leaf-rudiments of the 'eye,'
two very small leaf-rudiments belonging to each being
visible.

Thus the stem, leaves and axillary buds all become
very much modified during the formation of a tuber.
Nevertheless even the mature tuber cannot be mistaken
for a root, although it is found below ground. The 'eyes'
are clearly modified buds. Besides, the eyes and their
subtending ridges are closer together towards one end of
the tuber, and at the very end instead of ridges we find
leaf-rudiments, like those of the axillary buds, forming a
terminal bud. At the opposite end is a scar, where the
tuber broke off from the unthickened part of the branch.
By such features as these it is usually not difficult to tell
the nature of any organ after carefully examining it, even
though it has become very much modified; and more can
be told by a study of its development.

Modifications of the forms of organs cannot always
be explained as easily as in the example just considered.
It is, for instance, seldom possible to give satisfactory
reasons for the various forms assumed by the foliage-
leaves of different plants.

Many seedlings have their earlier leaves different in
form from their later ones. The first two foliage-leaves
of the Runner Bean, although large and fully expanded,
are simple, while the later leaves are compound. In
other plants the early leaves assume a succession of forms,
which show a transition between the earliest and that
characteristic of the mature plant. A leaf of a mature
Broad Bean plant has two pairs of leaflets and two leaf-
like outgrowths from the leaf-base, called stipules; the

leaf-axis bearing the leaflets ends in a little point. In the lowest leaves of the seedling only the stipules are well-grown, the rest of the leaf hardly developing at all and appearing merely as a little tooth between the stipules. Between these reduced scale-leaves and the leaves with two pairs of leaflets are leaves with only one pair of leaflets.

Another interesting example showing transition is the

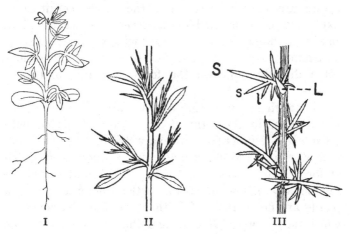

Fig. 43. I, young seedling of Gorse, showing simple cotyledons and first leaves with three leaflets (after Dale, from Ward); II, part of an older seedling, with the upper leaves simple and subtending branch-spines; III, part of older twig, with leaves (*L, l*) and branches (*S, s*) spiny.

seedling of the Gorse. The leaves of the mature plant, the Gorse bush, develop as green spines, but the first foliage-leaves are compound, having three little leaflets. Later leaves are simple, but still flat and leaf-like. Finally, the narrow leaf-spines are produced, with stiff branch-spines in their axils which are green and help in photosynthesis. Clearly these different forms of leaves are not to be explained fully by differences of function; for from this point of view it is difficult to understand why the compound leaves should be produced at all. It is probable that the remote ancestors of the Gorse had compound leaves and did not grow in dry exposed situations; since then the Gorse has become adapted, by the reduction of its leaves, to a dry habitat, and the compound leaves of its seedling are a relic of the ancestral type.

If Gorse seedlings be grown in moist air the leaf-spines are not produced, but the leaves continue to expand into little leaf-blades. Even the mature plant produces similar leaves in very moist weather. This is an example of the important fact that external conditions often directly influence the form of plants and their organs.

Although many of the most important modifications which occur in the form of organs are connected with differences of function, and every difference must affect their working, it would be a mistake to expect every detail of form and structure to 'have a use,' that is, to be functionally significant. Some of the differences between plants are no more 'useful' than are the characters in which the features of Negroes and Chinese differ from those of Europeans.

Forms of leaves.

Even foliage-leaves, having one and the same principal function, photosynthesis, exhibit a great variety of forms. Our knowledge of the reasons for these differences is far from complete, but one important factor has already been mentioned which affects the size and structure, and probably the shape, of leaves, namely, the need for regulating transpiration and evaporation according to the dryness of the habitat. In dry, sunny situations, plants have usually small, simple, tough, often hairy leaves;

in moist, shady situations often large, broad, thin and delicate leaves.

When many different leaves are compared certain parts can be distinguished, though not all are represented in every leaf: they are the leaf-base, leaf-stalk or petiole, leaf-blade, and stipules. The *leaf-base*, by which the leaf is attached to the stem, is not always distinct from the stalk; but in many leaves it is broad and flattened, clasping or sheathing the stem (e.g. Buttercups, most Monocotyledons, etc.). The sheath of a Grass leaf is a long sheathing leaf-base. In the Gooseberry bush the spines are outgrowths from the back of the leaf-bases.

Stipules occur one on each side of the leaf-base, more or less united with it, or quite separate from it (*free* stipules). They are sometimes green and leafy, sometimes membranous. Many leaves have none. When present they usually serve as an additional protection for the axillary bud, or for the younger leaves in the terminal bud while the leaf is expanding (as in the Broad Bean); the stipules of the Dock are fused and form a membranous sheath round the stem and the next younger leaf. In some plants stipules serve special functions (e.g. winter bud-scales of Beech).

The *leaf-stalk* or *petiole* varies in form and thickness and in strength with the weight of the blade and the position in which it must be supported. It may be cylindrical, or grooved on the upper surface, or flattened, sometimes with a thin green wing along either side. The leaves of the Aspen hang downwards and flutter in the slightest breeze because their petioles though tough are slender and flattened. On the other hand, large leaves, like those of the Horse-chestnut, are held up by stout petioles, while the stalk of the much larger leaf of Rhubarb is stouter still. Leaves which have no stalk are called *sessile*.

The *blade* varies very much in form, and in describing leaves it is useful to have special terms for various shapes, for different kinds of venation, margin, tip and base. Some of the most useful of these are illustrated in Figures 44 and 45.

In the first place a leaf may be *simple*, with a single blade; or *compound*, with the blade divided into separate *leaflets*, each of which might at first sight be mistaken for a leaf. The position of the axillary bud generally affords a guide: leaflets have no axillary buds. The leaflets may be arranged on either side along a common stalk (*pinnate*) or attached all at the end of the leaf-stalk (*palmate*). The blades of leaflets are described in the same terms as those of simple leaves.

The veining or *venation* of leaves is of two principal types, *parallel* and *reticulate*. In *parallel-veined leaves*, like the Grass leaf, the main veins enter the leaf-base separately and run nearly parallel to each other from one end to the other; while small veins cross at right angles and connect the main veins. This type, as we shall see later, is characteristic of a large class of plants, the Monocotyledons. In *net-veined* leaves, that is leaves with reticulate venation (Latin *reticulum* = net), the main veins

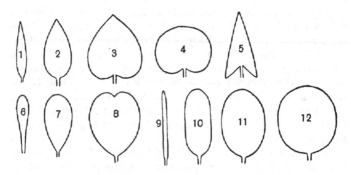

Fig. 44. Diagrams illustrating the principal terms used in describing the form of the blades of leaves or leaflets.

1–5, broadest near base; 6–8 broadest near apex; 9–12, middle broadest.

1, lanceolate; 2, ovate; 3, heart-shaped (cordate); 4, kidney-shaped (reniform); 5, arrow-head-shaped (sagittate).

6, spathulate (with blade merging into stalk; otherwise oblanceolate); 7, obovate; 8, obcordate.

9, linear; 10, oblong; 11, elliptical; 12, round (rotund).

branch either palmately (Garden Nasturtium, Sycamore) or pinnately (Beech, Sunflower), or in some cases run more or less parallel from base to tip (Plantain), but smaller veins run from them and branch repeatedly to form a fine and elaborate net-work. Most plants with reticulate venation belong to another large class, the Dicotyledons, though a few are Monocotyledons.

The edges of leaves or leaflets are sometimes even (entire) but more often indented in various ways (Fig. 45). When the edge is deeply indented the leaf is called either *lobed*, if the indentations extend about half-way to the midrib, or *divided*, if the indentations are deeper still. According to their venation leaves may be pinnately or palmately lobed (or divided).

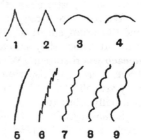

Sharp teeth enable rain to run quickly off a leaf, as drops readily form on and fall from them. Some leaves end in a tapering tip (Lime) which is still more effective and is called a *drip-tip*. In tropical rain forests many of the leaves have drip-tips, some of them very long ones. By these means the water is prevented from running under the

Fig. 45. Common forms of tips (1–4) and margins (5–9) of leaves: 1, tapering (acuminate); 2, pointed (acute); 3, blunt (obtuse); 4, notched: 5, entire; 6, serrate (with teeth pointing forward like those of a saw); 7 toothed (dentate); 8, crenate; 9, wavy (sinuate).

leaf and blocking up the stomata; or is drained quickly off so that, when the rain stops, the surface quickly dries again.

The upper surface of leaves, moreover, is often smooth and shiny, or waxy, so that water runs off as readily as from a duck's back. This is a marked feature of the floating leaves of water-plants (p. 433). Other leaves have their under surface closely covered with hairs, which prevent rain from coming in contact with the stomata and blocking them, as well as reducing the loss of water vapour in dry windy weather.

The following examples illustrate the use of terms in concise description of leaves: Privet—shortly stalked, elliptical, acute, entire, pinnately veined; Crack Willow (Fig. 69)—shortly stalked, oblanceolate, acute, finely serrate, etc.; Lesser Celandine (Fig. 57) —long stalked, with sheathing base, cordate, palmately veined, with margin slightly crenate; Hawthorn (Fig. 56)—stalked, ovate, pinnately divided and coarsely toothed or lobed, with ear-like stipules; Rose (Fig. 63)—stipulate, pinnate, general outline obovate, leaflets ovate, serrate, acuminate.

Arrangement of leaves. Leaves are borne on the stem either singly, or two or more together at the same node. The first arrangement is called *alternate ;* successive leaves are distributed

spirally round the stem. When two leaves are present at each
node, on opposite sides, they are called *opposite ;* usually successive pairs are arranged transversely to each other, crosswise,
and this arrangement is called *decussate.* If more than two leaves
occur at a node they are called *whorled* and the number of leaves
in each whorl stated. Where the stem remains short so that the
leaves appear to rise from the ground (from the 'root-stock')
they are called *radical,* though strictly speaking their arrangement
is usually alternate.

Some of the ways in which leaves are displayed in order to
catch as much light as possible have already been described.
The form of leaves should be considered in connexion with their
display, for form and arrangement must suit each other. Small
or narrow leaves, for instance, are naturally more closely arranged
than large or broad leaves. It should also be observed that
many leaf-mosaics are formed by the twisting of the leaf-stalks,
so that the leaves assume different positions from those which
they had in the bud.

Some leaves do not keep always the same position but are
capable of movement. The leaves of Sunflowers, for instance,
turn to face the light in the early morning, the movement being
brought about by the leaf-stalk curving during growth. Such
leaves only change their positions while they are still growing:
but they continue to grow slowly for a long time. The leaves of
Clover, Runner Bean, Acacia (and many other leguminous
plants) and of Wood Sorrel take up a changed position at night.
These movements are frequently described as *sleep-movements.*
They are brought about by the curvature of the specially modified
leaf-base or leaflet-base, called a *pulvinus,* which is usually
cylindrical, and has a thick cortex of soft parenchyma. In the
case of the Runner Bean (*Phaseolus multiflorus*), for instance,
there is a pulvinus at the base of the main petiole and one at the
base of each of the three leaflets. By the curvature of these
pulvini the leaf and leaflets can take up a suitable position in
relation to the light; but at night the leaflets bend down till
they hang nearly vertical from the stalk. This is due to the
parenchyma on the upper side of the pulvinus becoming more
turgid than that on the under side. Next morning this change
in turgidity is reversed and the leaflets are raised again from their
nocturnal to their diurnal position. The leaflets of Wood Sorrel
and Clover (Fig. 46) also fold downwards at night. One result of

these 'sleep-movements' is that the leaflets are less exposed to the air, so that transpiration is diminished.

Similar movements are performed also in other circumstances by many of these leaves that possess pulvini. If a plant of Wood Sorrel growing in a pot is jarred somewhat violently on a table a few times, the leaflets assume the nocturnal position in response to the shock.

The leaflets of Wood Sorrel and of some Clovers (e.g. the Field Clover, *Trifolium pratense*), when closely examined, are found to be in continual movement, whether in the dark or in the light. These movements are not brought about, like the sleep-movements, by changes in external conditions, but are *spontaneous*. To see them, a light bristle or glass fibre may be attached with a little thick gum or shellac varnish as an index to a leaflet; the position of the point on a vertical scale is then

Fig. 46. Leaves of Clover in the day and night positions. (After C. Darwin.)

recorded at short intervals. The spontaneous movements of this kind performed by the leaflets of the Field Clover are often considerable and fairly rapid, a complete up and down movement occupying only about four hours under favourable conditions.

Modified leaves. Scale-leaves are found especially on underground shoots. In many cases it is not certain if they have any function when mature, though they may afford some protection to the axillary buds. They are, however, very important while still in the bud, where they protect the growing point from friction with the soil, like a root-cap protects the growing point of a root. Others are clearly modified as protective organs, against injury or disease (like the outer tough brown scales of bulbs and corms). Similarly the scale-leaves of winter-buds act chiefly as a protection against loss of moisture. The thick fleshy scale-leaves of bulbs are modified for the storage of reserve food-materials (p. 171).

Much reduced leaves, called *bracts,* are found subtending the flowers in many inflorescences (p. 209). These may be green, or membranous.

Xeromorphic leaves are often very different in form and structure from ordinary foliage leaves and are often small. The very small grooved leaves of the Heathers (Fig. 23, p. 107), and the needles of Pines and Spruces, are examples. The leaves of the Stonecrop are short, thick and succulent, and nearly circular in section (Fig. 47). The small narrow leaves of the Gorse are pointed and somewhat spiny.

In the Barberry the leaves on the main shoots develop as spines, the foliage leaves being produced on short shoots and appearing therefore as axillary clusters. These spines, and also

Fig. 47.   A piece of Stonecrop (*Sedum album*), showing the succulent leaves.   (After Praeger.)

the spines on the leaves of Holly, doubtless to some extent restrain herbivorous animals from eating the plants.

In many climbing plants the leaves or parts of them are modified as tendrils. In the compound leaf of the Garden Pea tendrils appear in place of upper leaflets; while in the Yellow Vetchling the whole leaf develops as a slender unbranched tendril, the stipules growing large and acting as assimilating organs.

Other special forms of leaves are possessed by many water-plants (Chapter XXIV) and by *insectivorous plants.* The leaves of the Butterwort (*Pinguicula*) are covered with short, sticky hairs. When an insect alights on the leaf its legs stick to the hairs and, if it is not large and strong, it is unable to escape; meanwhile the leaf slowly rolls up from the edges and so encloses the insect. Digestive fluid is then secreted which dissolves the softer parts of the insect's body: the digested juices are absorbed and the

leaf afterwards unrolls, but a period of several days may elapse
between the capture of the insect
and the expansion of the leaf.

The Sundew is another British
insectivorous plant. It has a
rosette of stalked leaves, the blades
of which bear a number of
stalked glands, or 'tentacles,' the
gland at the top secreting a drop
of sticky fluid looking like a dew-
drop: the outer tentacles are long
and spreading, the inner ones
shorter and erect. If an insect
alights on one of the glands it
sticks, and, while it struggles, the
outer tentacles quickly bend inwards and fasten upon the insect
pouring over it an abundance of
a digestive fluid. They remain in
this position till the juices are absorbed, and then expand again.

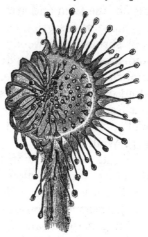

Fig. 48. Leaf of Sundew
(*Drosera rotundifolia*), with
tentacles on one side bent
inwards. (After C. Darwin.)

Shoots and stems.

The chief functions of shoots are photosynthesis and
transpiration. Both these functions are carried on
principally by the leaves. Stems bear the leaves in suitable positions, and conduct water to them, and carry
from them the food-material that they manufacture; but
young green stems, and even old stems, if they contain
green tissue, play also some part in photosynthesis, and
nearly all stems give off water vapour through stomata
or through lenticels (p. 390). Stems also store food-
materials when the supply from the leaves is greater
than is needed for immediate use. Ordinary shoots and
shoot-systems which carry on these functions are very
various, as we saw in Chapter I. Differences are found
in the forms of leaves and in the manner and the closeness

(or otherwise) of their arrangement on the stem (p. 159);
also in the form and arrangement of stems and branches.

Some stems grow erect and tall, and branch. Trees
and shrubs have woody stems which branch repeatedly
and grow in thickness from year to year: trees usually
branch high above the ground, the main stem forming a
stout cylindrical trunk; shrubs branch low down forming
bushes. Erect herbaceous stems may also branch either
high or low, in the latter case assuming a bushy habit
(e.g. Wallflower).

Other stems grow in a horizontal instead of a vertical
direction, *creeping* along the ground or sprawling over
neighbouring plants. They rest the weight of their leaves
on the ground or on other plants; thus they require less
building material and grow relatively weak, being quite
unable to support their leaves unaided in the erect position
(Creeping Jenny is an example). Most such creeping
shoots put out adventitious roots at intervals, chiefly
from the nodes. The leaves are often displayed horizontally in a close leaf-mosaic. The stems of *climbing*
plants may be compared with those of creeping plants:
these are also weak, but bear their leaves up into the
light and air by climbing up erect plants or similar
supports (see Chapter XXIII).

Modified shoots. There are many shoots or stems that
have special functions, and are modified in form.

By far the largest and most important class of modified
shoots are those which serve to spread and multiply the
plant (i.e. for *propagation*) or to enable it to survive
seasons unfavourable to growth (for *perennation*). Many
such are specially modified as storage organs, a store of
food-material being necessary in a resting shoot in order
that new leaves may be produced. Some leafy shoots
serve these as well as their ordinary functions. The stems
of trees and shrubs store large reserves of food-materials.

Creeping shoots that root at the nodes (Fig. 49) spread the plant and usually multiply it, for the old parts die, leaving the branches as separate plants. More usually the shoots which serve for propagation or perennation are not green and leafy. The Strawberry plant produces *runners* which arise in the axils of leaves at the base of the rosette. Each runner bears very few scale-leaves, far apart and closely pressed against the stem so

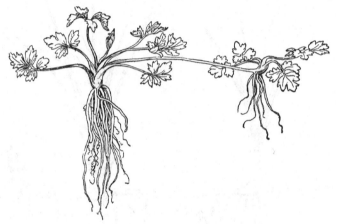

Fig. 49.　Creeping Buttercup (winter condition), showing creeping shoot arising in the axil of a leaf on the main plant. (After Praeger.)

that they offer no hindrance to its rapid progress over the soil or between other plants. After forming two or three long internodes the terminal bud turns up and forms a new cluster of radical foliage-leaves, puts out adventitious roots below, and so establishes itself as a new plant at a distance from its parent. From the axils of scale-leaves at the base of this plant grow other runners. In this way the Strawberry plant is rapidly spread and multiplied. The runners themselves die away sooner or later: it is the short stems of the rooted plants which last through the winter.

Numerous plants, on the other hand, have underground creeping shoots, or *rhizomes*, which not only spread the plant and multiply it, but also last through the winter, storing food in their tissues. The Couch Grass has thin, quickly growing rhizomes, bearing sheathing scale-leaves; after growing horizontally for some distance

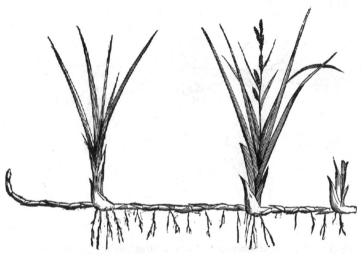

Fig. 50.   Rhizome of a Sedge, bearing scale-leaves; each portion ends in an aërial shoot, and an axillary branch continues the rhizome. (After Le Maout and Decaisne.)

they turn up at the end and grow into green aërial shoots. Buds in the axils of scale-leaves form branches of the rhizome at intervals, each of which produces an aërial shoot. It is these branching rhizomes that make Couch Grass so troublesome a weed, and so difficult to remove completely.

Many of the 'herbaceous perennials' of our gardens, like the small perennial Sunflowers, have long rhizomes,

by means of which they spread and last through the
winter. In other cases the rhizomes remain short, so
that all the new aërial shoots are close together: Michaelmas Daisies and Marguerites grow for this reason in
clumps.

Some rhizomes are very thick, and store large quantities
of food-materials. The rhizome of the Solomon's Seal is
an example. It is marked by many thin rings, the scars
of sheathing scale-leaves, and at intervals by circular
scars left by the annual aërial shoots. The upturned
end of the rhizome, which is enclosed in overlapping

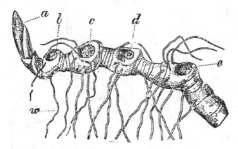

Fig. 51. Rhizome of Solomon's Seal: *a*, bud of next year's aërial
   shoot; *b*, *c*, *d*, *e*, scars of this and previous years' aërial shoots;
   w, adventitious roots. (After Strasburger.)

sheathing scale-leaves, grows in summer into an aërial
shoot, which bears leaves and flowers and dies down in
the autumn. A bud in the axil of one of the scale-leaves
grows out in the meantime to form a new piece of rhizome.
Each piece, from one circular scar to the next, is a lateral
branch of the piece behind it, although when it has become
swollen it appears like a continuation of the same stem
The rhizome of the Iris is very similar, but the segments
are shorter and more swollen.

The *stem-tubers* of the Potato have already been
described. They are the swollen ends of underground

branches; the unswollen portion of the branch, which
may be compared to the runner of the Strawberry, dies
away, leaving the tuber isolated at some little distance
from the parent plant. The tuber of the Artichoke
(Fig. 52) is a similar stem-tuber: the scale-leaves are

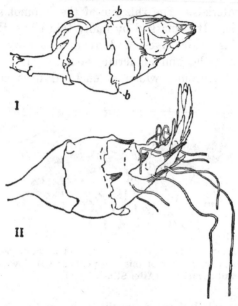

Fig. 52. Tubers of the Jerusalem Artichoke (*Helianthus tuberosus*).
I.   Resting tuber showing scale-leaves in whorls of three, with
     axillary buds, *b*, in one case expanded into a lateral tuber, *B*.
II. Tuber beginning to sprout. The terminal bud is forming an
     erect shoot, and adventitious roots are growing around the base
     of it.

better developed than in a Potato, and are arranged in
whorls of three. The terminal bud grows up to form an
aërial stem, just as in so many rhizomes.

Besides horizontal shoots, there are also vertical
underground shoots and stems. The stem must clearly

remain very short in such cases, and it is usually somewhat thickened, storing food and lasting through the
winter.   Such a short stem, bearing foliage-leaves, is
often called a 'root-stock.'   Meadow Plantains, Daisies
and other rosette-plants have short stems of this kind,
bearing adventitious roots at the base; while the similar
stem of the Dandelion is continuous with the long thick
tap-root.   These stems are kept below the level of the
soil as they grow by the contraction of the roots, which
are often seen to be covered with transverse wrinkles due
to this contraction.

When such a stem is much swollen it is a kind of tuber.
Two familiar examples are the tuberous
base of the stem of the Bulbous Buttercup (Fig. 53) and the *corm* of the Crocus.
The Crocus corm in its winter resting
condition is a hard swollen piece of
stem, packed with starch, covered by
scales and bearing horn-like buds at
the top.   When the scales are removed
one by one, a few outer scales are found,
attached at intervals, each completely
surrounding the stem and leaving a
narrow scar all round it when removed
(Fig. 54; 1, 2, 3).   In the axil of each
scale is a tiny bud (b_1, b_2, b_3), which
often becomes separated from it during
the swelling of the stem.   At the top of the stem the
scales are smaller: here a few of the axillary buds (B) have
grown large and pointed.   Each of these buds is covered
by several white sheathing scale-leaves, which completely
enclose the inner parts of the bud: when such a bud is
cut open longitudinally, yellowish young foliage-leaves
and tiny flower-buds are discovered.   The corm puts out
adventitious roots from its circular base, and in the early

Fig. 53.   Base of
stem of Bulbous
Buttercup. (After
Praeger.)

spring the buds expand and the flowers and leaves appear, growing at the expense of food-material stored in the corm. Later the stem of each bud begins to swell at the

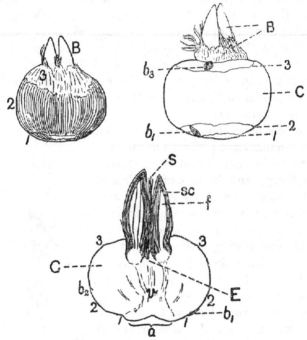

Fig. 54. Crocus corm, in winter.

C, swollen stem; E, its end, to which last season's aërial shoots have died down, leaving remains, S.

1, 2, 3, outer scale-leaves; b_1, b_2, b_3, dormant buds in their axils: B, large axillary buds at top of corm, with sheathing scale-leaves, sc, and young flower-buds, f: v, conducting tissue running from a, the base of the corm, where it was attached to the old corm, and where the adventitious roots will appear, to the new buds and the old aërial shoot.

base, storing the surplus food-material manufactured by the leaves and any that may be left in the old corm. In this way one or more new corms are formed, each a lateral branch of the old corm. When the old corm dies away it leaves a large scar at the bottom of the new corm: this is the flat base from which roots afterwards arise. The flowers soon wither as the summer passes, the fruits appear above the soil and shed their seeds, and the leaves die down. The white scale-leaves that covered the bud, and the bases of the outer foliage-leaves, become the brown scales which protect the new corm or subtend the new axillary buds at the top of it, in the midst of which can be seen the brown remnants of other leaf-bases and flower-stalks (S).

The Meadow Saffron also has a corm, but the bud which expands in spring arises near the base and is pressed close against one side of the corm, instead of being on the top of it.

Bulbs are erect shoots, usually underground, which store food in thick swollen scales, and not like corms and other stem-tubers in the stem. Many Monocotyledons form bulbs.

A Tulip bulb is covered by one thin brown protective scale, inside which are several concentric thick and fleshy scales containing food-materials. If the bulb be cut in half longitudinally it is seen that these scales are attached to a very short and broad flat piece of the stem, like a flat disc. The upper part of the stem, enclosed within the scales, is cylindrical; it bears young yellow foliage-leaves, and ends in a flower-bud, in which all the parts of a Tulip flower may be distinguished. This upper part of the shoot grows out in spring as the aërial shoot.

If the scales are carefully dissected off, an axillary bud can be found within each. In a Tulip bulb, dug up when the flower is faded, the scales are found to have lost much of their substance and become thinner, while one (sometimes more than one) of the axillary buds has grown and formed scale-leaves, which are swelling to store food now being made by the foliage-leaves: this is the new bulb. If a Tulip bulb is planted too near the surface of the soil, the new bulbs are carried out from the parent bulb and deeper into the soil on long stalks, forming *droppers*.

The Onion is another bulb, with several concentric fleshy scales storing food, and a few outer, thin and tough protective scales. They are not, however, scale-leaves, as in the Tulip bulb, but simply the modified bases of the foliage-leaves, from which the cylindrical green blades have withered: the ragged remains of these can be seen at the top of the bulb. The upper part of the stem within the scales grows during the second year of the life of the plant into a tall thick stalk ending in a spherical head of flowers. When the seeds have been shed the whole plant dies and does not form another bulb: the Onion is thus a *biennial* plant, the bulb merely storing food manufactured in the first year for the production of seed in the second year.

Other common bulbs are those of the Lily (with numerous overlapping, not concentric, scaleleaves), the Daffodil and the Snowdrop (with scales formed from leaf-bases).

Stems as well as leaves may assume a modified form in plants growing in dry situations —for example, the green axillary spines of the Gorse. In the Butcher's Broom (*Ruscus*), a small

Fig. 55. Twig of Butcher's Broom, showing leaf-like branches (*cld*) in the axils of scale-leaves (*a*) and bearing flowers (*fl*). (After Ward.)

shrub, the leaves are reduced to small membranous scales,

and in the axils of these scale-leaves are broad green pointed structures, looking just like leaves. On these leaf-like axillary branches the flowers are borne, subtended by small scale-leaves. In the Prickly Pear (*Opuntia*) and other Cacti and Cactus-like plants the leaves are small and soon drop off, or are represented by spines, while the whole of the green succulent part of the plant consists of stem and branches.

The sharp stiff points of the green branches of both Gorse and Butcher's Broom probably deter herbivorous animals from interfering with them. The thorns of Hawthorn (Fig. 56) and Blackthorn may have the like

Fig. 56.   Hawthorn, showing branch thorns.   (After Gregson.)

effect. These thorns are modified axillary branches, and usually bear buds and leaves.

Some climbing plants have branches modified as climbing organs (Chapter XXIII).

Dwarf-shoots. Several of the modified shoots just mentioned are examples of dwarf-shoots. Some of these are of limited growth, the growing point soon losing its power of growth, as in the case of branch thorns. In other cases they are simply branches that grow very slowly (e.g. dwarf twigs of Beech, p. 395), but can develop into long branches if the neighbouring long shoots are removed.

The clusters of leaves and flowers of the Cherry and of the Barberry are dwarf-shoots with very short stems.

A *flower* is a modified shoot of limited growth. We know that it is a shoot because it occurs in the axil of a leaf (or bract); but after the parts of the flower have been formed the growing point does not develop any further.

Buds.

The young and delicate tip of a shoot with all its parts folded together for protection is called a bud. In a *vegetative bud*, the young leaves fold over the growing point, and the outermost of them protect the rest of the bud until in the course of their development they bend outwards and begin rapidly to expand. If the leaves have stipules these usually play an important part in the protection of the bud while each leaf is expanding (e.g. Broad Bean). Special means of protection, in the form of scale-leaves, or stipular bud-scales, are possessed by buds which have to last through the winter (*winter buds*, Chapter XXII).

Buds which grow underground are also protected by scale-leaves, which fit closely and usually sheath the bud, making it conical and pointed. This shape is very suitable for the tip of a shoot which, like roots, pushes its way through the soil. As one scale-leaf is left behind by the growth of the next internode another is exposed and in its turn protects the delicate growing point and leaf-rudiments from friction with the soil: since the bud tapers and the scale-leaves fit closely they are not pushed back and rubbed off, but rather pressed more closely together.

Brussels sprouts are large vegetative buds, which

store much food in all their parts. When a sprout is
carefully dissected the leaves are found to have tiny buds
in their axils. The leaves on the outside are green, but
those within, where little or no light has reached them,
are pale green, yellow or, in the middle of the sprout,
white. The close 'heart' of a Cabbage or a Lettuce is a
similar very large bud, storing food formed by the outer
leaves till it expands to form the flowers and seed (runs
to seed). Bulbs are also large modified buds with scale-
leaves that store food.

Flower-buds enclose the organs of a flower during their
early development. Some buds enclose whole inflorescences (p. 389).

Roots.

The root-systems of plants are of different types, as
we saw in Chapter I, even those which serve the ordinary
functions of roots. Two main groups may be distinguished: (1) root-systems in which all the branches of
the system arise from a main axis, continuous with the
main stem; and (2) root-systems composed of numerous
adventitious roots, arising directly from the stem (usually
creeping or subterranean) and each forming, as it were, a
separate small root-system.

To the first group belong the root-systems of most
trees and shrubs and many annuals. Most other perennials, at least after their seedling stage, develop adventitious root-systems.

Within each group the character and general form
vary greatly. In the first, the relative importance and
length of the main root, and the depth at which the
lateral roots are developed most strongly, are the most
important points of difference. Adventitious roots, too,
may be long or relatively short, may spread near the
surface (Grasses) or grow down deep into the soil (Bracken,

Artichoke); they arise from any part of some stems (Bracken), only from the nodes of others, while they occur in a tuft at the base of an upturning aërial stem or root-stock (Strawberry, Creeping Buttercup), corm or bulb.

Modified roots. Roots, like shoots, are sometimes modified for the performance of special functions. Of these the commonest is perhaps the storage of food-materials.

Carrots, Turnips, and Radishes are examples of swollen tap-roots: a Carrot is conical and tapering, a Turnip round, suddenly narrowing below into the ordinary part of the root, while Radishes of both forms occur. But if we study the development of these swollen organs in the seedlings we find that not only the root but the base of the *stem* swells; so that the lower part of a Radish, for instance, is root, but the upper part stem. All three plants are *biennial*. During the year in which the seed is sown, the plant forms only leaves, and stores in the swollen axis the surplus food which they manufacture. The stem remains short, with its terminal bud close to the ground. In the autumn the leaves die down, and during the winter the plant remains dormant. In spring few leaves are produced but the food stored during the previous season is used up in the production of flowers and seed. These cultivated vegetables are much more swollen than the corresponding organs of the wild plants —for instance, the Wild Carrot.

The Dandelion is a *perennial* plant with a thickened tap-root: here too the stem is also thick, the tap-root and the root-stock merging into each other.

The Lesser Celandine, Peony and Dahlia have a number of swollen roots—*root-tubers*—which may become isolated from the parent and give rise each to a separate plant, or may yield up their food-reserves to the perennial

parent plant again for the renewal of its growth. In the
Dahlia, lateral roots swell at a little distance from their
point of origin to form more or less spindle-shaped tubers.
Each tuber tapers into a long thin portion like an ordinary
root. In the Celandine, each small blunt-ended club-
shaped tuber is an adventitious root, developed really
on an axillary bud which remains dormant. If, however,
the tuber becomes detached from the parent plant the
bud is carried with it and in the spring uses the food
stored in the tuber to form a new plant (Fig. 57).

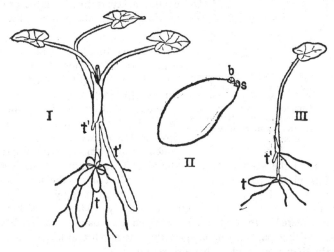

Fig. 57. Lesser Celandine (*Ranunculus Ficaria*) in early spring. I,
plant with cluster of old tubers (*t*) and two new tubers developing
(*t'*); II, detached tuber, enlarged, showing bud (*b*) and scar of
attachment (*s*); III, young plant growing from detached tuber (*t*)
with new tuber (*t'*) growing from a bud in the axil of a scale-leaf.

The swelling of the root-tubers of the Celandine is due entirely
to the expansion of the parenchymatous cells of the primary
cortex. In a transverse section the central cylinder occupies
only a small space in the centre, surrounded by a wide cortex,
and shows the ordinary structure of a root in which no secondary
growth has taken place.

On the other hand, in most cases (Carrot, etc.) the swelling is in part brought about by secondary growth in thickness. The cambium, however, instead of producing mainly vessels and other woody cells forms a large proportion of parenchyma and few water conducting elements.

In the root of the Beet additional cambiums appear, so that instead of only one zone of wood and phloëm, there are two or three such zones, one outside another (Fig. 58). These can readily be seen with the naked eye on cutting a beet-root across. Here also the cambium layers form a large proportion of parenchymatous cells.

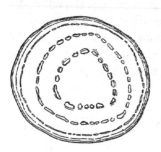

Fig. 58.   Cross-section of a Beet-root, about natural size, showing the concentric rings of bundles formed along with parenchyma by successive zones of cambium.

Some plants, such as the Ivy, have adventitious *climbing roots* which arise on the side of the stem away from the light and next the support (p. 429).

Many of the Orchids grown in hot-houses for their beautiful flowers have adventitious roots which hang down from the upper parts of the aërial stem into the air. These plants are natives of tropical forests where they live perched upon the branches of trees. In such situations their water supply, however abundant while it lasts, soon drains away. These *aërial roots* are covered with a spongy layer which absorbs water as readily as blotting paper; this enables them to catch and absorb more rain than they could otherwise retain.

The Mistletoe is an interesting *parasitic* plant which has no ordinary roots of its own but forms organs usually regarded as modified roots to enable it to absorb water from the tissues of a living tree—its *host* plant. These *haustoria* (i.e. suckers) spread in the inner cortex of a host branch, sinking frequent attachments to the wood of the branch so that water and mineral food pass from the host plant into the Mistletoe. The Mistletoe has green leaves and is able to make the rest of its own food by photosynthesis. But some other *parasites*, such as the Dodder, are not even green and are entirely dependent on their host for all their food and water. The Dodder grows on a variety of herbs, such as Clover, Nettle and Gorse, twining round them and putting out haustoria which penetrate the stem till they come to the vascular bundle: they then form connexions with the vessels and sieve-tubes of the host.

There are other plants—for instance, Eyebright, Yellow Rattle, and Bartsia—the roots of which attach themselves in the soil to the roots of Grasses or other plants, among which they grow, putting suckers into them and absorbing from them some of their nutriment. By digging up the roots of these plants and carefully washing away the soil, the attachment of the roots to those of other plants can be seen without great difficulty. Such plants have green leaves like the Mistletoe and make part of their own food; they are, therefore, called *partial parasites*, to distinguish them from *total parasites* like the Dodder.

SECTION III

REPRODUCTION

CHAPTER XIII

THE FLOWER

Hitherto we have dealt with the *vegetative organs* of plants; and we have discovered the functions of these organs in the nutrition and growth of the individual plant. When, however, we consider the whole life-history of a plant, we find that sooner or later it spends its energies lavishly in the production of flowers and seeds. We have already found that much food is stored up, for instance, in grains of wheat. We know also that seeds when sown produce new plants. In producing seed, therefore, the plant is bearing and providing for offspring—for a new generation of plants. The new plants which grow from the seed are just like the parent plant: they *reproduce* in every important feature the likeness of the parent. The whole process leading up to this, beginning with the formation of flowers, is therefore called *reproduction*. The flower contains the *reproductive organs* of the plant.

Great differences exist between plants in the extent to which seed production drains their resources. *Perennial* plants produce each year a crop of seed, but do not exhaust their stores of food-material completely in this process. They grow on from year to year, increasing in size, and storing food for their own growth. In trees and shrubs this continued growth of the parent plant is obvious; but in other perennials also, which have underground storage organs, the parent plant propagates itself vegetatively as well as producing an annual crop of seeds.

In marked contrast with perennials are those plants which only produce seeds once and then die, having spent all their energies for the new generation. *Annuals* like the Poppy and the Virginia Stock pass through their whole life in one season. Others (the *ephemerals*) pass through several generations within the year; such are the Chickweed, Shepherd's Purse, and other rapidly growing weeds. The Carrot, Turnip, Onion and other *biennial* plants store up food one year and flower the next. The Century Plant, or American Aloe, *Agave americana*, a slowly growing desert plant of Central America, stores up food in its thick leaves and stem for many years, and then expends it all in the production of one huge bunch of flowers like a candelabra, sets seed, and dies.

Flowering and seed production form thus a very important part of the life of plants. In seeking to understand this process of reproduction we must begin by examining a flower.

In an open Buttercup (Fig. 61) are visible on the outside
A Butter-   two rings or *whorls* of leaf-like parts, forming
cup.   the *perianth*. Just within the perianth are numerous yellow *stamens*, called collectively the *androecium*[1]; and in the middle of the flower are many green *carpels*, composing the *gynaecium*[1].

Of the leaves of the perianth the outer whorl of five are greenish; in the bud they overlap each other and cover the rest of the flower completely. These are the *sepals*, and the whole whorl is called the *calyx*. The inner whorl, the *corolla*, consists of five bright golden

[1] The meaning of these terms will be seen later (p. 186).

yellow leaves, the *petals*; each petal is opposite the space
between two sepals, so that the petals alternate with the
sepals.   Each petal is heart-shaped, and narrows towards
the base.   Near the base is a tiny yellow scale, covering
a small juicy spot (the *nectary*), on which a drop of liquid
may be found.   This liquid is sweet to the taste; it is
the nectar for which insects visit so incessantly the
flower, and which bees make into honey and store in
honey-combs for their young larvae.

The stamens are many; their number varies from

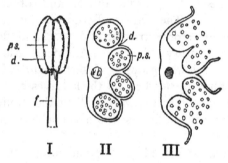

I   II   III

Fig. 59. I, stamen showing filament, *f*, and anther with four pollen
sacs, *ps*: *d*, lines of dehiscence. II, the anther cut across, showing
the four pollen sacs, containing pollen grains, and place of dehiscence; *vb*, vascular bundle. III, an anther dehisced. (Adapted
from Willis.)

flower to flower, and so is said to be *indefinite*. Each
stamen consists of a thin stalk, the *filament*, and the
anther above. The outer anthers, which like the outer
leaves in a bud are older, will probably have burst;
these disclose a mass of yellow *pollen*, which is powdery
and can be rubbed off with the finger, but is not dry
enough to fall very readily from the anther. Insects
visiting the flower, especially those with hairy bodies,
often become covered with yellow pollen in their efforts
to get at the nectar between the petals and the ripest

stamens. The anthers burst outwards, towards the petals,
and towards the heads of insects probing for nectar:
on this side a young anther is two lobed, and each lobe
is divided into two by a longitudinal groove. When it is
cut across four pollen-sacs full of pollen are found, in each
lobe two, separated by the groove. Each lobe of the
anther splits along the groove and the wall of each pollen-
sac curls back exposing the pollen, which very soon covers
the whole of this side of the anther so that its structure
can no longer be seen.

The carpels are, like the stamens, indefinite in number,
though not quite so numerous.
Examined under a lens each
carpel is seen to consist of a
round green body, the *ovary*, with
a very short tapering beak, the
style, bearing a yellow sticky
streak, called the *stigma*: when
the stigma is examined under
the microscope it is found to be
covered with short hairs or papillae,
which give it a velvety appearance.
If the carpel be torn open carefully
with a needle, a little round white

Fig. 60.   Carpel of a
Buttercup cut in half
longitudinally, show-
ing the stigma above
and the ovule inside.
(From Ward.)

object is found within it; this is an *ovule* (Fig. 60).

When all the parts of the flower are removed, the flower-
stalk is found to be a little swollen at the top where they
were attached: this swollen top is the *receptacle*. The
form of the receptacle and the arrangement of the parts
upon it can be seen very clearly if a flower is cut
longitudinally into two halves: this is best done by first
splitting the stalk with a sharp knife and then, holding
the halves of the stalk together, drawing the blade
upwards through the rest of the flower.

Except at the very beginning of the flowering season,

in any meadow where Buttercups grow, old flowers may
be found which have dropped their sepals, petals and
stamens; but the carpels have not only persisted but
have grown. Inside them are the ovules, which have
also grown. The wall of the carpel and the outer skin
of the ovule eventually become dry and hard. We
recognise the enlarged ovule as the *seed* of the plant:
it does not fall out of the carpel, but the whole ripe
carpel falls off, and, if it alights on a suitable spot, it
eventually germinates and produces a new plant. The
ripe carpels are called fruits; in the Buttercup each
fruit contains one seed.

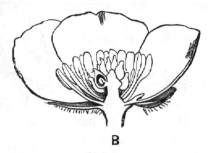

B

Fig. 61. Flower of a Buttercup, cut longitudinally to show the arrangement of the parts on the receptacle. (After Le Maout and Decaisne.)

So it is in the carpels that the seeds are produced.
All the other parts wither, dry up and disappear. What
then are their functions? What have they to do with
the production of seed? The sepals, we have seen,
protect the inner parts of the flower in the bud while
they are developing and are still young and delicate.
But what of the stamens and petals? Why does the
flower produce so much pollen, and of what value to the
plant is the nectar which the petals secrete?

In the first place we must ask whether the carpels
are able to produce seed by themselves. It is not difficult

to put this to the test of experiment. If a flower bud be carefully opened, the petals and stamens may all be removed with a small pair of forceps. This must be done before any of the stamens have burst: otherwise some of the pollen might be left behind and the experiment would not be completely trustworthy. Another precaution is necessary because of the insects which visit these flowers; if any came by chance to the flower which is being experimented with they might bring something from other flowers, and so spoil the experiment. We have seen, for instance, that insects get themselves covered with pollen. We therefore tie up the flower securely in a cover of fine muslin so that no insect can reach it.

Flowers prepared in this way produce no seed: the carpels shrivel up and die like the petals. This result might be due to the severe shock which the flower received when the stamens and carpels were removed. The experiment, therefore, is not by itself sufficient to prove that the carpels alone are incapable of producing seed. We must also show that the carpels of flowers treated in the same way can produce seed if they receive the right kind of help.

If we examine the carpels of ordinary open flowers very carefully with a lens, we find that the stigma is dusted with yellow pollen—in fact, its surface, covered with papillae, could hardly fail to have caught some pollen from the bodies of its numerous insect visitors. This suggests a further experiment. The petals and stamens are removed from several flower-buds which are each covered at once with muslin. After a day or two when the flowers would in the natural course have opened, the muslin bags are removed from half of them, and the stigmas are dusted with pollen from other flowers in which the stamens are ripe; they are then covered with muslin again.

In this case the carpels with *pollinated* stigmas produce seed, so that the failure of the others is not due to shock. Therefore the carpels alone are unable to set seed; pollen from the stamens must first be deposited on the stigma. In other words, *pollination* is essential to seed production. The function of the stamens is to produce the necessary pollen.

The two organs of the Buttercup, the carpel and the stamen, which we have now found to be essential to the production of seed, are called, by analogy with the sexes in animals, male and female. The carpels, in which the seeds are produced, are the female organs: the stamens, which bear the pollen that makes the carpels productive, are the male organs. The terms androecium and gynaecium by derivation mean 'male house' and 'female house' respectively (from the Greek *andros = man, oikos* = house; and *gunaikon* = the woman's part of a house, from *gune* = woman, and *oikos*).

The function of the petals also now becomes clearer. They bear the nectar for which insects visit the flower. We have only to watch the insects at work to convince ourselves that they must brush the stigmas many times with pollen as they crawl about the flower. We have already noticed that the pollen is shed outwards so as to fall between the petals and stamens, just where the insects probe for the nectar.

Besides providing nectar the petals also make the flowers conspicuous so that the insects can distinguish them more readily. If the petals of several flowers are removed these flowers will be found to receive fewer visits from insects than other complete Buttercups around them; and some of their carpels may not ripen.

The yellow stamens also help to make the flower conspicuous and thus to attract insects. Moreover the pollen itself is gathered by bees in special receptacles on

their hind legs, and by many other insects, as food for
their larvae; an abundance of pollen is thus an additional
attraction.

The presence of all these attractions would seem to
suggest that to Buttercups the help of insects is very
necessary. We can easily put this to the test of experiment by covering some unopened flowers with muslin,
without otherwise interfering with them. Under these
conditions fewer ripe fruits are produced; some Buttercups even produce none at all. As we have seen, the
anthers shed their pollen away from the carpels, so there
would seem to be little if any chance of it reaching the
stigmas of the same flower. If the flower were bent
over by the wind, some pollen might fall on to the stigmas;
some of the youngest anthers, too, may here and there
actually touch the stigmas of adjacent carpels. But the
pollination brought about in this way is uncertain and
meagre compared with the thorough dusting which
takes place during the frequent visits of pollen-covered
insects.

Further, why is it that the pollen is shed outwards,
and not inwards directly on to the stigmas? This question suggests another. Is the pollen of the same flower
suitable for its pollination? We may find the answer to
this latter question by carefully pollinating some flowers
each with its own pollen, taking great care that no pollen
from any other flower shall reach the stigmas. To
ensure this, the pollen left on the brush after pollinating
one flower must be destroyed before the same brush is
used on another flower: this can be done by dipping the
brush into alcohol, which is removed by wiping the brush
and allowing it to dry.

If the flowers are those of the Upright Meadow Buttercup (*Ranunculus acris*, see p. 279) even after careful
pollination in this way none of the carpels set seed. We

therefore come to the curious conclusion that the carpels of this kind of Buttercup only ripen if they receive pollen from other flowers—in other words, if they are *cross-pollinated*. To *self-pollinate* them, as in the experiment, is useless: they are *self-sterile*. Further experiments have shown that, in order that seed may be set and the flower be fertile, the pollen must not come from the flowers borne on the same individual plant, but from other plants of the same kind. It is now clear that the help of insects is essential to the successful pollination of this flower because they bring about *cross-pollination*. Insects must frequently bring about self-pollination as well in crawling over a flower (especially small insects, such as the small beetles so commonly found in flowers) but this has no result in this kind of Buttercup.

Other common Buttercups (e.g. the Bulbous Buttercup, *Ranunculus bulbosus*) are not self-sterile. Insects may self-pollinate as well as cross-pollinate these flowers also; so that some carpels may ripen as a result of self-pollination, others as a result of cross-pollination. But the help of insects must be important even to these flowers chiefly because they bring about cross-pollination, for the stigmas are not ready to receive the pollen till most of the stamens have already burst and insects have carried their pollen away. Moreover, there would be no need to attract insects in order to secure self-pollination, for, as we shall see in other flowers, this can be ensured much more simply; in the Buttercup it does occur to some extent even though insects are excluded, by actual contact of the outer stigmas with the youngest stamens, or when the flowers are shaken by the wind.

The blossom of the Cherry (or of the Plum) is another **The Cherry-** brightly coloured flower which is visited by **blossom.** bees and other insects. It has many stamens, though not as many as a Buttercup and in the

middle is a single carpel. Each stamen has a long thin
filament and a short anther which sheds its pollen towards
the centre of the flower. The erect carpel, which contains
a single seed, tapers into a long style with a knob-like
stigma at the top. Outside are five delicate pink petals,
and five green sepals, alternating with each other as in
a Buttercup; the calyx protects the flower in the bud
and the petals make the flower conspicuous. The petals,
however, bear no nectaries. The receptacle, too, is
hollow, like a cup: the carpel stands in the bottom of
the cup and the other parts of the flower are all attached

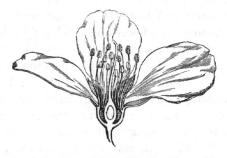

Fig. 62. Flower of the Almond (very like the Cherry-blossom) cut in
half longitudinally, showing the single carpel in a cup-like expanded
receptacle bearing sepals, petals and stamens on its edge. (After
Baillon.)

round its edge. The contrast with a Buttercup in these
respects is readily seen in longitudinal sections. The
curious form of the receptacle must clearly be due to the
outward and upward growth of the part just below the
carpel which has carried the other parts of the flower
with it. The inner surface of this cup, within and around
the bases of the filaments, is juicy with nectar. In this
flower, then, it is the receptacle that provides the nectar
for which insects visit it. Another feature of the flower
which probably helps to attract insects is its sweet scent;

and we may also notice that the flowers are the more conspicuous because they are borne in little clusters.

Many kinds of insects visit the flowers for nectar or for pollen. The nectar is more easily found than in a buttercup when the flower is fully open, for the stamens, which when young are curled inwards into the hollow of the receptacle, straighten as they grow and finally spread outwards away from the centre of the flower. The larger insects alight on the flower, and thrust their tongues down to the nectar. Smaller insects crawl over the flower and sooner or later find their way down between the style and stamens to the receptacle. Watch the insects at work and you will see that while turning or crawling about on the flower they can hardly help touching both the stigma and the anthers. Insects coming from other flowers bring from them pollen, some of which will probably be deposited on the stigma and bring about cross-pollination: on the other hand, as the stigma becomes ready to receive pollen at the same time as the anthers burst, it is clear that self-pollination may also result. Self-pollination can also occur even apart from insects.

Like a Buttercup, therefore, the flower of the Cherry is adapted to attract insects, and the importance to it of the visits of insects seems to be that they bring about cross-pollination which could hardly occur at all without their help.

The Dog-rose is a larger flower than either a Buttercup **The Dog-** or a Cherry-blossom. It has five large fringed **rose.** sepals, five large heart-shaped petals, and numerous stamens like those of the Cherry. All that can be seen in the centre of the flower, however, is a tuft of stigmas. Below the calyx is a swelling: when this is cut through longitudinally (Fig. 63, 2) it is found to be a hollow flask-shaped structure with a narrow neck, on the walls

of which are borne many carpels, each with a long style
that passes up through the opening at the top and ends
outside in a stigma in the midst of the stamens. In
this flower the receptacle has not merely expanded into
a disc but has grown up around the carpels and become
deeply hollowed, bearing the other parts of the flower to
a higher level than the carpels, which it nearly encloses.

Fig. 63. Dog-rose. 1, flowering shoot; 2, opening flower cut in
half longitudinally, showing the flask-shaped receptacle, with the
carpels attached inside it and the other parts of the flower to its
edge. (After Wossidlo.)

This receptacle does not, however, secrete any nectar,
nor do the petals. A Dog-rose is a honeyless flower,
which provides insects with pollen only; but on the other
hand it is large and so the more conspicuous, and it has
an even sweeter and stronger scent than the Cherry-
blossom. If such a flower is watched on a bright warm

day when the insects are on the move, many insects of different kinds may be seen to visit it. They usually alight on the stigmas, which afford a convenient landing-place in the centre of the flower, and therefore if they bring any pollen from other flowers on their bodies some of it is likely to be deposited at once on the stigmas. Cross-pollination thus often occurs first. But many of the smaller insects crawl about all over the flower, and so bring pollen from its stamens to its stigmas. Self-pollination is almost certain to happen in this flower in the absence of insects, for the stamens burst and the stigmas mature together: and, although the stamens spread at first a little away from the stigmas, they are slightly raised above them, so that the flower must often be swayed into such a position that pollen could readily fall on to the stigmas, while later on the stamens bend inwards again.

These three flowers all show remarkable adaptations to secure the visits of insects, which bring about cross-pollination. Many other flowers, however, are still more remarkable.

The Primrose has a tubular calyx, in one piece, with **The** five teeth at the top. The corolla is also in **Primrose.** one piece; within the calyx it forms a narrow tube but divides above into five broad, spreading lobes which alternate with the teeth of the calyx, just as the petals of the Buttercup alternate with its sepals. Anthers are seen inside the corolla-tube: they come off with the corolla, and when the tube is split open with a needle there are found to be only five of them, each anther attached almost without a filament (nearly sessile) to the corolla. Remaining behind in the bottom of the calyx-tube is an ovary, which unlike the carpels of the Buttercup or Cherry is round and symmetrical. When split open many tiny ovules are found within it, attached

to a central column, called the *placenta*. From the top
arises a long erect style ending in a knob-shaped stigma.
Nectar is found within the lower part of the corolla-tube
in the Primrose, and as the anthers block the way even
to very small insects only the larger insects with long
tongues can reach it.

When flowers from several different plants are examined
they are found to be of two kinds. In one, commonly
called a 'thrum-eyed' flower, the anthers are visible in

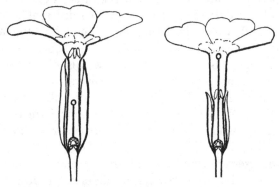

Fig. 64. Short-styled and long-styled Primroses cut in half longitudinally. The thick lines represent the actual section; the rest
of the flower is filled in more lightly.

the mouth of the corolla-tube, while the stigma is half
way down; in the other, 'pin-eyed' flower, the style is
long and the stigma occupies the mouth, while the anthers
are found half-way down the tube at the same level as
the stigma of a thrum-eyed, short-styled flower.

A bee visiting a thrum-eyed flower alights on the
flat spreading *limb* of the corolla. Round the mouth of
the tube the corolla is of a deeper yellow and this doubtless
helps the bee to find quickly the entrance to the tube.
It puts its proboscis straight down between the anthers
to the nectar at the bottom, and the part of the proboscis

at the level of the anthers is rubbed against them and well dusted with pollen. If the bee passes next to a pin-eyed flower, just this part of the proboscis will come in contact with the stigma, which will therefore be thoroughly pollinated. Meanwhile the proboscis is covered at a lower level with pollen by the anthers farther down the tube, and when the insect next flies to a thrum-eyed flower this part of the proboscis will be brought against the low stigma during the collection of the nectar.

There are thus two parts of the insect's tongue which get well dusted with pollen in the two different forms of flower, and which in turn rub round the stigmas of flowers of the *other* form. Most of the pollen deposited on the stigma of a long-styled flower is likely to come from the stamens of short-styled flowers, and vice versa, although there is also some chance of pollen from the same flower or from a flower of the same form reaching the stigma, while the insect is inserting or withdrawing its proboscis.

Primroses are thus adapted for a special kind of cross-pollination—between flowers of different forms, which are always found on different plants. Charles Darwin, who was the first to explain why there are two forms, pointed out another correspondence between the stamens and stigmas of the two forms. The papillae of the stigma are much larger in long-styled than in short-styled flowers: short-styled flowers on the other hand have much larger pollen-grains, so that large grains are mostly deposited on stigmas with large papillae, and small grains on stigmas with small papillae.

Darwin called the transfer of pollen from one form to the other 'legitimate' cross-pollination, as this is the mode for which the flowers are specially adapted, while cross-pollination between like flowers he called 'illegiti-mate.' He found by experiment that more and better seed resulted from legitimate than from illegitimate cross-pollination, and that the seed also yielded larger and healthier plants; while self-pollination yielded still less seed and still poorer plants than illegitimate cross-pollination.

It is important to observe how neat and precise is the method of pollination in the Primrose compared with the way in which it occurs in the other flowers we have examined. In these the whole under surface of the insect's body is covered with pollen as it crawls or turns about all over the flower. Only a little of the pollen ever reaches the stigmas, but this waste of pollen is necessary in order to ensure their pollination. In the Primrose, the nectar is enclosed, so that insects can only reach it in a particular way, and the stamens and stigmas are so placed that the pollen which is removed by an insect will be brought accurately to the stigmas of other flowers. Pollen is thus used very economically, and the stamens are few in number.

Notwithstanding this very effective mechanism, Primroses are not always cross-pollinated, for they open in spring when the weather is not always warm enough to tempt insects out; but if cross-pollination fails there is a chance of self-pollination. The flowers stand at first upright so that in short-styled flowers the stamens are above the stigma and pollen can be shaken down on to it; afterwards the flower stalks lengthen and the flowers gradually bend over, often to a pendant position, so that the stigmas of long-styled flowers are brought more or less under the anthers.

The White Deadnettle is another flower in which **The Dead-** cross-pollination is secured by the shape of **nettle.** the flower and the relative positions of the anthers and stigma. Calyx and corolla are each in one piece. The calyx has five teeth, which are not all equal in size. The corolla, which below is narrow and tubular, is divided into two lips, the upper, hood-like, standing erect, the lower spreading horizontally, forming a convenient landing-stage for insects. Under the hood and thus sheltered from rain are found four stamens, and a style ending in a stigma forked like an adder's tongue. The stamens have long filaments, attached low down to the corolla, bearing the anthers all close together. The style rises from an ovary within the bottom of the corolla-tube but quite free from it. This ovary has a curious

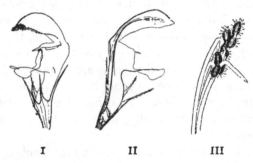

I   II   III

Fig. 65. Deadnettle. I, flower of White Deadnettle, showing platform and hood. II, the same cut in half lengthways, showing the style and stamens under the hood, the hanging lobe of the stigma, and the hairy ridge in the corolla-tube barring the way to the nectar. III, anthers and stigma from a garden Deadnettle.

form, being divided into four lobes which project beyond the point of attachment of the style: in each lobe is one seed. The forked stigma lies immediately under the anthers, with one lobe pointing forwards in a line with the style, and the other downwards. When the stigma is young its two lobes are close together. The whole of its outer surface is smooth like the style; only on the inner faces, which are not disclosed till the lobes separate, are papillae found, and only there will pollen readily adhere.

The meaning of these various points of structure and arrangement is seen when bees are watched at work on the flowers. They are visited chiefly by humble-bees for the nectar, which is present in sufficient quantity to be tasted if the corolla be removed and the base of the tube put in the mouth; it is efficiently protected from rain and small insects by a ring of hairs near the bottom of the corolla. A bee alights on the lower lip, puts its proboscis down the corolla-tube, and presses into it in the effort to reach the nectar, filling the mouth of the flower completely with its body. Its back thus comes against the anthers and stigma under the hood. It rubs past the stigma, pushing the lower fork backwards, and so must deposit on the stigmatic surface pollen from flowers which it has already visited. It also presses against the anthers and so removes more pollen; but on backing out of the flower it rubs the lower fork of the stigma outwards, and there is thus little chance of new pollen from the flower's own anthers being left on the stigmatic surface.

Here then is a mechanism by which cross-pollination is favoured and the chances of self-pollination at the same time diminished: on the other hand, as the stigma is so near the anthers, self-pollination is possible whether insects visit the flower or not, but cross-pollination is given the first and greatest chance: why the first chance is important we shall see later. As in the Primrose the pollen is used very economically and we find accordingly few stamens. The most conspicuous point of difference from the Primrose is in the form of the flower, which faces sideways, and is so formed that the insect must alight on the corolla in a particular way. Owing to this irregularity in the shape of the corolla the flower can only be divided into symmetrical halves in one way; it is therefore called *zygomorphic*. The Primrose, and other

flowers we have examined, can be divided symmetrically
in many ways, like a star, showing *radial* symmetry, and
are called *actinomorphic* [Greek *aktinos* = ray (of a star),
morphos = shape]. There are many flowers which face
sideways like the Deadnettle, and most of these are
zygomorphic.

Of the flowers we have so far examined, in the open
ones, Buttercups, Cherry-blossoms and Dog-roses, the
nectar, if present, is not difficult to get: numerous small
insects, including flies, little beetles and bees, therefore
visit these, and crawl about wasting the pollen, so that
pollen has to be abundant. In Primroses the nectar
can only be reached by insects with long tongues, and
their insect visitors are chiefly bees and some long-
tongued flies very like bees (Bee-flies). The Deadnettle has
a still longer tube and its principal visitors are restricted
to the larger humble-bees. Hive-bees, for instance, have
not tongues long enough to reach the nectar. The
Deadnettle is thus specially adapted to this very small
group of insects. Such flowers are called, for this reason,
highly specialised. They have remarkable floral mechanisms, being adapted some to bees, others to butterflies,
moths, hover-flies or wasps.

There are many plants which have small greenish or
brownish flowers very different in appearance from brightly coloured insect-pollinated
flowers. At haytime, Grasses, for instance,
send up tall green plumes or ears which bear
little resemblance to ordinary flowers. Yet,
when we walk through a field, our boots are covered
with a pale yellow dust which flies in clouds from these
plumes as we brush through them. When they are examined, stamens are to be seen hanging out of them,
though these stamens are very different from the

**Flowers
cross-pol-
linated
by wind :
Grasses.**

comparatively stout stamens of a Buttercup. They have a very thin flexible filament, and a large purple anther containing an abundance of pollen. The pollen is dry and dusty and falls from the anther at the slightest touch, or is blown from it by the slightest breath of wind.

If a plume be examined more minutely pairs of delicate white feathery structures may also be seen protruding here and there. A plume or ear consists of a number of tiny clusters, called *spikelets*, of green or

Fig. 66. Spikelets from the inflorescence of a Grass, each with two flowers. In the left spikelet the stigmas have protruded from the upper flower, and its three stamens are just appearing. In the lower flower of the other spikelet the swaying anthers have opened and the wind is blowing the pollen from their spoon-like tips, in which it collects. (After Kerner.)

brownish overlapping scales. One of these showing both stamens and white feathery organs should be removed and carefully dissected under a lens with a pair of needles. When the scales have been taken off, a tiny round pale green ovary is found, surmounted by the two feathery *stigmas*, which are very large in comparison with the size of the ovary. Below the ovary, three stamens are attached. No corolla or calyx is to be seen.

The spike or plume, as the case may be, is thus an *inflorescence*, bearing a number of flowers. Insects will seldom be observed to visit these inflorescences; the individual flowers especially are very small and inconspicuous. We have noticed, however, how readily the pollen is blown by the wind; it is so abundant, the plants growing so thickly together, that some can scarcely fail to be caught in the stigmas, which are, moreover, large and exposed to the wind. The pendent position of the stamens makes it almost impossible for their pollen to reach the stigmas of the same flower: the wind usually catches the plant sideways, and will therefore carry the pollen to other flowers, usually on other inflorescences. The flowers of the Grasses are thus cross-pollinated by the wind.

From the branches of the Hazel in spring long tassels called *catkins* dangle in large numbers. **The Hazel.** Each catkin consists of a central axis on which a number of tiny scales are spirally arranged, each

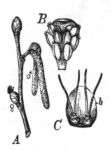

Fig. 67. Hazel. *A*, twig with male catkins, ♂, and female bud with stigmas protruding, ♀. *B*, scale from male catkin showing stamens. *C*, scale from the female inflorescence, bearing two ovaries each with two long stigmas. (After Eichler.)

bearing a number of stamens with short filaments and large anthers. No trace of stigma or ovary can be found. These catkins are already present in the winter, the scales closely overlapping each other; in spring they grow longer, the scales separate, and the anthers ripen and burst, disclosing an abundance of dusty pollen. If at this time a branch be shaken by hand or by the wind the pollen falls in clouds from the catkins.

At the same season small brown buds can be found on the twigs, which differ from the other buds in being tipped by a tiny brush of soft red stigmas. If we watch the later growth of these buds, we shall find that it is from them that the clusters of hazel-nuts come. When closely examined, it is found that the little red-tipped buds, like the catkins, contain a number of tiny flowers: each is an inflorescence, like a Grass plume.

In this plant, therefore, the ovaries and the stamens are not together in the same flower but are borne separately in different inflorescences, so that obviously there is no possibility of self-pollination. Pollen may very occasionally fall from the catkins on to the stigmas of female flowers beneath them, but in order to reach all the stigmas it must be spread abroad in clouds. The plant is, therefore, very dependent on the wind, which may also carry the pollen to stigmas of other Hazels.

The Alder, Oak, Birch and Poplar are other catkin-bearing trees with wind-pollinated flowers. The common Scotch Fir (Scots Pine) and other Pines are also wind-pollinated, and produce pollen in even greater abundance. All the Grasses and the Sedges, the Docks and the Stinging Nettles are wind-pollinated herbs.

Other wind-pollinated flowers.

Now all these plants are like the Grasses in having small and inconspicuous flowers which do not attract notice. In fact they are not what the ordinary person thinks of at all when he speaks of flowers. They stand in great contrast in this respect with the brightly coloured insect-pollinated flowers. Many of the wind-pollinated flowers do not possess petals or even sepals, but are merely protected by small dull-coloured scale-leaves known as bracts. Neither do they possess any scent or contain any nectar.

We have already noticed how very abundant is the supply of pollen, produced in the catkin-bearing plants by many stamens, in the Grasses and Sedges on the

other hand, where the stamens are few (2–6), by the large
anthers. It is evident that nearly all this extraordinary
quantity of pollen never reaches a stigma and is wasted.
Yet if less were produced there would be a greater chance
of some stigmas going unpollinated. A great waste of
pollen is therefore necessary where pollination is effected
by the wind in order that plenty of seed may be produced. The pollen is protected in various ways from
being washed away and spoiled by the rain. In catkins
like those of the Hazel, the bracts roof over the stamens.
In most cases the anthers only dehisce and expose the
pollen in dry air, and it is then soon dispersed.

Careful consideration of the various examples will
show that wind-pollinated plants are generally so situated
at the time of pollination that their flowers are well
exposed to the wind. In most of our catkin-bearing
trees the flowers ripen in early spring before the leaves
are out. This is a time of year when winds are prevalent,
and the wind sweeps easily through the bare branches,
carrying the pollen from one branch or tree to another.
The Grasses, Sedges and Docks too are found growing
on wide, open spaces such as moors, downs and commons,
prairies, steppes and pampas. All such places are wind
swept and unsheltered; and the inflorescences stand up
above the general level of the herbage so that they are
fully exposed to the wind.

The Study of Pollination.

It is impossible in a small book to describe more than
a very few of the great variety of ways in which pollination is secured in different flowers, especially in those
pollinated by insects. Some more examples will be
referred to in Section IV; but each student should study
as many of the common flowers of the district as possible,

carefully examining their structure, observing the arrangement of the parts, condition of anthers and stigma, etc., from the time of opening till they fade, and when possible watching pollination in progress. In the case of insect-pollinated flowers opportunities should be taken of watching, especially on warm, sunny days, to see what kinds of insects frequent them, and how each kind of insect behaves. Keen students will find special features of interest in almost every flower they meet, not solely with regard to their method of pollination, and it is well worth while examining at any rate a selection of flowers very fully, following their history right through until the seed is dispersed. Time would not allow of such an exhaustive study of all the interesting flowers that are found, but they may be examined and grouped in classes according to the kind of adaptations for pollination which they possess. The following summary is intended as a guide to such work: common examples are mentioned, some of which are sure to be met.

We have seen that flowers may be either insect-pollinated or wind-pollinated. The flowers of a few water-plants (e.g. *Elodea*, p. 438) are pollinated by water. These external agencies tend to bring about cross-pollination rather than self-pollination. There are many flowers, however, that are regularly automatically self-pollinated without the aid of any external agency.

In the great majority of flowers that are pollinated by external agencies cross-pollination is specially favoured in various ways, of which we have had examples. The structure of some flowers makes self-pollination impossible. On the other hand, many flowers that are adapted primarily for cross-pollination fall back on automatic self-pollination as a last resource.

Unisexual flowers. In some plants the pollen and

the ovules are produced in separate flowers which are therefore called *unisexual*. When the male flowers (with stamens) and the *female flowers* (with ovaries) are borne on different plants they are called *dioecious* (from Greek *di* = two, *oikos* = dwelling place): clearly in such cases self-pollination is impossible. When both kinds of flowers are borne on the same plant they are called *monoecious* (Greek *mono* = one, and *oikos*). Here, again, self-pollination in the strict sense is impossible, though pollen from male flowers may reach stigmas on the same plant.

Dioecious : Among plants with wind-pollinated flowers which are dioecious are the Poplars (p. 373), the Stinging Nettle, the

Fig. 68. Flowers of Sorrels, wind-pollinated, dioecious. In the female flower the stigmas are large and have many spreading branches: in the male flower the anthers hang from the filaments by slender threads and when they dehisce each lobe bends up at the tip forming a spoon-like receptacle for the pollen until the wind blows it away. (Female, of Common Sorrel: male, of Sheep Sorrel.)

Sorrel (Fig. 68), and the Dog's Mercury: among dioecious plants with insect-pollinated flowers are the Willows (p. 371), White Bryony, the White Campion (p. 310), and the Butterbur (p. 353).

Monoecious : The catkin-bearing trees, other than the Willows and Poplars, and many Sedges (*Carex*) are monoecious and wind-pollinated. Arrowhead (p. 442) and Cuckoo-pint (*Arum maculatum*, p. 216) are monoecious and pollinated by insects.

Dichogamous flowers. The great majority of flowers are *hermaphrodite*, having both stigmas and ovaries; but in very many of them the stamens and stigmas do not ripen at the same time. There are some cases in

which either the stamens are completely withered before
the stigma is ready to receive pollen, or vice versa, so that
a male and a female stage may be distinguished: in such
flowers self-pollination is impossible. In others the male

Fig. 69.   Male (1) and female (2) catkins of a Willow.
(After Wossidlo.)

and female stages overlap for a longer or shorter time,
during which self-pollination is possible unless prevented
by some special contrivance. This separation in time of
male and female stages is called *dichogamy*: if the male
comes first the flower is *protandrous*; if, on the other

hand, the stigma is the first to ripen the flower is *proto-gynous.*

Protogynous : Among wind-pollinated flowers protogyny is more commonly met with than protandry. The Wall Pellitory, Plantains, most Grasses, Rushes, Woodrush, and Sedges with hermaphrodite flowers (e.g. *Scirpus*), are also protogynous.

In the Plantains (e.g. the Greater Plantain, *Plantago major*, common on roadsides and waste places) the flowers are borne in a close spike, the oldest at the bottom. These spikes are often seen with the stamens hanging out of the lower flowers around withered stigmas, while in the upper flowers only the ripe stigmas are visible, the stamens still being enclosed within the scaly perianth (*Plantago media*, the Meadow Plantain, has brightly coloured stamens and its flowers are slightly scented; they are occasionally visited by insects).

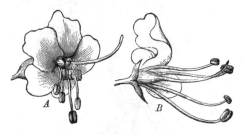

Fig. 70. Flowers of the Horse-chestnut: *A* is the female stage, the stamens still unripe, *B* is the male stage, the stigma withering. (After Engler and Prantl.)

The Christmas Rose (*Helleborus niger*), the Figwort and the Horse-chestnut (Fig. 70) are examples of the few plants with insect-pollinated flowers that are protogynous.

Protandrous : Protandry is very common in flowers pollinated by insects. The larger flowered Cranesbills (*Geranium*) and Willow-herbs are protandrous; so are Ivy, Harebells, many plants belonging to the Families Compositae, Umbelliferae, Labiatae and Caryophyllaceae, etc.

Dichogamy is often combined in insect-pollinated flowers with movement of the stamens and stigma to occupy successively the same spot in the flower. The Rosebay (*Epilobium angusti-folium*) is a beautiful example of this. When the flower opens the stamens stick out horizontally but the style hangs down; later the stamens bend downwards, while the style moves up and takes their place and the stigma, till then closed, spreads its four lobes and exposes their stigmatic surfaces. The style and stamens of the protogynous flowers of the Horse-chestnut behave in a similar way (Fig. 70). In the Garden Nasturtium (*Tropaeolum*) and in the Grass of Parnassus the stamens take up a definite position in turn, one by one, followed finally by the stigma. Other good examples are Honeysuckle, the Meadow Cranesbill (*Geranium pratense*), Red Campion, Pink, and Figwort (p. 339, Fig. 149); but there are many more.

Relation between insect-pollinated flowers and the insects which visit them.

The insects which visit flowers belong chiefly to four great groups:

(1) The *Coleoptera*, or beetles, have a front pair of horny wings (elytra) which act as covers to a second pair of delicate transparent flying wings, which are folded underneath them. As a rule beetles are only able to lick freely exposed nectar. The beetle most commonly found visiting flowers in Britain is a small bronze-black one called *Meligethes*.

(2) The *Diptera*, or flies, have only one pair of transparent wings. Many of them visit flowers both for honey and for pollen. Most of these have short tongues not more than 4 mm. in length. They are stupid creatures and, like the beetles, are only able to get nectar when it is freely exposed or only slightly concealed and easy to reach: many of them feed also on carrion. The hover-flies and the bee-flies, however, have longer tongues, from 4–12 mm., and are nearly as clever as bees in finding concealed nectar. They have usually the abdomen ringed with black and yellow or brown: the larger bee-flies are therefore easily mistaken for bees. The hover-flies can be distinguished by their remarkable habit of hovering, apparently stationary, in mid-air and suddenly darting to another spot and again hovering.

(3) The *Hymenoptera* are insects with two pairs of transparent wings, and include the ants (the workers are wingless), ichneumons (mostly black with very slender 'waist'), bees and wasps. The bees are the most important flower-visiting class; all the others are short-tongued. The bees have not all very

long tongues; some of the smaller ones have tongues less than 6 mm. in length.   The hive-bee (*Apis*) has a tongue about 6 mm. long, the large humble-bees (*Bombus*) 10 mm. or more.

(4)   The *Lepidoptera* include the butterflies and moths, which visit flowers for nectar.   Most of them have tongues about as long as those of bees, the larger ones up to 15 mm., closely coiled up under their heads: but hawk-moths have much longer tongues, even 30 mm., while that of the Convolvulus hawk-moth is of extraordinary length, reaching as much as 8 cms. Most moths fly at night, especially at dusk; a few fly in the daytime, like the butterflies, but can be distinguished by their antennae ('feelers'), which are usually tapering or feathery, not ending in knobs like those of butterflies.   From our point of view, however, all the day-fliers can be grouped together.   Flowers visited especially by these may be called butterfly-flowers; those visited by night-flying moths, moth-flowers.

Further information about insects, their structure and habits, can be obtained from textbooks of Entomology.

Means for attracting insect-visitors.

Besides providing nectar and pollen, flowers pollinated by insects are usually brightly coloured and have scent.

There can be no doubt that *scent* does attract insects.   Wind-pollinated flowers are not scented.   Small greenish but sweet-scented flowers, like Mignonette, receive numerous visitors. Moth-flowers are sweetly scented, often only at night.

The *colour* of flowers is related to some extent to the kinds of insects which visit them.   Moth-flowers are white or pale yellow, the colours most easily visible at night.   Butterfly-flowers are frequently red, bee-flowers blue or purple, while bright yellow is very common among flowers that are pollinated by shorter-tongued insects; but there are many exceptions.   It has been shown that bees can discriminate between different colours, and that, having found honey provided for them on paper of one colour, they will afterwards fly straight to paper of that colour among other pieces of different colours.   A bright colour makes a flower conspicuous against a background of green, so that it can be distinguished even from a distance.   In most flowers it is the corolla which is coloured.   In others, however,

conspicuousness is due to other parts of the flower. The whole
perianth is coloured in most bulbous plants, like the Tulip,
Crocus and Lily.   In the Narcissus and Daffodil there is an extra
coloured outgrowth, the corona.   In Fuchsias, Columbines, etc.,
sepals and petals are both bright but often differently coloured.
In Monkshood, Winter Aconite, Christmas Rose, etc., the coloured
leaves are the sepals.   The stamens are the brightly coloured
organs in Meadow Rue, in the male catkins of Willows, and in the
Meadow Plantain, and add to the conspicuousness of many other
flowers (as in the Buttercup), while the Iris in addition to its
coloured perianth has three stigmas that are broad and petal-like.

Honey-guides.   In numerous flowers the entrance by which
the nectar is approached is marked by a ring or flush, coloured
differently from the rest of the flower, or lines and streaks converge
upon it.   Often, as in the Pansy, these are in bright contrast with
the general ground colour.   Considering the evidence already
mentioned showing that some insects distinguish between different
colours, there can be little doubt that these 'honey-guides,' as
they are called, do help the more intelligent insects to reach the
nectar with directness and rapidity.

Inflorescences.   A large patch of colour is far more
conspicuous and attracts attention from a greater distance
than a small spot; solitary flowers that need to be conspicuous therefore succeed better if they are large.   The
great majority of flowers, however, do not occur singly
but in inflorescences, and the effect of the massing together
of colour thereby achieved is to increase very greatly the
chances of pollination.   When the plants are gregarious
and bear many inflorescences the effect is further increased.

Not only do inflorescences increase conspicuousness,
but flowers are so arranged in them that insects pass
more readily from one to another, and pollinate and get
nectar from a number of flowers without travelling far.

Some inflorescences are much branched and spreading
(*panicles*); being erect, they display the flowers above
the general level of vegetation.   In others the flowers are
borne on stalks, in the axils of bracts (p. 162), along a

central axis, the youngest at the top (*racemes*). Many
protandrous bee-flowers (e.g. Monkshood) are borne in
large erect racemes; the bees start at the bottom and
work upwards, beginning therefore with older flowers in
the female stage and ending with younger flowers in the
male stage. Smaller racemes are very common: some
are pendulous (Laburnum, Barberry). *Spikes* are similar
but with sessile flowers, usually more closely massed
together, as in the Meadow Plantain; the catkins of
Willows (Fig. 69) are pendulous spikes.

In a great number of inflorescences the flowers are all
arranged at nearly the same level. The Cow Parsley
and other Umbelliferae, Ivy, White Rock, Candytuft,
Elder, and Guelder-rose are examples. In many of these
the individual flowers are small. In the Guelder-rose the
outer flowers have a much larger corolla than the others,
but are barren, containing neither ovary nor stamens:
thus in addition to the aggregation of the flowers there
is a division of labour between them, the outer ones
serving to give additional conspicuousness to the whole
inflorescence.

Fig. 71. Inflorescence of a
Scabious. (After Baillon.)

There is a large class of
very interesting cases in which
the aggregation is carried still
further. The globular or
cylindrical heads of the Clover
and the Teasels are very
much condensed spikes. In
the large group called the
Compositae, including the
Daisy, Dandelion, Thistles,
etc., and in the Scabiouses,
the axis of the inflorescence
is expanded horizontally into
a flat or more or less convex or concave receptacle on

which the little flowers (or florets) are spread out nearly
on a level. These modified heads, or *capitula* (in the
single, *capitulum*), often show a division of labour, like
the Guelder-rose, the outer florets being large, sometimes
of a different colour from the central flowers, and at the
same time lacking either stamens or both stamens and
ovary (Daisy, Coltsfoot, Cornflower). The inflorescence
is protected as a whole when young, and is kept com-
pactly together later on, by a common *involucre* of
bracts, which is often mistaken by beginners for a
calyx, the inflorescence being regarded as a single flower.
Such inflorescences are often called *social flowers*. They
are very conspicuous and provide an abundance of nectar,
and are visited by all kinds of long-tongued insects in
large numbers.

Classes of insect-pollinated flowers.

Flowers may roughly be grouped according to the insects
which visit them and the way in which they reward their visitors.

Pollen-flowers, like the Dog-rose, contain no nectar. They
are visited chiefly by flies, but also by bees, for the pollen which
they provide in abundance. These flowers usually have numerous
stamens and are of simple structure, self-pollination occurring
about equally with cross-pollination. Common examples are
the St John's Worts, with tufts of branching stamens, Meadow-
sweet, Elder, Rock-rose, Wood Anemone, Old Man's Beard
(*Clematis*), and Poppy.

Many of the less specialised flowers yielding nectar provide
abundant pollen as well.

Flowers with nectar freely exposed are visited mainly by a
great variety of short-tongued insects of the various classes, such
as are only able to lick nectar from an open surface or a very
short tube (e.g. Ivy, Golden Saxifrage). Many such flowers are
small, like those of Bedstraw and most Umbelliferae (p. 328).
Cross-pollination is favoured in many of them by protandry and
by the spreading of the stamens away from the centre of the
flower, so that the insects often alight first on the stigma.

Flowers with nectar partly concealed. In these the nectar can

only be obtained by insects with some small degree of intelligence
and tongues a few millimetres long. They are largely visited
by flies with rather longer tongues, and receive more attention
from bees and Lepidoptera. The concealment is brought about
in various ways. The stamens hide the nectar in the Buttercup,
Stitchwort, etc.; in the smaller-flowered Cruciferae the erect
sepals form a short tube; the receptacle forms a shallow cup
in the Strawberry, Cinquefoil, etc.; while in the smaller-flowered
Compositae and Labiatae, the Guelder-rose, etc., there is a short
corolla-tube.

In other flowers *the nectar is fully concealed* and out of reach
of short-tongued insects. Only hover-flies and bee-flies, bees
and wasps, and Lepidoptera have tongues long enough to obtain
it, and these insects are their most frequent visitors.

Small insects which would rob them of nectar to no purpose

are even in most cases excluded. In some
flowers the tube has a narrow opening
(Forget-me-not, Garden Geranium); in
others the way is blocked by the anthers
(Woody Nightshade, Primrose, Narcissus),
or by the filaments (Daffodil), or by hairs
(Gooseberry, Deadnettle, Pansy). Other
flowers are closed (Snapdragon). All these
features, as well as the ordinary means
of concealment of the nectar, have the
additional effect that they prevent rain
from trickling down to the nectar and
spoiling it.

Fig. 72. Flower of
the Gooseberry,
showing hairs on
the style obstruct-
ing the tube. (After
Wossidlo.)

The greater depth of the nectar is in
most flowers due to the presence of a longer calyx-tube (Campions)
or corolla-tube (Primrose, Deadnettles), but in some either to
stiffly erect sepals (Wallflower) or to a more deeply hollowed
receptacle (Blackberry, Rose Bay). In other cases a pocket-like
spur is formed by some part of the flower (Toadflax, Larkspur,
Garden Nasturtium, Violet).

From flowers with longer tubes than 6 mm. only the longer-
tongued bees and Lepidoptera can obtain nectar. Bees especially
are clever and active, and, as they usually confine their attention
to the flowers of a few or even one kind of plant at a time, there
is greater certainty of cross-pollination and less waste of pollen

than in flowers that are visited by less intelligent and methodical insects. The contrast is clear if we first watch the desultory hap-hazard wanderings of flies and then the diligent way in which bees pass without waste of time from flower to flower. The habit of visiting one kind of flower at a time must be of advantage to the insects as well as to the plants, for they are able, so to speak, to 'get their hand in' for the one kind, and so do their work more quickly and visit more flowers. Many of these flowers are adapted to particular kinds of insects.

Bee-flowers. Flowers with deeply hidden nectar which assume a horizontal or hanging position (Daffodil, Canterbury Bell), more especially those that are zygomorphic, with a landing stage, and those in which the nectar is concealed in some special way so as to be difficult to find, are visited chiefly by bees,

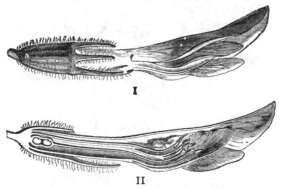

I

II

Fig. 73. I, flower of Red Clover; II, the same cut in half longitudinally. (After Baillon.)

which are by far the most intelligent of flower-visiting insects. Most interesting are the *humble-bee flowers*. Only the longest-tongued bees can reach the nectar in such flowers as the Deadnettle, Monkshood, Larkspur, Columbine, or Red Clover (Fig. 73). Only humble-bees are strong enough to open the *closed* flowers of the Snapdragon, Toadflax and many Leguminosae. The geographical distribution of plants with such highly specialised flowers might be expected to depend on that of the insects pollinating them. This appears to be true for the Monkshoods: they belong to the north temperate zone and there extend nearly but not quite as far as the humble-bees.

Lepidoptera flowers. There are many long-tubed flowers, like Knapweed and Thistles, that are visited by butterflies and moths as well as by bees. In flowers with tubes more than about 12 mm. long, however, the nectar is beyond the reach of most bees. Such flowers are pollinated by butterflies and moths. Most

Fig. 74. Twig of Perfoliate Honeysuckle, showing the clusters of long-tubed moth-flowers. (After Wossidlo.)

of them stand erect and are actinomorphic. Red Campion and Ragged Robin are examples of *butterfly-flowers*. Others are pollinated chiefly by night-flying moths. These *moth-flowers* are usually white or yellow: many of them, such as the Evening Primrose and the flower of the Tobacco-plant, emit their scent only at night. Honeysuckle, which is chiefly visited by the Privet Hawk-moth, first opens at night. Other moth-flowers are Privet, White Campion (*Lychnis vespertina*), and Bladder Campion. The large White Convolvulus is said to be pollinated by the exceptionally long-tongued Convolvulus Hawk-moth, now very rare in the British Isles: owing perhaps to this fact the flowers very seldom set seed in this country.

Many of the long-tubed flowers are robbed of their nectar by a common humble-bee, *Bombus terrestris*, which has a habit of biting through the tubes near the nectar and so reaching it in an illegitimate way without pollinating the flowers at all. The hole once made, other insects can obtain an entrance through it and continue the robbery. The Bell-Heathers, Monkshood, Deadnettle, Ground Ivy, and Toadflax are frequently found to have been treated in this way.

Among the very numerous mechanisms, most of which make for cross-pollination, but which show a bewildering variety in detail, there are some types which are sufficiently common or well-marked to be mentioned in a general way.

Special mechanisms in flowers with concealed nectar.

In flowers with anthers and stigmas ripening together or nearly together, the stigma is very commonly placed a little in front of the stamens.

In dichogamous flowers the anthers and stigma commonly assume the same position in succession.

The existence of two forms of flower in the Primrose has been mentioned: this is called *heterostyly*, from the differing lengths of the styles in the two forms. The Cowslip and Garden Primulas are very similar. The Purple Loosestrife and the Wood Sorrel have three forms of flower with different lengths of style—short, medium, and long—and two groups of stamens, the anthers of which occupy the positions in each flower corresponding to the positions of the stigmas in the other two forms.

In the Pansies and Violets, in the Heathers, Lousewort, Borage, Snowdrop and some other plants, the pollen is powdery like that of wind-pollinated flowers and is 'peppered' down on to the head of a visiting insect that shakes the flower or disturbs the anthers. The Violet, for example, has its five sessile anthers closely fitting together round the ovary, with projecting flaps around the crooked style: from the box so formed, the pollen, which is shed inwards, only escapes when the style is disturbed and moves the flaps. Each of the two lower stamens bears, as an appendage, a nectary, which projects into the tube, or *spur*,

formed by the lowest petal. The stigma is a small hollow, the under margin of which forms a little flap extending downwards, in the way of a bee probing for nectar. When a bee alights on the lower petal and puts its proboscis up into the spur it rubs pollen received from other flowers on to the receptive side of the flap; meanwhile pollen falls out of the pollen box on to the fore part of its head. As it withdraws, the flap is raised so that self-pollination is practically impossible.

Fig. 75. Flower of Sweet Violet cut in half longitudinally. (After Baillon.)

A few flowers have parts that are *sensitive to touch*. The stamens of the Barberry when touched at the base with a needle move sharply inwards: when the touch of an insect's proboscis induces the movement the anthers are as a rule brought firmly against some part of the insect. In the Musk, the forked stigma closes up when touched: an insect may therefore deposit pollen as it enters but not when it leaves, and so self-pollination is prevented. (Compare also Knapweed, p. 350.)

(For other special mechanisms see Section IV.)

Deceptive and trap-flowers. These are flowers which do not fall into any of the classes already mentioned. Some of them attract small carrion-loving flies by their disagreeable odour and entrap them. The Cuckoo-pint (*Arum maculatum*) is a British example. The flowers are unisexual, without any perianth, and are grouped upon a common axis, the *spadix*, surrounded by a large bract called a *spathe*. The ovaries are borne at the bottom of the spadix, a belt of male flowers is found a little above them, and above these again is a ring of downwardly directed hairs, at the narrowest part of the spathe. Above this part the spathe opens and discloses the red club-shaped end of the spadix, which probably helps, along with the odour, to attract the flies. These enter the spathe, eventually push their way down past the barrier of hairs, which is easily passed in this direction, but they cannot get back again. The stigmas are ripe first and receive any pollen which the flies have brought from other inflorescences. Later the anthers burst and cover them with pollen, the hairs which have imprisoned them wither, and they escape to find their way into another inflorescence.

The Grass of Parnassus is a deceptive fly-flower found in boggy places. Five of its stamens are modified as branched structures ending in shining knobs that look like drops of nectar, while the other five stamens are normal.

Arrangements for automatic self-pollination if cross-pollination fails. In many flowers self-pollination may happen during insect visits, or by chance at other times: in some, like the Scarlet Pimpernel, the closing of the flower at night may bring anthers and stigma into contact. In others, however, self-pollination is brought about automatically after insect-pollination has had its chance. In the Dandelion and many other Compositae, and in the Canterbury Bell, the lobes of the stigma eventually curl back and come in contact with their own pollen: in many others the same result is brought about by the inward movement of the corolla when it withers, or by the movement of the stamens themselves.

Self-pollinated flowers.

There are many coloured but small and not very conspicuous flowers which are seldom visited by insects. They regularly pollinate themselves automatically, the anthers and stigmas coming in contact with each other. Many of these have as close relations plants which are regularly insect-pollinated, and it is found that as a rule the small self-pollinated flowers belong to annuals, the larger insect-pollinated flowers to perennials. Examples of this occur among the Geraniums, and the Willow-herbs. The advantages of self-pollination to annual plants are obvious: it is more certain and more economical, both of pollen and of material (for corolla, etc.), than insect-pollination; and as their life is short and they depend for the continuance of the race entirely upon seed, economy and certainty are of the greatest importance. Perennials, which have a longer growing season and propagate themselves vegetatively, can afford to expend more reserves on the production of large flowers and abundant pollen, and may even risk an occasional scanty seed-crop owing to scarcity of insects in order to secure the advantages of cross-pollination.

Other examples of these small self-pollinated flowers are the Chickweed, Shepherd's Purse, and Groundsel, some of the commonest garden and wayside weeds.

These small, simple flowers, however, are not the only ones to be self-pollinated. Some annuals with flowers highly specialised as regards their structure, like the Sweet Pea and the Garden Pea, are almost invariably self-pollinated.

Violets, Wood Sorrel, and some other plants, which grow where they are apt to be overlooked by insects, or receive in any circumstances but few visits, bear in addition to normal flowers (in the summer) others called *cleistogamous* flowers (towards the end of the season) that never open but are self-pollinated.

CHAPTER XIV

FRUITS

As a consequence of pollination the carpels of the Buttercup, as we have seen, enlarge and finally become dry fruits; while within each carpel the single ovule also grows and becomes the seed. Pollination results therefore in changes both in the ovary and in the ovule.

We have already met with ovaries very different from simple one-seeded cárpels; they must clearly ripen into fruits very different from the one-seeded fruit of the Buttercup, which does not break open to allow the seed to escape, but falls off as it is. The ovary of the Primrose, for instance, is perfectly symmetrical, and contains many small ovules attached to the free central placenta that rises from its base (Fig. 64, p. 193). After successful pollination the ovary grows much longer, protected by the calyx, which does not fall off but persists around it. The ripe fruit projects a little beyond the calyx and gradually dries, forming a sort of box or capsule. In drying, it finally splits at the top into five teeth, which separate, leaving the capsule open at the top. The ripe seeds fall to the bottom when they drop from the placenta and, as the stalk is dry and stiff and the capsule is held erect, they cannot escape while the capsule is still. When a strong wind blows, however, the fruit is swayed to and fro and the seeds are thus jerked out a few at a time. As they are small they may be carried some yards by the wind before they reach the ground.

In the Cherry flower there is as a rule a single carpel, though occasionally two or three carpels are found in a flower. Each carpel, like the carpels of the Buttercup, contains only one seed. After pollination the petals, sepals and stamens all wither and fall, while the ovary grows until it becomes the familiar cherry. At the top of it, a point still marks the place from which the style fell. Within the edible flesh is the stone. When the stone is cracked open a kernel is found within it, covered by a brown skin: this is the seed. The ovary wall has thus increased greatly in thickness; the inner part of it has become hard and woody, the rest soft and juicy, except for the skin which bounds it. The flesh is eaten as food by birds as well as by man. Birds often carry away wild cherries, which are smaller than cultivated ones, before they eat them. They then peck away the flesh and drop the stone, in which the seed remains unharmed; or they swallow the fruits whole and the stones pass unharmed through their bodies. In this way seeds may often be dispersed to considerable distances. Other stones, particularly of cultivated cherries, are robbed of their flesh and left hanging on the tree; they then only reach the ground when they drop off.

Comparing the fate of cherry-stones with what happens to the seeds of the Primrose it seems clear that in both cases the peculiar structure of the fruits results in the distribution of a large proportion of the seeds to some distance from the parent plant, though in two very different ways.

Let us examine the fruits of a few more common plants. The pod of the Sweet Pea develops from the single carpel found in the flower, which, unlike that of the Buttercup, contains several seeds, attached in two rows to the placenta that runs up one side of it. After pollination the

Fruits that eject their seeds.

ovary grows to many times its original length; the seeds also swell and gradually fill out the pod. If it be left on the plant the pod dries, and eventually splits along both sides into two valves. This often happens suddenly with some violence, and as the valves twist up at the same time the seeds are flung a little distance away. This explosive splitting takes place in dry weather; as it happens very quickly it is not often seen, but the empty twisted valves can readily be found (Fig. 76).

Fig. 76. Fruits of Sweet Pea, one still closed, the other split showing the twisted valves. Below each is the persistent calyx, with five teeth, and the dried filaments of the stamens (one separate, the other nine joined, see p. 323).

The similar but short pod of the Gorse also explodes in the same way: on a hot dry day the 'popping' of the pods is often heard on heaths where Gorse bushes are abundant.

Another fruit which opens is that of the Pansy. This is formed from an ovary in which the ovules are arranged on *three* equidistant placentae running up the walls. After pollination the style and stigma shrivel, the white ovules gradually ripen to small golden brown

seeds with shiny coats, and the ovary grows greatly in size and finally begins to dry. At last it splits between the placentae into three valves which incline away from one another (Fig. 77) Then as each valve dries still further it gradually folds over the smooth hard seeds and squeezes them more and more tightly until it shoots them out.

Fig. 77. Capsule of the Pansy, split into three valves on which the seeds are attached, and surrounded by the persistent sepals. (After Le Maout and Decaisne.)

The ovary of the Poppy, with many large placentae

Wind-dispersal. projecting into it from its wall (Fig. 78) and bearing numerous ovules all over their surface, becomes a large capsule containing many very small seeds. Under a roof-like top, on which the sessile stigmas radiate, small openings or pores are formed by a three-cornered flap which is cut out and folded downwards. The flower-stalk becomes stiff and dry and holds the capsule erect. Just as in the Primrose, therefore, the seeds can only escape, a few at a time, when the capsule is

Fig. 78. Capsule of a Poppy, cut across. (After Engler and Prantl.)

jerked or swayed. Fruits like these clearly form very effective mechanisms for the dispersal of their seeds in all directions by the wind; for the seeds are small and usually escape only when a strong wind is blowing. *Censer mechanisms*, as they are called because the escape of the seeds depends upon the swaying of the fruit, are possessed by a very large number of plants.

In the Sycamore the wind brings about seed-dispersal in a different way. The ovary is two-chambered, each chamber containing a single ovule. Even when quite young, the wall of the ovary extends on either side as a flattened wing. After pollination these wings grow, and when the fruit is ripe each half consists of a round nut-like part, enclosing the seed, and a large thin wing, with a stout rib along one edge. These two halves

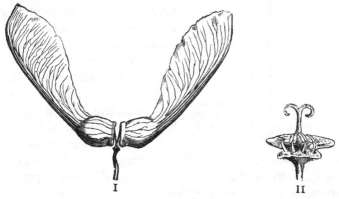

I   II

Fig. 79. Sycamore: I, fruit. (After Baillon.) II, flower from which the perianth has been removed, showing the stamens and winged ovary on the glandular disc. (After Willkomm.)

split away from one another, and remain attached to their common stalk merely by two thin fibres. They are then readily broken off by the wind, which catches the broad wing and may thus hurl the fruit to a distance. Even a light breeze may waft it away from the parent tree, for it is so constructed that it falls through the air slowly, spinning round and round. In this case it is the fruit itself which is dispersed by wind, though in two halves. Like the fruits of the Buttercup it does not

split open to let out the seeds, but the ovary wall remains
as a protective coat around them.

In the Willow-herbs dispersal by wind is secured
in still another way, even
more effective. The ovary is
here below the rest of the
flower, and is therefore called
inferior: as it is long and
narrow it appears like a part
of the flower-stalk. When
cut across it is found to be
divided into four chambers
by partitions, or *septa*. The
ovules are attached in each
chamber along the centre or
axis of the ovary where the
septa meet. The placentae

Fig. 80. Fruit of a Willow-
herb, splitting and dis-
closing the plumed seeds.
(After Baillon.)

are, therefore, called *axile*,
because they are not, like
the *parietal* placentae in the
one-chambered ovaries of the Pansy and the Poppy,
on the outer walls. After pollination the sepals, petals,
stamens and style fall from the top of the ovary,
which grows in length. When ripe it dries and gradually
splits between the septa from above downwards. Not
only does the outer wall separate into four valves, but the
septa also break so that as the valves bend back a central
column is left to which the seeds are attached. As
soon as the splitting has begun it can be seen that each
tiny seed bears a tuft of long silky hairs, which at first
cling to the valves so that they are spread out as these
bend away; but soon the seeds and their plumes dry and
become loosened from the fruit-wall and the placenta. Then
the lightest wind will waft them away. Such very light
plumed seeds may be carried long distances in dry weather.

Let us now examine the structure and development **Edible** of a few more *succulent* fruits—those which are **fruits.** called fruits in ordinary language because they are edible.

The flower of the Blackberry has many carpels, each with one seed, surmounted by a slender style and stigma. After pollination each carpel swells and forms a little fruit, like a very small cherry, with an outer skin, a fleshy part, and a pip, which corresponds to the cherry-stone, protecting the tiny seed inside it. The Blackberry is thus a cluster of fruits, in the botanical sense of the word: we may call it a *compound* fruit. Below the cluster the remains of the calyx, and usually also of the stamens, are found, at the edge of the disc formed by the outgrowth of the receptacle; while at the top of each little fruit may often be found a shrivelled style. Birds peck the little fruits off and swallow them. The pips are not digestible and so pass through a bird's body unharmed, to be dropped eventually, often far away from the parent bramble.

Strawberries are also compound fruits. The flower has many one-seeded carpels like the Blackberry flower; but after pollination each carpel becomes small and dry like those of the Buttercup. It is the receptacle which swells and becomes fleshy and bears the true fruits —the pips—on its surface: the shrivelled styles which often remain attached to the pips show that these are the true fruits in the botanical sense. The pips are not digestible and are distributed like those of the Blackberry.

In the Currants the ovary is below the rest of the flower, like that of the Willow-herb. This inferior ovary has only one chamber with two parietal placentae, bearing the many ovules. A short tube above it bears sepals, petals and stamens. After pollination the ovules become hard-coated seeds, while the ovary grows and becomes fleshy, and surrounds the seeds with a soft and juicy pulp. When it is ripe it changes from green to a bright red or black colour, either of which is very conspicuous against the green background of the foliage. These *berries* differ from the cherry not only because they contain many seeds but because each seed is protected by its own coat instead of by a hard layer of the ovary wall, the whole of which is here fleshy. The remains of the rest of the flower are found on the top of each fruit.

Seeds and fruits are thus, like pollen, dispersed by wind and animals, and their structure is adapted in various ways to either method. Those which are dispersed by wind are small or are provided with a wing or plume. Those carried by birds are fleshy, and when ripe have a bright or readily distinguishable colour.

Some fruits are distributed by animals in another way. The fruits of Cleavers (Goose-grass), for instance, attract notice because they stick to the clothes. This is a common plant, found scrambling amongst the vegetation in hedges and ditches, tall, thin and rough, with small leaves in whorls of six. The fruit is two-lobed, with one ovule in each lobe; when ripe the halves separate from each other very readily and come off as one-seeded nutlets. When they are examined with a lens they are found to be covered with hooked hairs, which adhere to the clothes or to the fur of animals or the feathers of birds so firmly that they are carried often for considerable distances before they are dislodged and fall to the ground.

These examples are sufficient to show that fruits are **Dispersal and the struggle for life.** adapted for dispersal, in some cases to a considerable distance, from the parent plant. Even the fruits of the Buttercup are more likely to break off the receptacle in windy weather or when brushed by a passing animal than to fall straight to the ground.

A little thought will make clear the importance of such dispersal. It is obvious that seeds are of no use unless they can germinate. In order to germinate they must find a suitable situation where there is moisture and soil to enable them to live. Otherwise all the seeds produced may be wasted. Now unless the parent plant is an annual and dies down before its seeds germinate, the soil below it is already occupied. Apart from this, if the seeds fell near together they would all be struggling in a restricted space, when they germinated, for the light, air and soil. Anyone who has grown Cabbages or Lettuce from seed knows that if the plants are left to grow thickly together they become weak and straggling, and only grow healthy and vigorous when planted out at a distance from one another. Thus if all the seeds fell together below the parent plant very few, if any, could survive. On the other hand, there may be unoccupied soil at some distance from the parent which would afford favourable spots for germination if only the seeds could reach them.

But even if there be no unoccupied ground, it is in most cases more likely that the seeds will germinate and produce successful plants away from the parent and from each other, even though crowded amongst other plants, for the chances are that these will have different requirements. Moreover, we shall see later that different plants can in many cases grow together because they can share soil and space between them, one being deep-rooted and another surface-rooted, one tall and another small and prostrate.

For all these reasons, therefore, it is of great advantage for a plant to scatter its seeds abroad in order that some of its offspring may survive and its race be successful.

THE STUDY OF SEED-DISPERSAL.

The following notes and classification of fruits is intended, like the previous classification of flowers according to their method of pollination, as a help to students in studying for themselves the dispersal of fruits and seeds. Each student should keep a classified list and add to it the name of each plant of which the fruit has been found and examined, with notes on any special features shown by its fruits and seed, illustrated by accurate drawings.

Wind, birds or other animals, or water may bear seeds or fruits away, or they may be forcibly and mechanically dispersed by the plant itself. The adaptations for these different methods of dispersal are very varied.

The form of the fruit also depends, however, upon the kind of ovary which the flower possessed. For the present we may distinguish:

(1) Carpels, each containing one seed (Buttercup) or a number of seeds on a single placenta along one side of the ovary wall (Pea); there may be several carpels in a single flower.

(2) Complex ovaries, usually symmetrical in form and with more than one placenta. They may be one-chambered, with parietal placentae (Pansy, Fig. 77) or a free-central placenta (Primrose), or they may have more than one chamber and then in most cases the placentation is axile (Willow-herb, Fig. 80). In some flowers the ovary is inferior (Willow-herb, Currant).

We have already met with examples, among the fleshy fruits, of ovaries of very different types that become adapted for dispersal in similar ways. In some cases other parts than the ovary share in the adaptation for dispersal, as in the Strawberry. The Blackberry, the Rose-hip, and the fruiting head of the Burdock are other examples. Such points as these must be noted in recording the structure of fruits, and wherever possible the fruit should be carefully compared with the flower, in order that the origin of each part may be determined. Even if both are not obtainable at the same time, if the flower has been examined and proper records made of its structure, the fruit can be compared with the drawings of the ovary, etc.

It is important always to distinguish carefully between fruits which open, so that the seeds are dispersed—*dehiscent* fruits—and *indehiscent* fruits, which do not open but are themselves dispersed. Some indehiscent fruits break up into portions, usually each enclosing one seed, and are called *schizocarps* (e.g. Sycamore).

A. *Wind-dispersed seeds and fruits.*

1. These are usually small, so that they present a large surface to the wind in proportion to their weight.

(a) *Small seeds.* The seeds of Orchids are nearly as small as pollen grains and so are easily carried; this is a great advantage especially to those Orchids of tropical forests that grow perched up on the highest branches of the trees.

Other seeds are not so small and in many plants are retained by the fruit until a wind, strong enough to carry them, shakes them out or dislodges them. One very common method is by *censer mechanisms*, already mentioned in the Poppy (Fig. 78) and the Primrose. It is not always the single capsule which is swayed: frequently, as in the Plantain and the Foxglove, the inflorescence bearing many capsules is swayed as a whole.

Censer mechanisms formed from complex ovaries may split at the top into teeth (Primrose, Campions (Fig. 81), Stitchworts; Foxglove) or open by pores (Poppy, Harebells). In some Harebells the capsule hangs downward and in these the pores are near the base of the capsule. The upper half of the capsule in the Scarlet Pimpernel and the Plantains comes off like a lid.

The several separate carpels of the Columbine, Monkshood (Fig. 82), etc., stand erect close together and, when ripe, split

along the inner margin (*follicles*) only for a short distance from
the tip, thus forming an efficient censer mechanism.

Fig. 81.　Capsule of Red Campion
opening by teeth at the top,
surrounded by the persistent
calyx. (After Le Maout and
Decaisne.)

Fig. 82.　Compound fruit of
Monkshood, of three erect
follicles. (After Baillon.)

In yet other plants the seeds remain attached to the opened
fruit. The wall of the two-chambered pods of the Wallflower,
Cabbage, Honesty, Shepherd's Purse, and
most other Cruciferae (*siliques* and *silicules*)
splits from the partition between the
chambers as two valves. The seeds remain
attached to the placentae round the edge of
the partition and are gradually dislodged
by the wind: they are in many cases
flattened and their surface thereby increased.

(b)　*Small fruits.* Many one-seeded dry
indehiscent fruits (achenes) are small and
light. Grass 'seed' consists really of the
fruits of grasses.

Larger achenes are in a few cases distributed by a censer mechanism. In the
Sunflower the 'censer' is formed by the
involucre surrounding the whole head of
achenes: while in many Labiatae the four
nutlets (p. 342) lie in the bottom of the

Fig. 83.　Fruit of
Wallflower. (After
Baillon.)

cup-like persistent calyx. Other achenes (Buttercup, Potentilla)
or one-seeded portions of schizocarps (Umbelliferae) may remain
attached to the plant until a strong wind breaks them off.

2. The buoyancy of many small seeds and fruits is greatly
increased by the presence of a tuft or plume of hairs: so that
they float in the air and catch the slightest breeze.   Such seeds
and fruits are often carried great distances, and those kinds of
plants which bear them are usually found very widely distributed
over the earth.

(a)   *Plumed seeds* with tufts of fine, silky hairs are possessed
by Willows and Poplars (p. 371) as well as by the Willow-herbs
(p. 224).   The Cotton plant also has plumed seeds: the hairs of
the plume are very long and are spun into cotton thread and
woven into cloth.

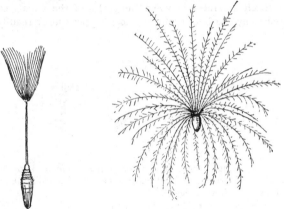

Fig. 84.   Plumed fruits of Dandelion and Spear Thistle.
(After Praeger.)

(b)   *Plumed achenes* with tufts of hairs are possessed by the
Bulrush and the Cotton-grass.

In the flower of the Traveller's Joy (*Clematis*), which has many
one-seeded carpels, the styles grow in length and become covered
with hairs, forming long feathery plumes to the achenes (Fig. 114,
p. 287).   In the Thistles the achene, which is formed from an
inferior ovary, is crowned with a ring of hairs, called a pappus,
which acts like a parachute: these hairs represent a much modified
calyx (see p. 349).   The achenes of the Dandelion, Goat's-beard
and Sow-thistle are also furnished with a pappus which is
separated from the fruit itself by a stalk (Fig. 84).

3. Other seeds and fruits, usually of larger size, are provided

with thin wings, which catch the wind and in some cases, as in the Sycamore already described, act like aeroplanes.

(a) *Winged seeds* are possessed by Pines (Fig. 85, II), Firs, Larches and other Conifers: in these plants the ovules are not enclosed in carpels but are borne on the scales of the *cone* which separate to allow the ripe seeds to be blown away (Pine and Larch), or drop off in order to free them (Silver Fir). In some cases the wings are so arranged that the seeds spin in falling, like the fruits of the Sycamore.

(b) *Winged fruits.* In some of these the wing develops from the wall of the ovary itself. The small and light winged fruits of the Birch are released when the scales of the female catkins in which they are borne fall off: this happens most readily in a

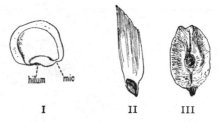

hilum   mic

I   II   III

Fig. 85. I, winged seed of Yellow Rattle (*mic.* = micropyle; see p. 254); II, winged seed of Scots Pine; III, winged fruit of Elm. (II and III after Praeger.)

wind. The Elm and Ash have winged fruits, while those of the Sycamore (p. 223) and Maple separate into two one-seeded portions each with a wing which makes it spin when it falls. In the Dock the perianth persists around the three-cornered achene and forms three wings. Each fruit of the Hornbeam is partly surrounded by a three-lobed leafy structure which acts as a wing, formed from three bracts; while in the case of the Lime the stalk which bears a cluster of fruits is joined for some distance with the midrib of the bract in the axil of which it arose, and this bract acts as an aeroplane to the whole cluster.

B.   *Fruits and seeds adapted for dispersal by birds and other animals.*

Among these we have to distinguish those that are carried on the outside of an animal attached to feathers or fur, usually without the knowledge of the animal itself, from those which have an edible part by which the animal is tempted to carry them off as food.

Very few seeds belong to this group. The Yew berry is a seed with a bright fleshy cup (the *aril*) partly enclosing it, which is attractive to birds. The seeds of the Gorse have a small fleshy appendage which is used as food by ants: they are therefore often carried farther by ants after they have been already scattered by the explosive dehiscence of the pod above referred to (p. 221).

1. *Hooked fruits.* The minute hooked bristles on the fruits of Cleavers or Goose-grass have been mentioned: the fruit splits into two round achenes. The fruit of the Wild Carrot also breaks into two achenes covered with stouter hooks, while in the Hound's-tongue the four-lobed ovary forms four hooked nutlets. In the

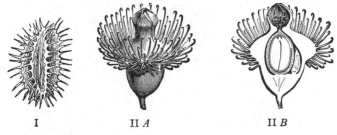

I   II *A*   II *B*

Fig. 86. I, Spiny achene of Wild Carrot (after Praeger); II, fruit of Agrimony, in *B* cut in half, showing the receptacle furnished with hooked bristles and enclosing the achene and also an abortive carpel (after Baillon).

Avens each of the many one-seeded carpels of the flower has a long style, hooked just below the stigma which eventually withers: the flower thus produces a cluster of hooked achenes. Hooks are borne in the Agrimony on the outside of the hollow receptacle, which encloses two achenes: the compound fruit is thus dispersed as a whole. In the Burdock a whole head of fruits, formed from an inflorescence, is dispersed: each of the narrow bracts composing the involucre that surrounds the achenes ends in a hook. Sometimes the head does not come off when it catches on the fur of a passing animal but breaks away again and the achenes are jerked violently out (compare the censer mechanism of the Sunflower).

2. *Fleshy fruits.* The wild fleshy fruits are eaten chiefly by birds. They are usually green and inconspicuous and often unpleasant to the taste until ripe, when they become attractively coloured. In some cases the seeds are swallowed and pass unharmed through the alimentary canal of the bird, in others the flesh is removed and the seed, either protected by a hard seed-coat or enclosed in a hard covering derived from part of the ovary wall, is rejected. Fruits like wild Cherries are swallowed whole by some birds, but have the flesh pecked from them by

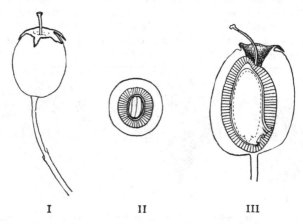

I   II   III

Fig. 87. I, a fruit of the Hawthorn—a haw—showing the persistent style and calyx on the top. II, a haw cut across showing the fleshy outer and hard inner (shaded) parts of the wall of the fruit, enclosing the seed in which are seen the two cotyledons. III, a haw cut in half longitudinally, but with the seed left untouched: the stalk of the seed and the micropyle are shown, and the position of the embryo is indicated by the dotted outline.

others. In all cases the seed is protected from digestion or mechanical injury by a hard resistant covering.

(a) *Stone fruits.* These include the fruits of the Cherry, Plum and Blackthorn (or Sloe), all of which are formed from single carpels, the inner part of the carpel wall becoming hard and protecting the soft-coated seed, the kernel, inside: they are called *drupes*. The haw of the Hawthorn is a somewhat similar fruit formed from an inferior ovary (Fig. 87): remains of the calyx and style, and often of the stamens also, may be seen at the top.

(b) *Berries.* In these the seeds themselves are hard-coated and the whole ovary wall becomes fleshy. Grapes are berries formed from superior ovaries and so are those of the Cuckoo-pint (*Arum*, Lords and Ladies): they contain several pips, the seeds, and a point at the top marks the former position of the style. Another berry is the date, the fruit of the Date Palm; this berry contains a single seed, the date-stone, and at its base can some-times be found the leathery persistent perianth. The fruits of the Gooseberry, Currant, Bilberry and Honeysuckle are berries formed from inferior ovaries: the remains or scars of the calyx, etc., are seen on the top of these. Oranges and Lemons are large

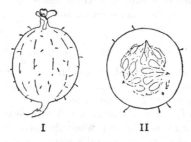

I   II

Fig. 88.   Gooseberry: I, fruit, with remains of sepals and stamens at the top; II, the same, cut across, showing the attachment of the seeds to three parietal placentae.

berries, the segments representing the chambers of the superior ovary; the juice is contained in enlarged bladder-like hairs that grow out from the ovary wall. Bananas are berries which under cultivation no longer produce seeds.

(c) *Compound or aggregate fleshy fruits.* Succulent fruits formed from dense inflorescences are aggregated together and thus, like the flowers from which they arise, gain in conspicuousness. The clusters of berries of Honeysuckle, Cuckoo-pint, Currant, etc., are examples. The flowers of the Bramble and the Raspberry contain many carpels each of which becomes a small succulent fruit similar in structure to a Cherry or Plum and therefore called a *drupelet*. The Blackberry and Raspberry are thus compound fruits: below the cluster of drupelets are found the remains of the calyx and stamens attached to the expanded receptacle. A

Strawberry is a compound fruit in which the receptacle becomes fleshy and bears the numerous achenes on its surface: on each achene can usually be found a withered style. In the hip of the Rose, too, the bright red fleshy part is the receptacle which here is flask-shaped and encloses several achenes: the latter are covered with rather stiff hairs, and are rejected by birds when they peck away the receptacle.

A Fig is formed from a whole inflorescence, in which numerous male and female flowers are borne on the inner surface of a hollow pear-shaped receptacle which becomes fleshy: the pips are achenes. In the Mulberry the flowers are borne in a small cluster and the perianth becomes fleshy, while a Pineapple is formed from a spike-like inflorescence, in which the axis, bracts and fruits swell and fuse into one large fleshy mass: above the inflorescence the main axis grows on and forms a crown of green leaves.

C. *Fruits and seeds dispersed by water.*

A few fruits and seeds of water-plants or plants growing by the water-side are capable of floating and may therefore be carried by a flowing stream or by surface currents.

Floating seeds. The seeds of the White Water-lily have a spongy covering filled with air: when the seeds escape from the fruit, which ripens and opens under water, they float to the top and only sink when the water has soaked in and displaced the air.

Floating fruits. The Yellow Water-lily has spongy fruits that float about and only let the seeds escape when they decay. The Coconut is a fruit in which the ovary wall has formed a fibrous layer (which yields 'Coconut fibre'), and a hard inner shell, within which the true seed is enclosed: the fibrous air-containing layer is covered by a water-tight skin. These fruits can float on the sea for weeks and are sometimes borne from one island to another many hundreds of miles distant.

D. *Explosive and sling fruits.*

Many seeds are dispersed mechanically by the plant itself, usually by the explosive dehiscence or movement of the fruit.

1. The expulsion of the hard shiny seeds from the split capsule of the Violets and Pansies has been described (Fig. 77). The seeds of the Wood Sorrel are shot out in a somewhat similar

way. The fruits in ripening do not dry up but remain green and
become very turgid: when quite ripe
they can be caused to split by a touch,
and then the seeds may be watched as
they are shot out, each by the pressure
of a surrounding turgid mass of white
tissue.

2. The pods of the Vetches and
some other Leguminosae (p. 322) like
those of the Gorse and Sweet Pea (called
legumes) split explosively into two valves
which twist up. Another explosive fruit
of a different kind is that of Herb Robert
and other Cranesbills. The fruit has
five one-seeded lobes, from the midst of
which rises a long beak (whence the
name Cranesbill). When the fruit is

Fig. 89. Fruit of Herb
Robert. (After Baillon.)

ripe the outer wall of each lobe springs up, along with a strip
from the beak, as if on a spring, and flings out its seed, which
leaves the pocket in which it lay as a stone leaves a sling.

E. *Sowing mechanisms.*

A few plants may be said to sow their own seeds. The fruits
of the Ivy-leaved Toadflax turn
away from the light before they
dehisce and, as it were, seek the
cracks and crannies in which the
plant usually grows: the seeds
themselves are rough so that
they obtain a hold on the soil
more readily. In the fruits of
the Storksbills, which are very
similar in structure to the sling
fruits of the Cranesbills, five one-
seeded portions enclosing the five
seeds split off, each carrying a
strip of the beak as an *awn*,
which twists up as it dries. When
the fruit has fallen to the ground,
naturally with the heavy achene

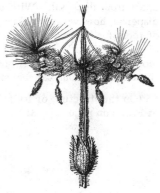

Fig. 90. Fruit of Garden
Geranium, similar to that of
Storksbill. (After Baillon.)

downwards, the awn continues alternately to twist and untwist

according as the air becomes drier or damper, but as it catches
against the surrounding blades of grass or other obstacles the
achene is turned round and round and gradually bores its way
into the soil.   The lower end of the achene is hard and pointed
and is furnished with a barb of hairs which prevent it from being
withdrawn again.

F. *Miscellaneous.*

Many fruits and seeds appear to have no special adaptations
for dispersal. There are many dehiscent fruits that allow the
seeds to fall under the parent plant, and the seeds are spread abroad
apparently only by accident.   In some cases the seeds are large
(e.g. Horse-chestnut) and may at times roll or bounce away for a
short distance when they fall.   In other cases dry seeds or achenes
are carried off by birds as food, and some of these are doubtless
dropped.   Some kinds can even if swallowed resist the action
of the digestive juices and pass unharmed through the alimentary
canal of a bird: seeds of Pondweeds, Sedges, and Bur-reeds,
for instance, have been found, quite capable of germination, in
the intestines of wild ducks.   Nuts (like Beech nuts, Hazel
nuts and acorns) are collected by squirrels for food, and must
occasionally be dropped. We may not conclude, however, that
large size or a store of food material are adaptations for dispersal,
for these characters are obviously of advantage in other ways, quite
apart from dispersal.   With regard to many of the problems of
dispersal, however, our information is scanty, and there is plenty
of room for observation of dehiscence and of the exact method
of escape of seeds, or of the fall and dispersal of fruits. It is,
moreover, only by careful observation that the effectiveness of
special adaptations can be correctly appreciated.

(On the structure of fruits see also Notes on the Morphology
of Fruits on pages 378–81.)

CHAPTER XV

SEEDS AND SEEDLINGS

We have seen that ovules develop into seeds and that these are dispersed in a variety of ways either still enclosed in the fruit or after they have escaped from it. These seeds when sown produce new plants. We must, therefore, examine the structure of seeds more closely and observe the changes that from so small a beginning lead up to plants with leaves, stems and roots.

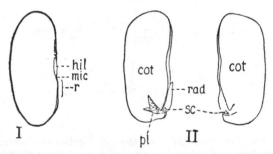

Fig. 91. Seed of French Bean: I, external view showing hilum (*hil*), micropyle (*mic*) and position of radicle (*r*). II, after removing the seed-coats and parting the cotyledons to show the plumule (*pl*), and radicle (*rad*); *sc*, scar left when second cotyledon was broken off.

The seed of the French or Kidney Bean is fairly large **The** and therefore easily examined. It is kidney-**Kidney** shaped, with a small white oval scar on **Bean.** one side where it broke off the short stalk that attached it to the placenta. This scar is called the *hilum*. If a bean which has been soaked for several hours be wiped dry and then squeezed, water is seen to exude from a tiny hole, the *micropyle*, near one end of the scar. Water can also, of course, enter the seed through this pore. The tough brown outer layer of the seed is easily stripped off. This is the *seed-coat*. Within it the

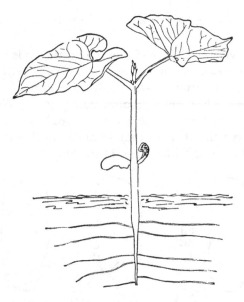

Fig. 92. Seedling of French Bean, showing the two cotyledons, now above ground and green, but wrinkled on the outside, the first two foliage-leaves, which were already visible in the seed as part of the plumule, and the apical bud. The radicle has grown into a long primary root with lateral secondary rootlets.

seed is white: the greater part splits readily into two halves, called the *cotyledons*, which remain attached at one end to a tiny structure, part of which lies between them, part alongside and visible before they are separated. The part between the cotyledons, which ends in what appears like a pair of tiny leaves, is called the *plumule* (i.e. little feather), the part outside is called the *radicle* (i.e. little root).

The nature of these structures is shown when the seed germinates. Beans soaked overnight and then planted in soil in pots will readily germinate. In two days the seed-coat is found split near the micropyle[1] and the radicle has grown out and turned downwards. In a week it is clear that the radicle becomes the primary or main root of the new plant; it has grown down much farther into the soil. The cotyledons are now beginning to appear above the ground. The stem below them arches itself and grows up, drawing the cotyledons after it with the plumule still between them. Later the cotyledons turn green, while the little plumule grows out and shows itself clearly as a bud with young leaves. The first pair are opposite and simple, but the later leaves are alternate and compound, each with three leaflets. The cotyledons remain fleshy and soon fall off but probably help for a time in the work of photosynthesis; their position on the stem and their green colour show them to be leaves, and as they were present in the seed they are sometimes called the *seed-leaves*.

Inside the seed therefore was a tiny plant; the plumule was a minute bud enwrapping the stem apex, the radicle was the main axis, the cotyledons were leaves. This very young plant, resting dormant in the seed, ready to appear when germination begins, is called an *embryo*. The thick, tough cotyledons contain an abundant supply of food, mainly proteid (p. 44); it is this store which supports the embryo in its early growth while the root is establishing itself in the soil. As the food supply is used up the cotyledons get softer and thinner. Later they turn green, as we have seen, and

---

[1] The radicle does *not* emerge *through* the micropyle, which is far too small.

then take part in photosynthesis, forming fresh supplies of food for the seedling.

The Date. The Date-stone, the seed of the Date Palm, has a very different structure. Along one side is a groove, and in the middle of the opposite side is a minute hollow: this is the micropyle. If a seed

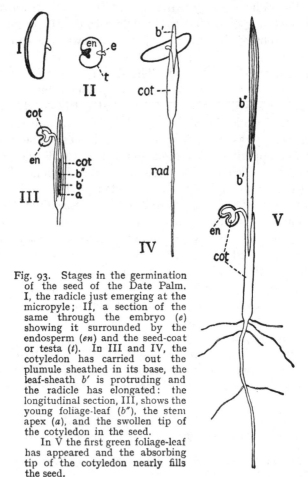

Fig. 93. Stages in the germination of the seed of the Date Palm. I, the radicle just emerging at the micropyle; II, a section of the same through the embryo (*e*) showing it surrounded by the endosperm (*en*) and the seed-coat or testa (*t*). In III and IV, the cotyledon has carried out the plumule sheathed in its base, the leaf-sheath *b'* is protruding and the radicle has elongated: the longitudinal section, III, shows the young foliage-leaf (*b"*), the stem apex (*a*), and the swollen tip of the cotyledon in the seed.

In V the first green foliage-leaf has appeared and the absorbing tip of the cotyledon nearly fills the seed.

be cut in half through the micropyle, a very small and soft object is discovered just behind it which can be picked out with a needle: it is embedded in the hard horny mass that composes the rest of the seed within the brown seed-coat. In a seed which has begun to germinate a tiny radicle is seen coming from the seed at the micropyle and turning down into the soil (Fig. 93, I). The tiny structure to which it belongs must, therefore, be the embryo, although little can be made out of the rest of it which is still in the seed.

Later, however, a thicker object follows the radicle downwards. Sometimes its growth forces the seed out of the ground. It splits near the seed and a little spike appears (Fig. 93, IV; b'). This grows up and from it emerges another spike, bright green and folded like a fan, which expands into a narrow leaf-blade with parallel veins (V; b''). Meantime the seed has become soft. When it is cut open the tip of the embryo is found to have expanded and now nearly fills the seed, while the hard horny mass has largely disappeared; what is left of it has become quite soft. We infer that the enlarging tip of the embryo has been absorbing the substance of the seed and has passed it on to the growing embryo.

To what part of the embryo this absorbing tip belongs may be discovered by cutting the seedling longitudinally (Fig. 93, III) or carefully dissecting it. The stem is sheathed by the bases of the foliage-leaves, one of which has already expanded. Outside these are the two sheaths already mentioned, which are elongated leaf-bases on which no blades have developed. It is the outer sheath which ends in the seed; its tip is the absorbing organ. It was this sheath which grew and carried the young shoot, the plumule, out of the seed after the radicle had found its way into the soil. Clearly this is a leaf with special functions, and may be called the seed-leaf or cotyledon. But its functions are very different from those which the two cotyledons of the Bean perform. It does not store food in the seed, nor does it expand after germination and help in the work of photosynthesis. The food supply is stored outside the embryo. The food-storing tissue is called the *endosperm*: the walls of the cells are very thick, the carbonaceous food being in the form of *cellulose*; this explains the hardness of the dry seed. The embryo itself, as we have seen, is very small.

The seeds of the Bean and the Date thus differ in the following important respects. The embryo of the Bean is large and fills the seed, which contains no endosperm; the Date embryo is very small, embedded in endosperm which forms the bulk of the seed. The Bean embryo has two opposite cotyledons, the Date embryo only one. The cotyledons of the Bean store the food-reserves in the seed, are afterwards carried above the ground out of the seed-coat, and form the first green leaves of the seedling. The cotyledon of the Date does not store the food-reserves, but its tip remains within the seed and absorbs the food stored in the endosperm, while its base carries the plumule out of the seed, but does not always come above ground, and never forms a leaf-blade. The ways in which the plumule is protected and brought out of the seed above ground also differ. In the Bean it is protected and carried above ground between the cotyledons; these are carried up by the growth of the portion of the stem between them and the radicle, called the *hypocotyl* as it is *below* the cotyledons. In the Date the plumule is carried out of the seed by the growth of the sheathing base of the cotyledon; but it comes above ground by its own growth, protected from injury as it forces its way through the soil by the second pointed leaf-sheath.

Dicotyledonous exalbuminous seeds.

There are many seeds which are like the Bean, differing from it only in details. Most large seeds and many small ones in which the embryo has two cotyledons (*dicotyledonous*) have no endosperm and are therefore called *exalbuminous*. The seeds of Lupine and Mustard, and

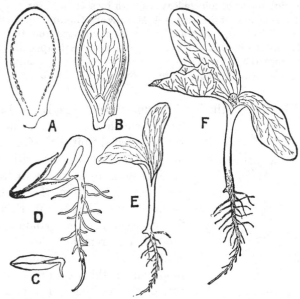

Fig. 94.   Germination of Gourd: *A*, the seed; *B*, seed laid open, showing one cotyledon and the radicle; *C*, the radicle emerging; *D*, the peg has caught on the seed-coat, and the growth of the arched hypocotyl is freeing the cotyledons; *E*, the cotyledons free and erect; *F*, the first foliage-leaf has developed from the plumule. (After Willis.)

those contained in the fruits of the Sycamore and Sunflower, are common examples of exalbuminous seeds with two cotyledons which come above ground during germination.

The hypocotyl of the Mustard seedling does not arch but remains erect: the cotyledons are pressed close together and make their way edgeways through the soil. The cotyledons of the Lupine, Mustard, etc., expand and assimilate for much longer than those of the French Bean; those of the Sunflower are still more leaf-like. In the Sycamore the cotyledons are long and narrow and are folded in the seed; in the seedling they become long narrow leaves. However leaf-like cotyledons may become, there always remain marked differences between them and the later foliage-leaves (e.g. Fig. 94) in form and texture. (On transition forms of foliage-leaves shown by seedlings, see p. 154.)

The seedlings of the Vegetable Marrow and Gourd (Fig. 94) show a remarkable adaptation for freeing the cotyledons from the seed-coat. At the base of the hypocotyl an outgrowth appears which pegs down the lower half of the seed-coat while the cotyledons are withdrawn by the growth of the arched hypocotyl.

The seeds of the Scarlet Runner Bean, Pea, Broad Bean,

Fig. 95. Seedling of Evergreen Oak: the cotyledons remain in the fruit and do not come above ground. (After Dale from Ward.)

and Horse-chestnut, and the acorn, the one-seeded fruit of the Oak tree, are also exalbuminous, with two large cotyledons stored with food; but they differ from the Kidney Bean in one important respect. The cotyledons do not come above ground, but remain below in the seed-coat and merely hand on the food which they contain to the rest of the seedling. The plumule makes its own way out of the seed and up through the ground. To prevent its delicate young leaves from being rubbed back and torn off and its growing tip damaged as it pushes through the ground, the stem is bent over; the top of the arched stem thus bears the brunt of the friction and pressure, and the leaves are drawn gently through the ground after it. In these seedlings, therefore, the hypocotyl does not grow in length as it does

in the French Bean: it is the stem of the plumule, *above*
the cotyledons (the *epicotyl*), which grows, arches itself
and brings the shoot above the ground.

Dicotyledonous albuminous seeds.

Many dicotyledonous seeds, however, are *albuminous*,
that is they do contain endosperm, like the monocotyledonous Date seed.   Most of these are small, like the seeds
of Violet and Pansy, Pinks and Campions, Plantain,
Buckwheat, and the achenes of Buttercup, Carrot, Dock,
Beet and Spinach.   The Castor-oil bean is larger and
therefore more easily examined.   The seed-coat is hard
and brittle.   At one end is a soft outgrowth which soaks
up water very readily.   Just by this is the micropyle; its
position is shown, if the seed be dipped into hot water,
by bubbles of air which emerge through it.   The hilum
is next the micropyle but is hidden by the absorbent
outgrowth.

When the seed-coat is removed, a thin white inner
membrane is found, and within this is a uniform white
mass.   If pressure be carefully applied to the edges of
this between the fingers, it is often possible to split it into
two parts, revealing the embryo which is enclosed within
it.   This has two large but flat and thin cotyledons, and
a small radicle which points towards the micropyle.
Between the two cotyledons lies the tiny plumule.   The
embryo is much more readily removed whole from a
soaked seed, as the cotyledons stick less to the endosperm.
When germination begins the radicle and hypocotyl
grow and turn downwards.   After the radicle has fastened
itself by root-hairs in the soil, the hypocotyl grows
upwards; it arches itself like the hypocotyl of the French
Bean, and eventually draws the cotyledons out of the soil.
By this time they have grown much larger and turned
yellow; the endosperm tissue has also expanded in breadth

though it is much softer and thinner because the cotyledons
have been drawing on the food supply contained in it.
It is often brought up from the soil out of the seed-coat
with the cotyledons and for a little time continues to
diminish. Then the cotyledons separate and expand and
act as the first green leaves of the seedling. The later
leaves are not entire like the cotyledons but palmately
lobed.

Seedlings of Pines and Firs.

Another interesting group of albuminous seeds are
those of the Pines and Firs, our cone-bearing trees. In

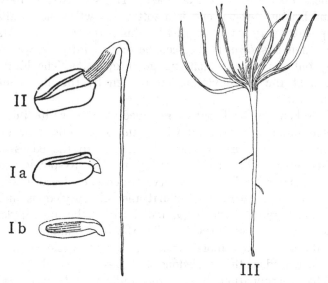

Fig. 96. Stone Pine. I*a*, seed with hard shell (the seed-coat) split
showing the endosperm from which the radicle is emerging:
I*b*, the same after removal of the seed-coat and half the endosperm,
showing the embryo with several cotyledons. II, a later stage of
germination, showing the elongated radicle, arched hypocotyl, and
cotyledons with their tips still in endosperm.

III, older seedling with expanded cotyledons and plumule
beginning to grow. (After Dale from Ward.)

these, however, the embryo has several narrow cotyledons
arranged in a whorl.   The Stone Pine, for instance, has
a large seed with a thick woody coat, which splits into
two halves when the seed begins to germinate (Fig. 96, II).
The seed is often carried up by the growth of the hypocotyl
and cotyledons, and the tips of the cotyledons remain
inside it till the food-material is exhausted.

The embryo is readily dissected out of the endosperm.
The cotyledons turn green while still in the seed, even
when grown in complete darkness; they are thus an
exception to the general rule that chlorophyll is only
formed in the light.

Monocotyledonous Seeds.

Seeds with *monocotyledonous* embryos like the Date
are usually also albuminous; but the cotyledon does not
behave in all of them as it behaves in the Date.

In the germination of the
small Onion seed the cotyledon
carries the plumule out of the
seed into the ground, as in the
Date, but then itself grows up,
piercing the soil with a sharp
knee-like bend, and turns green.
It afterwards straightens itself,
usually carrying up the seed
on its tip, by which it continues to absorb any food left
in the endosperm.   Its base
sheaths the plumule and successive leaves emerge each from
the sheathing base of the preceding one.   They all have, like
the cotyledon, a hollow cylindrical
'blade.'

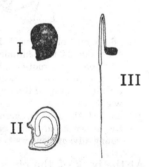

Fig. 97. Onion. I, external
appearance of seed; II,
seed in section showing
the curved embryo em-
bedded in endosperm, the
radicle pointing towards
the micropyle.

III, seedling showing
the elongated radicle, and
the bent cotyledon with
its tip still in the seed.

The embryo of Maize lies at one end of the grain, and is not embedded in the starchy endosperm. On one side of the grain is a patch, whitish in colour, in which two marks may be seen, one towards the narrow, the other towards the broad, end of the grain. If a soaked grain be cut in half through these marks (Fig. 98, I) the embryo can be seen, the radicle behind one mark, the plumule, with several sheathing leaves, behind the other.

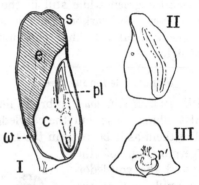

Fig. 98. Grain of Maize: I, longitudinal section showing the embryo consisting of the plumule, *pl*, radicle, *r*, with a special sheath in addition to its root-cap, and the very much enlarged cotyledon, *c*; *e*, endosperm; *s*, scar of the style, showing the grain to be a fruit; *w*, wall of the grain formed from the seed-coat and ovary-wall joined together. II, an embryo dissected from a soaked grain, and III, the same cut across the middle, showing how the cotyledon has grown around and enwrapped the rest of the embryo; *r'*, a young adventitious root.

At the back of the plumule and radicle is a flat expanded part, in contact with the endosperm, attached to the rest of the embryo at its centre. If the embryo be dissected out of a soaked grain this part is seen as a shield-like organ (the *scutellum*, or 'little shield'), presenting a large surface to the endosperm.

When the grain germinates the radicle emerges first; later the plumule grows straight up as a little pointed

spike, covered by a leaf-sheath without any blade, like
that which protects the Date plumule. Later the green
blades appear. The scutellum remains in the grain,
where it produces enzymes, which pass into and dissolve
the endosperm. The food so digested it then absorbs.
It thus plays the same part as the tip of the cotyledon in
the Date and Onion; we may, in fact, regard it as the
cotyledon of the Maize embryo, which does not grow out
of the grain, nor protect and carry out the plumule, but
merely digests and absorbs the food reserves of the
endosperm for the use of the radicle and plumule.

Meanwhile new roots have appeared, one from either
side of the embryo at the point where the cotyledon
is attached, that is from the stem; they are, therefore,
adventitious roots. Later others arise from the same
node and from other nodes higher on the stem; and
eventually they usurp entirely the function of the radicle.

This is the way in which the characteristic root-
systems of Grasses always arise. Grass fruits are all
very similar to that of the Maize in structure and in
method of germination. Wheat is another example. The
grain is grooved down one side, and on the other side at
one end is the embryo. A longitudinal section of the
grain through the groove shows the radicle, plumule and
scutellum, or cotyledon, arranged just as in the Maize
grain. During germination the radicle never becomes
as prominent, nor lasts as long as in the Maize; the
adventitious roots arise sooner, and some of them force
their way through the radicle and split it till it can hardly
be recognised.

CHAPTER XVI

FERTILISATION AND THE DEVELOPMENT OF THE SEED

Having seen what a seed really is, we must now go back and consider how such a structure develops from a tiny ovule This development is preceded by pollination, without which no seed is formed. How is it that the presence of a little yellow dust on the stigma can influence so greatly the ovules in the ovary below?

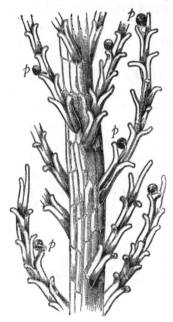

Fig. 99. Pollen grains (*p*) germinating on the stigma of a Grass. (After Kerner.)

If a stigma of rather loose texture, like that of a Grass, be examined in a drop of water under the microscope after pollination, it is often possible to see that the pollen grains which were deposited upon it have not remained unaltered; a delicate hair, like a root-hair, has grown from each grain into the stigma and may sometimes be followed for a long distance down towards the ovary. These hairs are called *pollen-tubes*. Pollen grains can easily be induced to form pollen-tubes by putting them into a drop of sugar solution, on a glass slide, keeping it from drying up by covering with an inverted tumbler lined with damp blotting paper. The tubes grow very long compared with the size of the pollen grains. They can only grow so long by feeding on the sugar in the solution, for if the grains are sown in water they only produce short thin tubes even if they germinate at all.

The best way of studying the germination of pollen grains is by growing them in drops of solution hanging from the under side of a cover-glass—hanging-drop cultures. The cover-glass is supported on a ring of glass, or on a ring built up with plasticine. so that a little closed chamber is formed in which the air remains moist and the drop does not evaporate. With such an arrangement it is easy to observe a grain at short intervals under the microscope and watch the growth of the pollen-tube.

The best strength of sugar solution varies for the pollen of different flowers. For that of the Garden Pea or Broad Bean about 10% is suitable.

If germinating grains are stained in iodine, it is sometimes possible to make out their nuclei. When the tube begins to grow there are two nuclei, one of which, the *tube-nucleus*, enters the tube, keeping as a rule near the tip, and probably influences its growth. The other divides into two, both of which enter the pollen-tube later.

When pollen grains germinate on the stigma the food that is necessary for the growth of the long pollen-tube must come from the stigma and style. We have already noticed that the stigma appears sticky. The cells of its surface, moreover, project as separate papillae which help to hold the pollen grain; the pollen-tubes grow down between the cells of the style, absorbing food from them, till eventually they reach the ovary. At this time the ovules are very small and delicate. They differ in details

of form in different plants, but on the whole they are
remarkably similar in structure in most flowering plants.

Ovules can be dissected out and examined under a

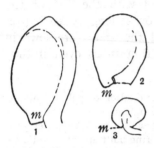

lens or a microscope. They
then appear as illustrated in
Fig. 100. Three main features
are usually visible: the stalk
by which the ovule was
attached to the placenta,
the main body of the ovule,
and the micropyle, which
appears as a small projection.

Fig. 100. Ovules as seen when
dissected from the ovaries and
examined under the micro-
scope. 1 and 2, anatropous
ovules of Narcissus and Dog
Violet; 3, curved ovule of
Wallflower: *m* = micropyle.

In a few ovules (for instance,
those of the Docks) the micropyle
and the stalk are found at opposite
ends: these are called *orthotropous*
ovules. But in very many the
body of the ovule turns over
during its development, so that the micropyle is brought next
the stalk which is joined to the side of the ovule from one
end to the other: such ovules are called *anatropous*. Others
turn half over, so that the stalk appears to be attached to one
side of the ovule, or are curved as in the Wallflower.

In order to study the structure of the ovules, transverse sections of the inferior ovary of a Narcissus or
Daffodil are convenient (Fig. 101); the ovary is large, the
ovules are numerous and they lie across the chamber in
such a way that horizontal sections of the ovary pass
through the ovules longitudinally. A median section of
an ovule looks very like that shown in Fig. 102. The
stalk can be traced along the side of the ovule to which
it is joined. If the section is accurately median, the
micropyle is seen to be a small opening leading between
the two coverings or *integuments* of the ovule to the main
body of the ovule inside, called the *nucellus*. Inside

the nucellus is a hollow cavity (the *embryo-sac*), which contains a special cell called the *egg-cell*, without any cell-wall, situated near the micropyle. It also contains another important nucleus, usually near the centre, as well as several other nuclei.

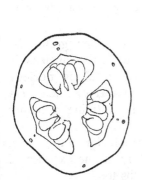

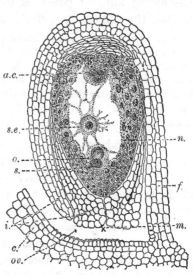

Fig. 101.   Ovary of Daffodil cut across, showing the three chambers, the axile placentation of the ovules, and some of the vascular bundles in the ovary-wall.

Fig. 102.   Longitudinal section through an ovule of Marsh Marigold; *i*, the two integuments; *m*, micropyle; *f*, stalk of the ovule attaching it to the wall of the ovary, *c*; *s.e*, central nucleus of the embryo-sac; *o*, egg-cell; *s* and *a.c*, other cells of the embryo-sac. (After Darwin.)

A pollen-tube eventually reaches such an ovule, passes through the micropyle, and makes its way through the nucellus to the cavity containing the egg-cell. By this time the tube-nucleus has usually disappeared. The other two nuclei pass down the tube, the end of the tube opens, and they enter the cavity of the nucellus. One of them joins with the nucleus of the egg-cell, fusing

completely with it to form a new nucleus; as a result
of this fusion the egg-cell forms a cell-wall, begins to
grow and becomes the embryo. This fusion of the female
egg-cell with the male nucleus of the pollen grain is called
fertilisation. Pollination is merely a means to this end.
The wrong kind of pollen is useless, because it cannot
fertilise the egg-cell (i.e. make it fertile or productive).

We have noticed that in many flowers a mixture of pollen
from the same flower and pollen from other flowers is usually
deposited on the stigmas: it has been shown in a few cases, and
is probably true in most, that this is sufficient to ensure *cross-
fertilisation*, because the pollen grains from other flowers of the
same kind germinate more quickly and their pollen-tubes grow
faster than those from the same flower.

The second male nucleus of the pollen grain also
enters the embryo-sac and fuses with the nucleus in
the middle. By the activity of the resulting nucleus,
which divides and forms a large number of nuclei, the
embryo-sac becomes filled with cells, forming a tissue:
this is the young *endosperm*.

The young embryo remains small for some time.
Later it begins to grow more rapidly at the expense of
the endosperm, from which it absorbs food.

When the fertilised egg-cell begins to grow it forms first a row
of cells. The end cell farthest from the micropyle becomes the
embryo proper, the rest form the *suspensor* which elongates and
pushes the embryo down into the endosperm. Stages in the
growth of the embryo may be obtained very easily from young
pods of Shepherd's Purse by dissecting out the ovules, soaking
them for half an hour in a 2% solution of caustic potash,
mounting them in the solution and squeezing out the embryos
by gentle pressure of the cover-slip.

The endosperm in the embryo-sac also grows at the
expense of the nucellus around it, so that in most cases
it comes to fill the mature seed, the remains of the nucellus
forming merely a thin papery membrane of crushed cells,

often difficult to detect, just within the seed-coat. Even those seeds, which when ripe are exalbuminous, have an endosperm when young; but, before they become mature and enter on their period of rest, the embryo absorbs all the food-material of the endosperm, and grows till in its turn it fills the whole seed. In albuminous seeds, on the other hand, the embryo stops growing and becomes dormant while a large part of the endosperm yet remains.

CHAPTER XVII

THE CONDITIONS OF GERMINATION AND GROWTH

When a seed is ripe the embryo has ceased to grow
Properties and has become dormant: when the seed
of seeds. germinates the embryo re-awakens and begins
active growth once more. In the dormant condition it
has remarkable properties which do not belong to it
when germination has begun, nor to the mature plant.
In the first place the seed becomes dry, without the embryo
dying, whereas a growing plant soon dies if not supplied
with water. When dry, moreover, seeds can endure
great extremes of cold and heat without losing the power
of germination. This can be shown very simply by
Exp. 40 putting (1) six dry peas, and (2) six peas
that have been soaking in water for about 12 hours, into
separate test-tubes, dipping these for half-an-hour into
water heated to 60° or 70° C., and then sowing each lot
of peas in a separate labelled pot. After a week the dry
peas will have germinated, but not those that had been
soaked: the latter were killed by the high temperature.

Exp. 41 A similar experiment in which a freezing
mixture of ice and salt is used in place of hot water will
show that the seeds are capable of withstanding very low
as well as very high temperatures while they are dry.

Conditions necessary for germination.

(1) *Moisture.* Seeds remain dormant so long as
they are dry. They may be sown in dry earth or dry

sawdust without any change appearing, and only germinate if supplied with water. The presence of water is thus one of the essential conditions for the re-awakening of the embryo.

(2) *A suitable temperature.* There are other conditions which affect germination. If lots of six peas are sown in separate pots, and kept at different temperatures —for instance, one pot in a cold cellar, another in an ordinary room (about 15–18° C.) and a third near a radiator—it will be found on examining the seeds after a few days that germination is very slow at the low temperature, rapid at the high temperature. This experiment should be performed with several different kinds of seeds. At a temperature below 9° C. French Beans, for instance, will not germinate at all. Many other seeds will germinate slowly at lower temperatures, but at temperatures approaching the freezing-point of water few seeds are able to germinate. On the other hand, very high temperatures also prevent germination.

(3) *Oxygen.* Another essential condition for germination, as for all growth, is a supply of oxygen for respiration. That seeds will not germinate in the absence of oxygen, even though the other necessary conditions exist, can be shown by the following simple experiment: Two lots of
Exp. 42   a dozen peas are placed, one in shallow water which just covers them, the other in a flask nearly full of water which has recently been thoroughly boiled to drive off all air and then cooled. The flask is corked to prevent the entry of air, and both lots are set aside together. In a few days the difference is already apparent.

One other condition is necessary for the successful
Effects of   development of a seedling, although it is not
darkness.   as a rule necessary for germination: this is light. Light does not penetrate soil, and most seeds will germinate in total darkness (a few require some exposure

to light after soaking, e.g. Tobacco, Willow Herb, Purple Loosestrife).

In order to observe the results which follow if the shoots are still deprived of light after they have emerged from the soil, seedlings must be grown, some in the dark, others in the light, under conditions identical in all other respects. If we attempt to do this, and study the conditions very carefully, we find that it is not easy to shut out light without enclosing the plant in air that soon becomes saturated with moisture, or introducing differences of temperature between the darkened plants and the controls. Perhaps Exp. 43 the best plan is to grow three pots of seedlings, one under an inverted light-tight box or tin, with sawdust round the edge to prevent light entering below it, the second under a bell-jar, and therefore in damp air, the third without a cover, all three side by side in bright diffuse light, not within reach of direct sunlight which would heat up the covered plants too much. Compared with the uncovered plants, those under glass grow taller and weaker, while those in the dark grow very tall and weak and remain yellow. The condition of these darkened seedlings is called *etiolation*.

Etiolated seedlings of various plants should be grown and compared with normal healthy seedlings: Wheat or Barley, the Garden Pea, Mustard, and Vegetable Marrow are useful kinds. In all these cases the plants remain yellow, and grow tall and weak. In different plants the greater height is due, however, to the more rapid growth of different parts. The leaf-sheaths and blades of Wheat grow long; in the Pea, on the other hand, it is the stem that grows faster, while the leaves remain very small. In Mustard and Marrow seedlings the hypocotyl grows long and thin, and the cotyledons are smaller than those of normal seedlings.

Similar though less extreme differences are shown by

seedlings grown in dull light, as compared with seedlings fully exposed to the light. Such behaviour is clearly of advantage to seedlings germinating under natural conditions, for they are frequently overshadowed by other plants crowded closely around them. In consequence of the deep shade and the moisture of the air with which they are surrounded they grow tall and may thus find their way up into the brighter light above, and get an opportunity of establishing themselves. The store of food-material at their disposal is thus employed by them as far as possible in increasing their height, this being their most vital and pressing need until they find enough light to keep themselves supplied by photosynthesis: then they set about strengthening their weak stems.

Etiolated seedlings illustrate a very important truth,

Effects of conditions on the form and growth of plants. other examples of which are frequently met with, namely, that external conditions influence not only the rate of germination and growth but also the form and dimensions of plants. This applies not only to seedlings, but to sprouting shoots of tubers, bulbs, etc., as well as to growing branches of older plants. As we saw, moisture, and still more dull light or complete darkness, favour faster but weaker growth. Bright, hot sunshine, poor water supply, exposure to dry wind, all tend to have a stunting effect on plants.

Examples can always be found of plants of the same kind, growing in different situations, that differ very greatly from one another. Chickweed, Plantain, and other common plants that grow in a great variety of situations should be collected from this point of view, and a selection of dried specimens with roots, with full notes of the conditions under which each was found growing, should be placed in the School Museum.

Some examples of especial interest are well worth growing. Old seedlings of Gorse if placed in moist air begin to produce broad instead of spiny leaves. Another interesting case is that of the House-leek, which in moist air forms an erect stem instead of remaining close to the ground, while when grown in darkness it is quite unrecognisable, with very small leaves. Very easy examples to obtain are sprouts of potatoes, left (1) in the dark, (2) in the light, in moist air under a bell-jar, and (3) uncovered on a table.

Experiments on the absorption of water by seeds.

The amount of water absorbed by a seed during germination is very great. A French Bean, for instance, if soaked in water, increases considerably in size and weight: at first the seed-coat absorbs water and becomes wrinkled, then as the rest of the seed swells the seed-coat is stretched smooth again.

We saw in examining the bean that the micropyle in this seed becomes an open pore through which water can be squeezed. Water could therefore also enter by it. How much water enters by the micropyle and how much passes directly through the seed-coat can be found by carefully weighing seeds at intervals, some with the micropyle open and some with it closed. Broad Beans, in which the micropyle is also open, are larger, and are therefore more convenient for this experiment. Two equal lots of six or more beans should be selected. In one lot the micropyles should be closed with a spot of rubber solution, and a similar spot applied to some other part of the seed-coats in the other (control) lot. Each lot is then weighed and put to soak in water. After three to six hours they should be weighed again, all unabsorbed moisture being first removed with blotting paper from the surface of every seed. Another method is to support the beans on end, half-submerged in shallow water, some with the micropyle in the water, some with it out.

Exp. 44 *a*

If a good balance is not available, the volume of the seeds can be measured instead of their weight, by means of a small graduated cylinder. Sufficient water to cover the seeds is first poured into the cylinder, and its volume noted: the seeds are then carefully dropped into the water, and the level of the water again recorded. The rise of level gives the volume of the seeds. In this way the volume of each lot of seeds is determined, before and after soaking for six or more hours.

Exp. 44 *b*

It is not every kind of seed that has an open micropyle. In some the hilum is specially pervious: the part it plays can be determined, as in the case of the micropyle, by covering the hilum in some seeds, and an equal area of the coat in others, with rubber solution and comparing the percentage increase in weight of the two lots after a few hours' soaking.

The seed-coat of many seeds allows water to pass through it, but as a rule not with great readiness. By comparing the increase of weight of seeds with and without the seed-coat after, say, an hour in water, it can easily be shown that it is a hindrance to the entry of water.

Exp. 45

In some seeds the seed-coat is especially impervious, and may prevent the entry of water for a long time. In consequence of this, germination is postponed, in some cases for one or more years. Some of the seeds of the Lupine have coats that hardly allow any water to pass through them. If a dozen or more seeds are soaked in water for some hours, a few will be found at the end of that time still scarcely at all swollen. If the seed-coats of the unswollen seeds are scratched these seeds then swell rapidly.

In most seeds the seed-coat hinders the absorption of water, not only because it does not allow the water to pass through readily but also because it resists the expansion of the embryo or endosperm. It is only necessary to consider the hard woody seed-coats of the Stone Pine or the Brazil Nut, or the shells of Plum-stones, or Hazel-nuts, in order to realise what enormous resistance the inner parts of seeds may have to overcome. A simple method of demonstrating that seeds can absorb water against very great pressure is to fill a glass bottle, such as a medicine bottle, with dry peas. packing them in tightly, and afterwards immersing the bottle in a pail of water, so that the water enters it. The peas in swelling break the bottle.

Exp. 46

We have already found that living cells absorb water by osmosis, and absorb it against considerable resistance. If the swelling of seeds be also due to osmosis we should expect to find less water absorbed from a solution of salt. This can readily be tested by soaking a dozen beans in a 5 % solution of salt, and another dozen, for comparison, in water, weighing both lots before soaking, and afterwards at intervals. Much less water is absorbed by the seeds in salt-solution.

Observations and experiments on the growth of seedlings.

When seeds germinate, the radicle emerges first and very quickly makes its way down into the ground: only afterwards does the shoot begin to expand. The importance of this order of development is clear. The root must establish itself in a position to absorb water before the shoot can safely grow up into the air, where it will lose water in transpiration. The root must also obtain a firm grip of the soil.

Growth in the soil must often be against considerable pressure. It is remarkable in what hard soil some plants will succeed in growing. The roots of Plantains and other plants growing on waysides in trodden ground must exert considerable force in order to make any headway at all. This force is developed by the osmotic absorption of water by the living cells.

We have seen that the root-tip, which must bear the brunt of the pressure and friction, is specially adapted for the work of boring through the soil by the presence of a root-cap, continually renewed from below as fast as it is worn away, like human skin.

There is another feature that is important mechanically, as well as in other ways: the growing region of a root is short. If a part of the root at some distance from the tip were to grow in length it would push upon the parts next in front of it, but these would give way, and the pushing force exerted by the growing part would be wasted in bending the root. Similarly, if one tries to push the point of a long hat-pin into a piece of wood, holding it by the head, the pin bends from side to side: to be successful one must firmly grip the pin near the point.

In a seedling we can easily find out how much of
Exp. 47   the root is growing in length. Broad Beans
germinating in damp sawdust are very good for this
purpose. A few are selected with straight healthy
radicles 2–3 centimetres long, and after carefully removing
all surface moisture with filter paper, a series of fine
marks are made with waterproof Indian ink on each
radicle at intervals of 2 mm., beginning from the tip.

The marks can be made with a very small camel-hair brush,
or with a fine fibre of cotton or silk stretched between the ends
of a curved piece of stout wire and moistened with the ink. The
marks should be accurately placed with the help of a millimetre
scale, or else the distance from each mark to the next should be
measured and recorded. The beans are then pinned to sheet
cork inside the lid of a tin box (or a piece of wood covering a jar),
in such a way that the radicle will hang vertically in the box
when the lid is replaced. The air in the box is kept moist by wet
blotting paper or sand. After a day or two the distances between
the marks are measured again.

For comparison with the result of this experiment the
Exp. 48   shoots of seedlings about three weeks old
should be marked in a similar way. The contrast is
very striking: a great part of the stem is found to be
still growing in length, even many centimetres distant
from the tip; whereas the marks on the radicle beyond
about 12 mm. from the tip remain at their original
distance apart. The root grips the soil firmly by its
root-hairs a short distance from the tip, and so is the
better able to push its tip through the soil. The shoot,
growing in the air, has not the same resistance to overcome.

The short growing region of the root is not solely of
mechanical advantage to it. For the safe growth of the
root-hairs it is necessary that the part bearing them
should have ceased growing; and the radicle of a seedling
must form its root-hairs early, in order to absorb water

from the soil as soon as possible to supply the shoot.
Each new portion of root added by the growing point
grows quickly to its full length, so that it is soon ready
to bear root-hairs. This can be observed by making
marks as before and measuring daily the distance between
the first two (i.e. the part of the root lying at first
between 2 mm. and 4 mm. from the tip).

How the direction of growth is determined.

If rightly considered, it is very remarkable that the
shoots of seedlings should always find their way upwards
into the light and air, and their roots as invariably turn
downwards into the soil. If they were human beings we
should ask how they could tell which way to grow. As
it is, we must inquire what causes them to grow in one
direction rather than in another

When seeds, such as beans, in which the position of
the radicle is easily recognised, are sown in various positions

Exp. 49　　and examined after a few days, it is found that
in whatever direction a radicle may at first have pointed,
it has, in every case, where necessary, curved until its
tip points downwards and then grown straight down.
Similarly, the shoot turns upwards from any position.
Moreover, if at any time after the root has begun to grow
downwards or the shoot upwards it is inverted or placed
on its side, it will again curve until it is growing once
more straight up or down.

The way in which the curvature takes place may be

Exp. 50　　studied by pinning germinating beans to
the cover of a bottle as in Experiment 47, so that the
radicles are horizontal. If the radicles are marked as in
that experiment it will be seen that only those parts
curve that are still growing: clearly, the upper side must
grow faster than the under side. The curvature is,
therefore, called a *growth-curvature*.

But *why* does a radicle that is pointing vertically
downwards grow straight, while one that is not curves
in growing? It is no answer to this question to say that
a root grows downwards in order to be better able to
absorb water, fix the plant in the ground, and perform its
other functions, any more than the fact that the captain
of a ship wishes to reach a certain port, in order to fetch
or deliver merchandise, explains how he gets there. He
would be helpless without a compass or the stars to tell
him in what direction he is going. And a seedling just
beginning to germinate in the soil is surrounded, not only
on all sides, but above and below, by a trackless ocean of
earth. By what stars or compass can it distinguish the
upward or downward direction from any other?

This problem can only be solved by careful experiment. In the first place we must ask whether there are
any differences between the conditions above and below
the seed which might affect it. We know that above
the soil are light and air, which are needed by the shoot,
in the soil is moisture, needed by the root. Is it possible
that some light penetrates the soil, or that the air supply
is better in the upper than in the lower layers, or that the
water supply increases in abundance the deeper the soil
is penetrated, so that the shoot is enabled to distinguish
in which direction the light or the atmosphere is to be
found, the root the direction in which to reach a better
supply of water? The first two of these suggestions
seem very improbable. Soil appears quite opaque;
and roots require air as well as shoots, so that we should
rather expect them to grow towards it than away
from it. We can, however, put all three suggestions to
the test in two simple experiments.

If seeds be sown in soil and put in a cupboard from
Exp. 51   which all light is excluded there can be no
question of light entering the soil and reaching the

seedlings. Nevertheless, the roots grow downwards, and, as we have already seen in our experiments on the effect of darkness on seedlings, the shoots emerge into the air. Clearly, light can have had nothing whatever to do with this result. We must next sow seeds in soil so arranged that moisture cannot be more abundant below nor air more abundant above. The bottom of a wooden box is Exp. 52 replaced by a piece of string netting. The box is then filled with soil, in which the seeds are sown. On the surface of the soil is laid blotting paper which is kept wet. By this arrangement the upper layers of soil are kept quite moist while air enters most freely from below. The roots nevertheless grow downwards, and their tips eventually appear below the netting, while the shoots lift the blotting paper above. From this experiment we infer that, in moist soil, neither differences in the amount of moisture nor in the supply of air above and below determine the direction in which roots or shoots grow, for in either case the roots would have grown up and the shoots down.

A similar result follows from Experiment 50. There the seedling is surrounded by moist air on all sides, and if all light is excluded neither moisture, air, nor light can influence the direction of growth. Some other influence must be at work, apart from any of these. What is it?

If a stone is dropped it falls to the ground. So long as it is held it presses upon the hand and it is therefore said to have weight. This means that the stone is continually pulled downwards, towards the earth. It has been found by very delicate experiments that every piece of matter attracts every other piece, but that the attraction depends on the *mass* of the piece (i.e. the amount of matter in it) ; so that the attraction of the earth itself is very great, while that of one stone for another is wholly

insignificant in comparison. This attraction of the earth
for objects near it is called the *force of gravity.*

Now the force of gravity tends to pull every part of a
seedling downwards. The growth of a root downwards
might be due to a simple yielding to this pull; but not
the growth of a shoot upwards, against the attraction of
the earth. Such an explanation can be shown, however,
not to apply even to roots. In whatever position they
are growing they are supported by the soil itself against
the downward pull tending to bend them. Moreover,
if a germinating seed be fixed so that the radicle lies
horizontal in a layer of water on the surface of mercury,
Exp. 53   it turns and pushes its way down although it
has to exert considerable force in displacing the mercury.

How can we find out whether gravity does affect the
direction of growth? We cannot eliminate it: but it is
possible (1) to arrange so that it affects the seedling
equally on all sides, and (2) to replace it by or add to it
another force of a similar kind in a different direction.

(1) If seedlings are placed with their radicles and
shoots horizontally, and are then slowly rotated, the side
which is underneath is continually being changed. By
this plan we ensure that gravity shall pull the radicle and
shoot as much towards one side as towards another, just
as we ensured in previous experiments that air or moisture
should affect them equally in all directions.

The seedlings are rotated by clockwork. An apparatus
can easily be constructed with a cheap clock. A long steel
knitting needle is heated in a flame an inch or two from one end
and bent sharply at right angles. Two holes are bored at the same
level in opposite sides of a tin box to take the long arm of the
needle. The short arm of the needle is firmly wired to the
minute hand of the clock, and the long arm is passed first through
one side of the box, then through the middle of two or three
large corks covered with wet blotting paper, and finally through
the hole in the other side of the box, so that the corks are inside

the box. Germinating seeds are fastened by pins to the corks, with their radicles pointing in various directions, and the clock set going: obviously the seedlings will be rotated once every hour by this apparatus. The lid of the box keeps the air moist. Such an instrument as this is called a *klinostat*.

Grown in the dark on a klinostat neither roots nor shoots curve in any particular direction, but continue their growth in any direction in which they may happen to have been placed.

This experiment does not, however, solve the problem. The position of the plant is merely being continually changed before it has time to curve and we have still not proved that the influence which is thus prevented from inducing curvature is gravity and not some other vertical

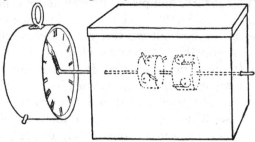

Fig. 103. A simple klinostat.

influence of unknown nature. For positive evidence we must resort to another method.

(2) If a weight be swung round on the end of a piece of string it pulls on it, and if the string is let go the weight flies off at a tangent. The faster the weight is revolved, the stronger is the force pulling it away from the hand, the centre of rotation. This force is called a *centrifugal force*. If a seedling were exposed to such a force it would be pulled away from the centre of rotation in just the same way as gravity pulls it downwards, and if gravity affects the direction of growth of its root and shoot so also should the centrifugal force: the root should turn in the direction of the pull, away from the centre of rotation, the shoot in the opposite direction towards the centre.

A simple apparatus for carrying out this experiment is a
little water-wheel made from a disc of galvanised
Exp. 54   iron about a foot across, with a cork fitted firmly
in a hole in the middle of it.

A knitting needle is passed through the cork to serve as an
axle, and a series of discs of cork are fixed on the circumference
of the wheel, as shown in the figure. The axle can be made to
turn in two short pieces of glass tube, firmly fixed on wooden
uprights. Small corks fixed on the axle near enough to the

Fig. 104. Simple water-wheel for exposing seedlings to a
centrifugal force. (After Osterhout.)

uprights will stop the wheel from moving too much from side to
side. Seedlings are firmly pinned to one side of the cork discs,
and then a jet of water is directed on to these discs in such a way
that, as each disc comes within reach, the water hits the side
opposite to the seedling. In this way the wheel is driven rapidly
round, while the seedlings are kept moist but not too wet.

On such an apparatus, when the wheel is rotated on
a horizontal axis, the force of gravity acts equally in all
directions, but the seedlings are fixed in relation to the
centrifugal force which pulls them continually towards

one side. Radicles under these conditions *are* found to turn outwards, in the direction of the force, and shoots to turn inwards in opposition to it.

If the wheel is set up horizontally and rotated on a vertical axis, gravity also pulls the seedlings continually towards one side. With gravity pulling downwards and the centrifugal force also pulling outwards, the radicles grow obliquely downwards and outwards, the shoots upwards and inwards, thus answering to both forces.

Now what is the meaning of these experiments? **Geo-tropism.** The force of gravity and a centrifugal force affect roots and shoots in such a way that they grow in definite directions. The direction depends partly upon the direction in which the force acts, but it also depends upon the plant itself, for a shoot grows in the opposite direction to a root, although the same force, in the same direction, is acting on both. We describe the result, therefore, by saying that the root and shoot *respond*, each in its own way, to the influence which the force exerts upon them. The force *does not bring the movement about* in either case, it merely *causes the root or the shoot itself to curve by its own active growth.* The force is, therefore, said to act as a *stimulus*, calling forth a response in shoots or roots that are sensitive to it. The capacity for being affected in this way by a stimulus is called *irritability* or *sensitivity*, and the organ is said to be *excited* by the stimulus.

The response of a root or shoot to the stimulus of gravity is a growth-curvature, and the position assumed as a result of it bears a definite relation to the direction of the stimulus. A *directive* response of this kind is called a *tropism*; the response of an organ to the force of gravity or a centrifugal force is called *geotropism* (Greek *ge* = earth; *tropos* = turning). A radicle, which turns in the same direction as the force, is called *positively geotropic*;

a shoot, which turns in the opposite direction, is *negatively geotropic*.

Leaves, lateral branches, and secondary roots take up horizontal positions in response to gravity, and are therefore called transversely geotropic. Many flowers that face sideways also take up their position in response to gravity. This can easily be shown in the case of the Narcissus, for instance. If a young and fresh flower is fixed with its stalk in a test-tube of water, so that it

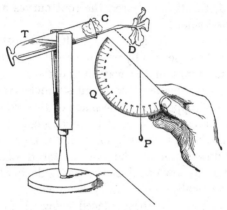

Fig. 105. Flower of Narcissus set up to show geotropic curvature, in test-tube (*T*) with plug of cotton wool (*C*); *D*, original direction of the perianth-tube: *Q*, a protractor with a plummet (*P*) attached, for determining accurately the direction in which the flower points. (After Darwin and Acton.)

points downwards, its position may be recorded on a piece of card fixed behind it, by lines drawn parallel to the perianth-tube, and will be found to alter until the flower is again directed horizontally.

Although it is in response to gravity that roots grow **The response** downwards they are sensitive to other in-**of roots to** fluences besides. If Experiment 52, in **moisture and** which the seedlings are grown in a box **contact with** **obstacles.** with a net bottom, be continued with the box tilted, the roots emerging from the soil into the air turn and grow along next the soil. This has usually been taken to show that roots can distinguish between moist soil and dry air. Recent experiments, however, throw doubt on this. Root tips are very sensitive to touch, so that roots tend to keep in contact with objects. In the soil they are stimulated more or less equally all round; but emerging from the tilted box they curve towards the lower side where the soil touches them longest. If, on the other hand, the tip is injured the root curves away from the wounded side.

How roots behave in contact with hard obstacles, can Ex. 55 be observed by growing seedlings behind glass, and placing stones in the way of the roots. One side is removed from a wooden box, about six inches deep, and a piece of glass supported in the place of this side by a few small nails in such a way that it slopes slightly outwards. The box is then filled with soil, with a layer of broken pieces of flower pot at the bottom for drainage. The radicles of seeds sown against the glass side, in attempting to grow vertically downwards, keep close against the glass. If stones have been placed below them they can be watched as they make their way round these obstacles.

So long as they are buried in the soil, shoots are not The re- reached nor affected by light: they turn sponse of upwards and so make their way eventually shoots to into the light, not in response to light but light. to gravity. It is, however, common knowledge that shoots do turn towards the light. Plants grown in a window have to be frequently turned in order that they may not lean permanently in the direction of the window. This means that when the light comes from one side the shoots respond to light rather than to gravity. This power of responding to light is called

heliotropism (Gr. *helios* = sun) or *phototropism* (cf.
photosynthesis). The following experiments enable us to
obtain more accurate information about the phototropism
of shoots.

A long, narrow box is made with an overlapping lid,
Exp. 56   and a round hole is cut in the middle of the
bottom. This box is turned upside down. Mustard
seeds are sown in the lid, on wet blotting paper, and covered
with the box, so that light only reaches the seedlings
through the hole. The result after about a week, during
which the blotting paper is kept wet, is shown in Fig. 106.

Fig. 106.

The middle seedlings grow upright and are stout and
green; those on either side bend towards the hole in the
cover, leaning over more and more and becoming taller,
weaker, and paler green the farther they are from it:
towards the ends of the box the seedlings are sometimes
quite yellow. If the box is long enough the seedlings at
the ends farthest from the light may lean less than those
rather nearer: this shows that the light reaching the ends
is too weak to induce the seedlings to respond very
vigorously, and they take up a position nearer to the
erect position that they assume in complete darkness.

A pot of healthy seedlings, such as Peas, with the
Exp. 57   shoots erect, is placed so that light only
falls on them horizontally. This may be done by putting

them into a deep box, laid on its side, and reflecting light upon them by means of a mirror a few yards away: the direct light from a window falls obliquely downwards and will, therefore, not do for an accurate experiment. Light from the sides should be screened off by brown paper extensions to the sides of the box. After a time the shoots will take up an oblique position. The angle they make with the ground varies very much with different plants. Some turn over till they are almost horizontal, others bend over comparatively little, but all show a phototropic (heliotropic) curvature.

Another experiment will make the results of the last Exp. 58 quite clear. The same apparatus is used, but the seedlings are laid on their side, so that the shoots are pointing straight towards the light. Under these conditions they bend *upwards*, in response this time to gravity, and take up a position similar to that taken up by seedlings of the same kind in the last experiment. In these two experiments, then, the shoots would in response to gravity alone assume or retain a vertical position, in response to light coming from one side a horizontal position; but, as both stimuli are acting together, the result is a compromise[1]. The slope of the shoot shows whether it is more sensitive to light (of the strength used in the experiment) or to gravity.

These experiments show how in nature the positions of shoots and roots may be influenced by more than one kind of stimulus acting together. The main direction is determined in the soil by gravity. In response, however, to irregularities in the distribution of moisture or to the presence of obstacles, the direction of growth of roots is modified; while shoots, when they have come above

[1] Compare Experiment 54, in which the positions assumed lie between those which would be taken up in response to gravity and centrifugal force respectively if each acted alone.

ground, are influenced by light as well as by gravity. The response is of the same nature in every case—a growth-curvature—but the direction depends on the combined influence of the different stimuli. The response to one of these stimuli alone can only be studied if the others are made to act equally in all directions or if their influence is entirely eliminated.

Leaves as well as stems respond to light, and their position is very largely determined by the direction from which most light reaches them. As they place themselves at right angles to the direction of the light they are *transversely* phototropic.

Some leaves, like those of the Garden Nasturtium (*Tropaeolum*) are capable at any time of altering their position: this can be shown by placing a plant in a pot so that light from a window falls upon it, when the leaves soon bend over till their blades face the light. Clovers and most Leguminosae move their leaves or their leaflets at the pulvinus (see p. 160) with a similar result. The younger leaves of erect plants, like the Sunflower, are brought facing the light by the phototropic curvature of the stem itself, but the leaves will often be found to move also, by the curving or twisting of their stalks. Not all leaves, however, remain capable of response to changes in the direction of the light; on a great many plants the leaves take up while young a definite position in which they become fixed as they grow older.

SECTION IV

THE CLASSIFICATION OF PLANTS

CHAPTER XVIII

EVOLUTION AND THE PRINCIPLES OF CLASSIFICA-
TION AS ILLUSTRATED BY THE BUTTERCUP
FAMILY

We may go into a meadow and find many plants so
like each other that we give them all the
same name, Daisy. Their flower-heads are
similar; their leaves are alike in form and are arranged
in rosettes; when the plants are dug up, they are found
to correspond also in the character of their root-systems,
in the way in which one rosette branches from another
—in fact, in their whole build or *habit*. Individual plants
differ from each other, it is true, in size, and in the number
of their leaves and flower-heads; such differences depend
partly on their age. But, besides, plants taken from a
very moist spot may have exceptionally large leaves
and flowers, or those found in a dry place may be small
and stunted. Yet we do not hesitate to call them all
Daisies.

When the achenes of one of these plants are sown,
there grow from them other Daisy plants, showing a
similar range of differences, according to the conditions
in which they are placed . So we come to think of Daisies
as all related to one another, in much the same kind of

way as the people who belonged to an old Highland tribe
were all relations.

Such a group of like plants is called a *species*.

If, however, we collect a number of Buttercup plants
from different places we shall find it possible to arrange
them in several groups, each forming a species like the
Daisies, but distinguished quite sharply from each of the
other groups by certain features of its own.

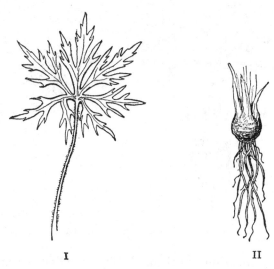

I   II

Fig. 107   I, leaf of Upright Buttercup; II, base of stem of
Bulbous Buttercup.   (After Praeger.)

Plants belonging to three of these species are very
common in the British Isles.   One is the Upright Meadow
Buttercup, a tall, strong plant, with large flowers and a
short perennial underground stem from which a number of
long, white roots penetrate the ground to a great depth.
A second is a rather smaller plant, which puts out runners
from its stouter and shorter perennial stem, and is called

the Creeping Buttercup: its leaves are usually smaller with broader segments, but its flowers are equal in size and very similar to those of the first. The third is the Bulbous Buttercup: it is erect like the first, and has no runners, but the stem is much swollen at the base; the flowers, too, are smaller, and the sepals are bent back against the flower-stalk (reflexed) instead of spreading as in the other two species.

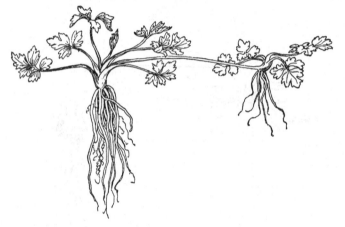

Fig. 108. Creeping Buttercup in winter. (After Praeger.)

Until we examined them closely, including the underground parts as well as those above ground, all these plants might perhaps have been thought to belong to the same species. **The genus Ranunculus.** Being so like each other in many of their characters, they are classed together in a larger group, called a *genus*, and each species receives two names, the first to show its genus, the second to distinguish it from the other species of the same genus. The Buttercup genus is called *Ranunculus*, and the three species we have referred to are *Ranunculus acris*, *R. repens*, and *R. bulbosus*.

There are many other species of Buttercups which differ from these three more than these differ from each other. Most of them are, however, so similar that they are without hesitation put into the same genus *Ranunculus*.

The Cornfield Buttercup, *Ranunculus arvensis*, for instance, is a small, nearly glabrous (i.e. hairless) annual plant, and its one seeded carpels ripen to achenes, which instead of being smooth and small are large and covered with curved spines (Fig. 109). Otherwise it agrees with the other three species in having similar divided leaves, with broad bases partly

Fig. 109.   Spiny achenes of *Ranunculus arvensis*. (After Baillon.)

sheathing the stem, and flowers with five sepals, five bright yellow petals having honey-glands protected by tiny scales, and numerous stamens and carpels.

The Spearworts are Buttercups, found in wet places, which differ from the Crowfoots, as the others are commonly called, in having undivided lanceolate leaves.

The Lesser Celandine, *Ranunculus Ficaria*, is another common plant which is classed in the same genus. It has a variable number of rather narrow yellow petals, usually eight or nine, and from three to five yellowish sepals, instead of having regularly five petals and five sepals like the other species; its leaves are smooth and undivided, and it stores its reserve food in a cluster of club-shaped root-tubers; but the petals have a honey-gland covered by a scale, the stamens and carpels are numerous, the fruits are achenes, and the leaves are palmately veined and have sheathing bases.

The Water Crowfoots have white flowers, but these flowers are in other respects very like Buttercups. Several species are distinguishable, differing from each other in size of flowers and form of leaves. Some grow in the water and have two forms of leaves (Fig. 201, p. 444), submerged leaves, which are subdivided into very narrow segments, and floating leaves, which are not divided but at most slightly lobed, and may be compared with the leaves of *R. Ficaria*. Others have submerged leaves only, while others again, growing in mud, only have undivided leaves. The flowers come above the water, and have five sepals, five white petals, with honey-glands not protected by scales as in other species, many stamens, and many one-seeded carpels, which ripen to achenes.

All these plants are more like each other than they are like any other plants, such as those which we shall study shortly; but within the genus, as we have seen, the true Buttercups have more points in common with each other than with the Lesser Celandine or the Water Crowfoots, and the Water Crowfoots most nearly resemble each other. It is possible to subdivide the genus into *sub-genera* in order to express the different degrees of likeness, and some botanists call these subdivisions the genera. The Water Crowfoots have been separated in this way and given the name *Batrachium* instead of *Ranunculus*. Even the Lesser Celandine is by some botanists put into a separate genus, *Ficaria*, and that is why this name, when used as its specific name, is spelt with a capital letter. So it is a matter for agreement among botanists what species shall be included in the same genus; it is partly a question of convenience and need not concern us further here, so long as we remember that our object is to classify plants which are most like each other together.

Just as species are grouped together into genera so **Allied** genera are assigned to larger classes.
genera. Let us examine a few plants belonging

to other genera which show points of similarity with plants of the genus *Ranunculus*.

Take, for instance, the Winter Aconite, *Eranthis hiemalis*, which is one of the earliest flowers to appear in gardens in spring. Below ground is a thick perennial rhizome, from which arise leaves, palmately divided, though less so than in the common Buttercups, and flowers with a green leafy collar just below each, called an *involucre*, which helps to protect the flower-bud. The flower itself consists of six or seven yellow leaves, without

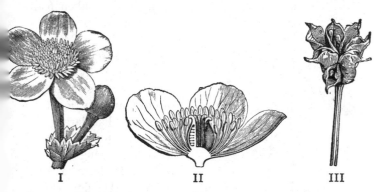

I   II   III

Fig. 110.   Marsh Marigold; I, flowering shoot; II, flower cut longi-
    tudinally; III, fruit.   (I and II after Baillon, III after Le Maout
    and Decaisne.)

any trace of honey-glands, a similar number of little yellow pocket-like structures containing nectar, numerous yellow stamens and several carpels. The little 'honey-leaves' are two-lipped, the outer lip being larger and yellower than the inner; we have only to imagine the outer lip to be expanded and the inner to become smaller, in order to see that these 'honey-leaves' correspond to the petals of a Buttercup. The conspicuous parts of this flower are the sepals: they serve to attract insects as well as to protect the inner parts of the flower in the bud. The carpels are larger than in the species of *Ranunculus*: each contains a number of seeds, and ripens, not to an achene, but to a small pod, which splits open along its inner rib, and so lets out the seeds. Such pods, splitting only down one side, are called *follicles*. In *Eranthis* they stand straight up and close together, and as they open inwards the seeds are only dislodged when the head of fruits is shaken (compare Fig. 82, p. 230).

The Marsh Marigold, *Caltha palustris*, which grows in marshy places by the banks of streams and ditches and in low-lying meadows, has large long-stalked heart-shaped leaves, growing together in a tuft from a thick rhizome. The flowers (Fig. 110) are large, often three or four centimetres in diameter; they have only one set of five or more yellow leaves, which serve both protective and attractive functions, like the sepals of *Eranthis*; there are no nectar-bearing leaves— nectar is produced by the carpels, which have small glandular areas on either side. The stamens are very numerous, and there are a number of many-seeded carpels, which ripen to follicles

Fig. 111.   A Wood Anemone.
(After Baillon.)

like those of *Eranthis*, but spread more when ripe
(Fig. 110, III).

The Wood Anemone, *Anemone nemorosa*, is like the
Marsh Marigold in having only one set of petaloid leaves;
these are pale pink, usually six in number and arranged
in two alternating whorls. The carpels, however, are
one-seeded, ripening to achenes as in the Buttercups.
The flower produces no nectar at all, but is visited by
small flies for pollen. Like *Eranthis*, the Anemone has a
leafy involucre below the flower, but it is half way down
the long stalk. There are other species of Anemone which
are pollinated by bees, and do yield nectar; *A. pulsatilla*,
the Pasque Flower, which grows in chalky districts, has
purple flowers in which nectar is secreted by the outer
modified stamens, which produce no pollen and are therefore called *staminodes*.

There is another group of genera in which the flowers
have honey-leaves or petals, as well as five large and
conspicuous sepals, and many stamens, but are greatly
modified in form in ways which adapt
them to pollination by humble-bees.
Their nectar is concealed, out of reach
of small insects. In the Columbine,
Aquilegia, both the petals and sepals
are coloured and conspicuous; the
petals grow out between the sepals as
long narrow pockets or spurs, and the
nectar is secreted at the bottom of
these pockets. The flower hangs
down, so that the pollen and nectar
are protected from the rain, and the
spurs are hooked over at the end
so that the nectar does not run out.

Fig. 112. Flower of
Columbine showing
spurred petals.
(After Baillon.)

A curious feature is the presence of staminodes in the
form of petaloid scales immediately surrounding the

group of carpels. There are only about five or six many-seeded carpels. When they are ripe the flower-stalk is no longer pendulous but erect, and the follicles, standing up close together, split open on the inside only for a short distance from the top, and so form a very efficient censer mechanism.

The Larkspur, *Delphinium*, has a zygomorphic flower. Two of the petals are spurred; besides these not more than two develop as tiny scales, sometimes none at all. One of the five sepals, lying between the two spurred petals, is also spurred, and the spurs of the petals fit into the spur of the sepal. Like most zygomorphic flowers the Larkspur faces sideways, and so the pollen is sheltered, while the lower sepals serve as a landing-stage for bees.

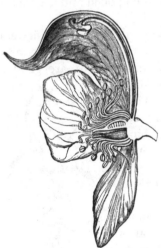

Though there may be as many as five many-seeded carpels, often only one is present, and the stamens, while still indefinite in number, are fewer than in such flowers as the Marsh Marigold or the Buttercups.

In the Monkshood, *Aconitum Napellus*, the sepals alone are conspicuous. They are blue, and of different sizes, the back sepals, the largest, being arched into a hood over two tall honey-leaves, which end at the top in hooked pockets containing nectar. Besides these two

Fig. 113. Flower of Monkshood cut longitudinally. (After Baillon.)

honey-leaves are only a few small and narrow scales. The flower is thus zygomorphic, and faces sideways. The stamens are fairly numerous, protected by the hooded sepal; but there are usually only three carpels. The stalk becomes erect when the follicles ripen, and, as in the Columbine, they open only at the top.

The Traveller's Joy (*Clematis Vitalba*), unlike other plants of the family, is a climbing shrub (see p. 425) with opposite pinnate leaves. The flowers have no honey-leaves, but four large white spreading sepals, and many stamens and carpels. They are pollen-flowers (p. 211), producing no nectar. The carpels are one-seeded, and ripen to achenes, each with a long feathery appendage formed by the style, serving for wind dispersal. Other species of Clematis are cultivated in gardens, especially one with large blue flowers.

Fig. 114. Achenes of Clematis with feathery styles. (After Baillon.)

The Meadow-Rues (genus *Thalictrum*) are peculiar in having a very inconspicuous perianth. There are no honey-leaves, and only four or five small thin yellowish sepals which soon drop off. The numerous pale yellow stamens are the only conspicuous feature of the flower. There is no nectar: *Thalictrum minus*, which grows in chalk pastures, is wind-pollinated, while the rather larger flowers of *T. flavum*, found in moist meadows, ditches and fens, are visited for pollen by flies. The carpels are one-seeded, and comparatively few, sometimes only three in the small flowers of *T. minus*. The leaves of these plants are much-branched compound leaves with numerous small, usually wedge-shaped, leaflets.

The Christmas Rose (genus *Helleborus*) has large drooping flowers with sepals that are white when the flower is open, but after fertilisation turn green. Within the calyx are several (8–10) small greenish tubular honey-leaves, similar to those of the Winter Aconite. The stamens are numerous, but there are few carpels, usually only three. The carpels are many-seeded: they ripen to follicles, which not only stand close together, as in the Monkshood, Columbine, etc., but are slightly joined together at the base. The flowers are markedly protogynous.

The Globe Flower (genus *Trollius*) has large yellow flowers with an indefinite, often large number of yellow sepals, which remain folded over the flower, nearly closing it, and giving it its characteristic spherical form. Within are a similar number of small narrow flat honey-leaves, numerous stamens, and several carpels that ripen to follicles. Although the sepals and honey-leaves are quite distinct, they are not arranged in true whorls, but spirally, attached not at the same level on the receptacle but one a little above another.

In the large red flowers of the Paeony (*Paeonia*) there are no honey-leaves. Surrounding the numerous stamens are five red leaves, and outside these, alternating with them, are five green leaves. Nectar is secreted by a disc around the base of the carpels, between them and the numerous stamens. The carpels are few in number, often only three, and ripen to large follicles, with thick fleshy walls covered with woolly hairs, opening down the inner margin to expose several bright red seeds.

Love-in-a-Mist (*Nigella*), often grown in gardens, bears flowers surrounded by an involucre of five much-divided leaves with very narrow segments. The flower has five coloured sepals, several small tubular honey-leaves, with curious lids which prevent small insects from reaching the nectar, and numerous stamens. Instead of several separate carpels, however, there is a complex ovary in the centre of the flower, with five chambers, each ending at the top in a free tip bearing a style; the fruit is a capsule in which each chamber opens at the top.

It is clear that few statements can be made about the plants we have just considered, taken as a whole, to which one plant or another would not form an exception. Some of the plants, belonging to different genera, for instance the Wood Anemone and the Monkshood, may even appear at first sight to have very little in common with each other. The Monkshood has an erect habit and zygomorphic, highly specialised bee-flowers, with two honey-leaves and only three many-seeded carpels; whereas the Wood Anemone has a creeping rhizome and its flowers are pollen-flowers, with no honey-leaves at all and many carpels, which are one-seeded. Yet when they are carefully examined and compared with plants of the other genera, each is found to be linked to these other

plants by some similar characters. The flower of the
Monkshood is similar in construction to the actinomorphic
flower of the Columbine, which has five larger honey-leaves
and a variable number of carpels. This again has points
of resemblance to the Winter Aconite on the one hand,
and the Buttercup on the other, and so has the Anemone,
with its numerous one-seeded carpels, involucre, and
conspicuous calyx, in spite of the total absence of honey-
leaves. Thus no satisfactory line can be drawn between
these genera. Even those which are quite exceptional
in certain of their characters (such as Love-in-a-Mist,
Nigella, with its five-chambered ovary) are linked with
the rest by other features. We have therefore, in classifying plants, to arrange all these genera near together.

Such a group of allied genera is called a Family[1].
This particular Family is called, after the genus *Ranunculus*, the Ranunculaceae.

If we do attempt to make general statements about
the characteristics of the Ranunculaceae as a whole, we
may mention the following points: The plants are mostly
herbs with alternate leaves (except *Clematis*); most of
them are poisonous (from the Monkshood, for instance,
is prepared aconite, which is used in very small doses in
medicine). The flowers are hermaphrodite, and with few
exceptions actinomorphic: in most genera they are borne
in branching cymose inflorescences, but in a few (e.g. *Aconitum*) in racemes. The stamens are indefinite in number,
and usually numerous. The gynaecium consists in nearly
every case of separate carpels, variable in number, often
numerous, but sometimes only one (*Delphinium; Actaea*).
As a rule, at least the stamens and carpels are arranged
spirally on the axis, and in many cases also the leaves of
the perianth (e.g. *Trollius*, in which they are indefinite

---

[1] By international agreement 'Natural Orders' are now called
Families.

in number). An interesting feature of the Family is the presence in so many genera of nectar-bearing leaves. Where these 'honey-leaves' are absent or inconspicuous, and also in some cases (e.g. *Aquilegia*) where they form elaborate petals, the sepals are petaloid, and serve the function ordinarily performed by petals as well as the protective function characteristic of sepals.

We have now seen how species are grouped into genera and sub-genera, and genera into a **Evolution.** Family, in order to express the varying degrees of resemblance that exist between them. What do these degrees of resemblance mean?

Charles Darwin, in his great book, *The Origin of Species*, showed that just as the individual plants forming a single species are related to one another, having common parents or ancestors, so species and genera may also be related, only more distantly, and that close resemblance probably means close relationship.

He pointed out, in the first place, that species do alter. Gardeners are continually bringing out new plants with more beautiful flowers or better fruit. These are, of course, all the offspring of other plants. In some cases we can still compare the cultivated plant with the wild species from which it was originally derived, and the differences are very striking. Compare, for example, double flowers like Double Stocks or Garden Roses with the ordinary Single Stock or a Wild Rose, and cultivated Apples with wild Crab-Apples. In the colours of flowers the variety is extraordinary: from the wild Snapdragon, for instance, with purplish-red flowers, have been derived numerous garden varieties, with purple, red, orange, yellow, cream or white flowers, some too with striped flowers, some with the corolla-tube and lips differently coloured.

Thus in many cases we know that different forms have arisen from common ancestors, and, in other words, that some species have altered under cultivation. This alteration is called *variation*. As a result of variation new forms appear which, though different, are related to each other just in the same way as individuals of a species which are all alike.

In nature, species seem at first sight to remain constant in their characters, except for the differences in size and vigour that depend on the kind of spot in which the individual plants may be growing (p. 261). The handing on from a parent to offspring of its own characters is called *heredity*. It is due to heredity that acorns always produce Oaks for generation after generation. Just as, however, under cultivation, variation steps in and modifies heredity, so that some of the offspring are not exactly like their parents, so, it is believed, has variation occurred among wild plants, and is probably occurring still.

Those who study wild plants minutely, frequently meet with difficulty in deciding whether a particular plant belongs to a certain species, or whether it should be placed in a separate species because of small differences which it shows from other plants of the species. It is necessary to decide, in the first place, whether these differences are due in any way to conditions, and often this is not easy. Among the Water Crowfoots, for instance, some of the species are very variable, having only much divided leaves in deeper water, only slightly lobed leaves in mud, and in shallow water leaves of both forms. The extreme forms of the same species are thus very different in habit from one another, and are easily mistaken for other species growing always in mud or always in deep water, as the case may be. The result is that some botanists have regarded all the Water Crowfoots as forming a single species, others have seen in them a very large number of separate species. Just how many species there really are can only be decided by growing seed from plants of the various kinds in different situations and carefully comparing the range of forms assumed by the offspring. This is an example of species with different *habitat forms*; to distinguish such species may be laborious, but the difficulty is not primarily due to the effects of true variation.

On the other hand what appears at first to be a single species turns out in many cases to be separable into a number of groups, differing very slightly but quite constantly one from another. Sometimes, the more minute the examination, the larger the number of distinct forms that have been detected (e.g. *Draba verna*).

The species must in such cases really consist of smaller groups, just as genera are made up of species. In practice, such species are subdivided into sub-species and varieties.

Many of these distinct forms differ only in very insignificant points from each other, and are very difficult for any but experts to distinguish. Yet they undoubtedly exist, breed true, flourish, and hold their own.

A third source of difficulty is the occurrence of plants intermediate in their characters between two species. In some cases they are only found where both species occur together, and in some such cases it has been actually shown by experiment that plants of these intermediate types are obtained by 'crossing' one species with the other, i.e. by cross-pollinating the stigmas of one with pollen from the other. Such intermediate plants are called *hybrids*. The Common Avens and the Water Avens (*Geum urbanum* and *G. rivale*), the first a common hedge plant, the other found in marshy places, are two species between which intermediate forms are found in nature. When these two species are crossed experimentally the offspring are like the commonest wild hybrid (Fig. 115): its flowers

are large like those of *G. rivale*, but have spreading sepals,
and are, therefore, more open like those of *G. urbanum*;
moreover, they combine the red colour of the one with the
yellow colour of the other, and are neither pendent nor
erect but take up a position intermediate between those
characteristic of the flowers of the parent species. When
these hybrid plants are crossed with the parents numerous

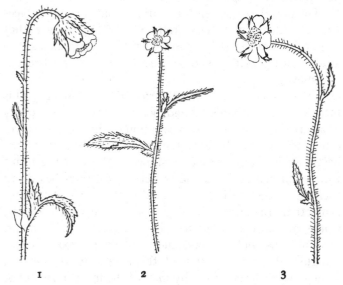

Fig. 115. 1, Water Avens, *Geum rivale*; 2, Common Avens, or Herb
Bennet, *Geum urbanum*; 3, the hybrid, *Geum intermedium*, formed
by crossing 1 and 2.

other intermediate forms are obtained, some of which have
been found in nature, and these combine in various ways
some of the characters of the one species with other
characters from the other species. Thus by crossing, or
hybridisation, the characters of two species are re-sorted
into different combinations. Plants can be bred, showing
many of these re-combinations, which will yield only off-

spring like themselves so long as they are not crossed with a different form: in other words, like the plants of the original species, they will 'breed true.'

In such cases as this, new varieties are being produced in nature in the same way as many new varieties are bred by horticulturists—by crossing plants of two different kinds, and so obtaining various re-combinations of the characters of the parents.

The fact that these hybrid varieties only occur in places where both parent species are found shows, however, that in nature they do not succeed in establishing themselves and spreading as independent species. Often this is because they never set good seed: they are sterile. Others for some reason are not successful in the *struggle for existence*. This brings us to Darwin's second point, that, whereas many new forms may arise, only a selection of them can survive: others will die out because they are not well fitted for the battle against adverse circumstances. Many of the forms which man has selected and preserved, for their beauty or their usefulness to himself, would soon perish, if left unprotected. In nature those plants survive which, on the whole, are best adapted to their particular environment. Darwin thus looked upon the struggle for existence as bringing about a selection of the fittest (in this sense), analogous to the selection by man of the most beautiful or most useful. These principles he summed up in his phrases 'the survival of the fittest' and 'natural selection.'

Selection alone can produce nothing new. It works upon the results of variation and hybridisation. Since Darwin's day much knowledge of these two phenomena has been accumulated, an account of which can be found in books on Genetics.

We may picture evolution therefore as the result of a continual production of new forms by variation and hybridisation. In the struggle for existence some of these forms have died out. Those which remain are separated from one another by gaps which have widened with time. Of the species we recognise to-day those that have arisen more recently from a common stock resemble one another, in general, more closely than those whose common ancestry is more distant.

So it may be that all the Buttercups were once represented on the earth by a single species, probably not quite like any of those which we know; and the more closely similar the different genera appear to one another, the more closely related do we suppose them. If we could go far enough back we might be able to trace all the plants which we put in the Family Ranunculaceae to a small ancestral group or even one species, though what this may have been like we can only guess.

It is nevertheless instructive to compare, for instance, the different forms of flowers which occur in a family like the Ranunculaceae, and consider in what features they can be regarded with probability as very different from or similar to the flowers of their ancestors.

Evolution of complex ovaries.

We have seen that most of the Ranunculaceae have a number of separate carpels. *Nigella*, however, has a complex five-chambered ovary, with axile placentation. This exceptional character was, therefore, probably evolved from separate carpels. A first stage in this process is seen in the Christmas Rose, where the erect carpels are united at the base. If we imagine five such carpels joined all the way up instead of just at the base, we have a five-chambered structure like the ovary of *Nigella*, with axile placentation, since in each separate carpel the ovules were attached on the side towards the centre of the flower. In *Nigella* the carpels still remain separate at the top, and each has its own style.

A complex chambered ovary is thus to be regarded as made up of as many joined carpels as it has chambers: it is therefore called a *syncarpous* ovary (*syn* = together).

The highly specialised flowers of Monkshood, Larkspur, and Columbine must also be regarded as having departed more from the ancestral type than relatively simple flowers like those of the Hellebore and Winter Aconite. Of the three, the actinomorphic Columbine is nearest to

Evolution of special- ised flowers.

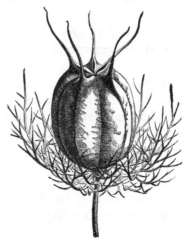

Fig. 116.　Capsule of *Nigella*.　(After Baillon.)

the simpler types, and has all its honey-leaves elaborated. The other two are zygomorphic: this feature is, as we have already seen (p. 197), an advance from the point of view of pollination, as the insect visitor must always approach such a flower in the same way; it involves a one-sided development of the young flower which is in all probability of comparatively recent origin. In these flowers we also find only two honey-leaves reaching full development; the others remain insignificant or do not develop at all.

Methods of recording the structure of flowers.

In comparing plants for purposes of classification it is necessary to make careful records of their characters, because they cannot all be obtained at the same time, and dried specimens are not easy to dissect and examine. We have seen that in the Ranunculaceae the flower is most important for indicating relationships, and records of flower-structure are therefore essential. There are three kinds of shorthand records which are very useful: they are the floral formula, the floral diagram, and the median longitudinal section. We have already made frequent use of the last.

Floral formulae. A floral formula is primarily a record of the number of the different parts of the flower, but several other points can be recorded by means of symbols. The conventional signs used in the formulae are as follows:

P = perianth
K = calyx
C = corolla
A = androecium (stamens)
G = gynaecium (carpels)

↓ before the formula = *zygomorphic*
( ) round a number means that the parts are *joined*
∞ = *indefinite*

When the parts, of the perianth for instance, are arranged in whorls, the numbers in the different whorls are given separately and connected by a +.

A line under the number of carpels (thus G5) means that the gynaecium is superior: a line above the number (thus $\overline{G(3)}$) that the ovary is *inferior* and therefore that the flower is *epigynous*.

Examples.

Buttercup, $K_5 C_5 A\infty \underline{G\infty}$.

Wood Anemone, $K_{3+3} C_0 A\infty \underline{G\infty}$ or $P_{3+3} A\infty \underline{G\infty}$.

Clematis, $K_4 C_0$ or $P_4 A\infty \underline{G\infty}$.

Monkshood, ↓ $K_5 C8 A\infty \underline{G_3}$.

Nigella, $K_5 C_{5-8} A\infty \underline{G(5)}$.

Floral diagrams. A floral diagram is a drawing of an imaginary section across a flower, passing through all its organs, and so represents how they are arranged in relation to one another. It shows, for instance, whether the parts of the perianth are arranged in whorls, and how they overlap in each whorl. In the gynaecium the placentation of the ovules is represented as seen in cross-section. It is often possible, especially with zygomorphic flowers, to tell that a particular side of the flower is towards the main axis, on which it was borne as a branch in the axil of a bract; the position of the axis and the bract are then indicated, and if there are any little bracts (bracteoles) on the flower-stalk these are also represented in their correct positions.

The series of diagrams in Figure 117 will show how floral diagrams are drawn and the way the different parts are represented. It is well in practice to make them fairly large and either to plan out the spaces to be occupied by the different parts before putting

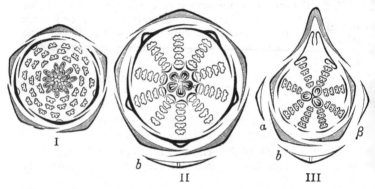

Fig. 117. Floral diagrams: I of a Spearwort, *Ranunculus lingua*; II of Columbine; III of Monkshood: *b*, bract; *a*, *β*, bracteoles. (After Baillon.)

them in, or to begin in the middle with the gynaecium, so that the inner parts will not be overcrowded.

In order to represent more accurately the general form of the flower and the mode of attachment of its parts to the receptacle a drawing of a median longitudinal section is extremely valuable. It can also be made to represent as a rule just those features of structure that are essential parts of the pollination-mechanism.

Longitudinal sections.

In order to draw such a section the flower must first be cut accurately into two similar halves. As a rule the knife must cut some perianth leaves in half and pass between others. A good way of making a successful cut is to start with the stalk and cut upwards (see p. 183), taking care that the section is in the right plane. Zygomorphic flowers may, of course, only be cut in one direction, but actinomorphic flowers have many 'planes of symmetry' and can be cut into symmetrical halves along any of these.

In making the drawing it is very important to distinguish carefully the section itself from the rest of the half-flower. The cut surface is seldom perfectly median, but it is easy to tell what parts would have come into a median section and what would not. When the section has been drawn with a firm clear outline, showing exactly how and where each part that comes into the section is attached, the rest of the half-flower can be drawn in more faintly, so as not to obscure the section. Drawings made in this way are really sketches of halves of flowers in which the *actual section* is drawn very carefully and distinctly in order to show the attachment of the parts. As examples, study Figs. 133 *A* and 164–6.

CHAPTER XIX

OTHER FAMILIES

A. *Families in which the flowers have separate petals.*

The Papaveraceae.

The flower of the common Field Poppy (*Papaver Rhoeas*) has many stamens, like a Buttercup; but it has two sepals, and four petals in two whorls, the three whorls

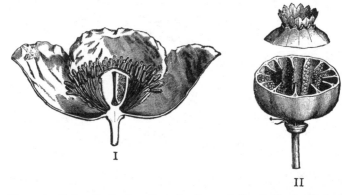

I

II

Fig. 118. I, longitudinal section of a Poppy (after Baillon); II, fruit of Opium Poppy, cut across to show the placentae (after Engler and Prantl).

alternating with each other; and instead of separate carpels it has a one-chambered complex ovary, from the walls of which project many placentae (Fig. 118) bearing over their expanded surface a very large number of tiny ovules. The flat top of the ovary is marked by a number of rays, the stigmas, equal in number to the placentae. The petals are large, and are crumpled in the bud. When the flower opens the sepals fall off. The flowers of the Poppy are pollen-flowers, and produce no nectar. The plant itself is an erect annual, with stiff spreading hairs and alternate pinnately divided leaves.

There are many other species of Poppy, all very similar to the common Field Poppy, differing only in such characters as the size and colour of the flowers, the shape of the ovary and the number of placentae and

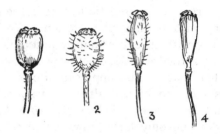

Fig. 119. Capsules of different British Poppies. 1, *Papaver Rhoeas*; 2, *P. hybridum*; 3, *P. Argemone*; 4, *P. dubium*. (After Praeger.)

stigma-rays, the size of the plant and the exact shape of the leaves; and so they all belong to the genus *Papaver*.

The Celandine, *Chelidonium majus*, has a smaller flower, with two sepals which fall off very early, four yellow petals crumpled in the bud, and numerous stamens; its ovary is one-chambered like those of the Poppies, but is long and narrow with only two placentae.

Besides the resemblances that the Celandine and the Poppies show to one another in the structure of their flowers and also in general habit, they all yield, when wounded, coloured milky sap, or latex, which rapidly sets on exposure to the air. This latex is poisonous; that of the Opium Poppy yields opium.

There is another group of plants, among which are the Fumitories (e.g. *Fumaria officinalis*), in which the flowers have only two (branched) stamens. Thus petals, sepals, and stamens are here all in whorls of two; while the ovary is like that of the Celandine. *Corydalis* and Dutchman's Breeches (*Dicentra*), often grown in gardens, belong to this group.

All these genera are put together in the Family Papaveraceae. They form a connected series, from the Poppies, with numerous stamens and placentae, to the Fumitory group. The Family thus includes a range of forms, like the Ranunculaceae, but there are more characters which are common to them all.

One important characteristic feature is the one-chambered ovary with parietal placentation. We have seen that an ovary with several chambers and axile placentation is to be regarded as made up of carpels joined together. An ovary like that of the Poppy can also be regarded as syncarpous. A free carpel, although very early in its development it becomes a hollow structure, begins as a flat rudiment and only later grows round till the edges meet. It is on the edges that the ovules are borne. In the development of complex ovaries, whether they have one or several chambers, separate rudiments appear at first, like the rudiments of carpels. In the case of one-chambered ovaries these rudiments grow together, edge to edge, and on these joined edges the ovules are borne. Thus the number of placentae is the same as the number of separate rudiments, and therefore the same as the number of carpels of which it is to be regarded as made up.

It follows that in the Poppy, as in the Buttercup, the carpels as well as the stamens are numerous. The floral formula is $K2\,C2+2\,A\infty\,\underline{G(\infty)}$. In the other genera mentioned there are only two carpels.

The Cruciferae.

The Wallflower is an example of a very large number of plants which are strikingly similar to each other in the structure and arrangement of their flowers. The species are grouped for convenience into many genera, but throughout the whole Family the flowers hardly differ from each other more than those of the single genus *Ranunculus*.

The Wallflower (*Cheiranthus cheiri*) is an erect shrubby plant with entire lanceolate alternate leaves. The flowers

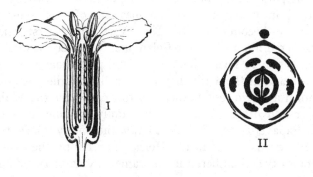

Fig. 120. I, longitudinal section of Wallflower; II, floral diagram representing the flowers of most Cruciferae.

are borne in racemose inflorescences, in which at first all the flowers are on the same level. As the older flowers become pollinated the stem grows and lifts the younger flowers above the pollinated ones, and so when the fruits are ripe they are exposed at intervals up a tall stalk. The flower has four erect narrow sepals, two of which tightly overlap the edges of the others and so form a sort of tube. The four petals are narrow and erect within the sepals, but above are broader and spread out horizontally. There are two short stamens, and two pairs of

longer ones, making six stamens in all. They stand erect
and the anthers block the way for small insects or rain
down to the nectar, which is secreted by green nectaries
round the bases of the shorter stamens and accumulates
within the sepals; the two outer sepals are hollowed at
the bottom into shallow pouches for its better reception.
The flower is thus so constructed as to secure the visits
chiefly of long-tongued insects.

The ovary is narrow and upright, about as long as
the longer stamens; it is two-celled, but the ovules are
arranged along the edges of the partition, i.e. the placentation is parietal. When the development of the ovary is
studied in young flowers the meaning of this unusual
structure becomes clear. The ovary is at first one-
chambered like that of the Celandine, but after the ovules
have begun to form, the two parietal placentae grow out
from between them till they meet in the middle and join
to form a single membrane. In this character the Wallflower and other plants of this Family are like some of
the Papaveraceae: in the Poppy the many placentae
grow inwards, while in the Horned Sea Poppy the ovary
becomes two-chambered in the same way as in the Wallflower.

A study of the development of the flower also suggests
an explanation of the presence of six stamens. The two
pairs of longer stamens arise as two outgrowths only, and
each of these then branches into two. It is possible,
therefore, that the longer stamens represent two stamens
that have branched, which with the two shorter ones
make up a whorl of four stamens alternating in the
ordinary way with the four petals. (The Fumitories
also, among the Papaveraceae, have their two stamens
branched.)

When the fruit ripens, the carpel walls split away
from the middle membrane, beginning at the base, and

may remain for some time hanging from the top. The seeds are left attached to the membrane and are gradually dislodged; they are thin and light, so that a strong wind will carry them some distance.

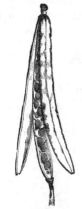

Except for differences of size and colour the flowers of Stocks, Watercress, Cabbage, Mustard are very like those of the Wallflower. The same floral diagram does for them all. Their fruits also are of just the same type. In fact, there are so many plants with similar fruits that it becomes convenient to have a special name by which to describe them; they are called *siliques*.

Fig. 121. Silique of Wallflower, dehisced. (After Baillon.)

Still other plants, usually with small flowers, have a shorter ovary, which is broad in proportion to its length. One of these is the Shepherd's Purse, *Capsella Bursa-pastoris*, a very common weed, growing in waste places and by waysides. It is a small plant with dark green pinnately divided leaves and small white flowers, which are self-pollinated. The ovary ripens to a flat heart-shaped fruit about 5 mm. long. The two valves split away from the septum just as in a silique; such a fruit is called a *silicule* (i.e. little silique). Honesty is a plant with a large silicule: the racemes bearing round white septa with a satin-like surface, which are left exposed when the valves fall off, are often used for decoration.

The Radish and a few other plants, with flowers of the same type as these, have fruits of a different form. The ovary of the Radish is at first one-chambered with parietal placentation, but ripens to a long pod which is jointed

between each of its few seeds and breaks up into one-
seeded portions.

There are also a few species with one-seeded fruits.
One of these is the Sea Rocket, *Cakile maritima*, common

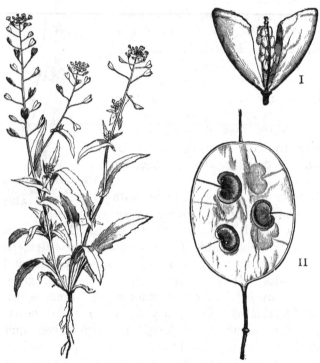

Fig. 122. Shepherd's Purse.   Fig. 123. I, silicule of Shepherd's
      (After Baillon.)   Purse,   dehisced   (enlarged).
                                            II, septum from silicule of
                                            Honesty. (After Baillon.)

in salt marshes all round the coast; its leaves are thick
and fleshy with few narrow lobes. The fruit is two-
jointed, but usually only the upper portion, which breaks
off, contains a seed. Sea-kale (*Crambe maritima*), also

a sea-shore plant belonging to this Family, has globular one-seeded fruits.

Thus, even in this Family, in which the majority of the plants have flowers so similar in type, there are genera which are exceptional in one of the features most characteristic of the Family, namely, the structure of the fruit.

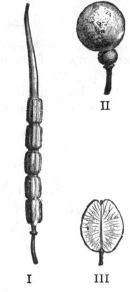

The Cruciferae include a number of important cultivated plants. To the genus *Brassica* belong the Wild Cabbage (*B. oleracea*), from which our cultivated cabbages, Brussels Sprouts, Savoys, Kohl Rabi, Cauliflower, Broccoli, etc., have been derived: the Wild Turnip (*B. campestris*) with the cultivated varieties, Turnips and Swedes; also the White and the Black Mustards (*B. alba* and *B. nigra*, often called *Sinapis alba* and *S. nigra*). The common garden Cress (*Lepidium*

Fig. 124. Exceptional fruits of Cruciferae: I, Radish; II, Sea-kale—two-jointed, the lower joint barren; III, Cress—winged, two-seeded.

sativum) and Watercress (*Nasturtium officinale*) also belong to this Family.

Other interesting members of the family are White Rock (*Arabis*), the grey green xerophytic leaves of which have a woolly covering of star-shaped hairs; Ladies' Smock or Cuckoo Flower (*Cardamine pratensis*) with pink flowers, common in moist meadows; and the Hedge Garlic (*Sisymbrium Alliaria*), with its masses of white flowers and large, thin, heart-shaped leaves. In the Candytuft (*Iberis*), often grown in gardens, the flowers are zygomorphic; in each flower the

Fig. 125. Zygomorphic flower of Candytuft. (After Baillon.)

petals that are towards the outside of the inflorescence are large, and so make it more conspicuous.

The Caryophyllaceae.

The Pinks, Campions, Stitchworts and Chickweeds are examples of another Family of plants, which are in many respects very like each other, though on the whole not so closely similar that the genera and species are as difficult to distinguish as they are in the Cruciferae.

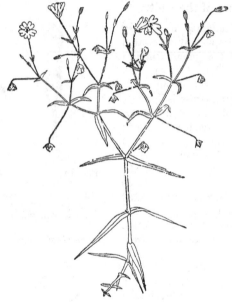

Fig. 126. Stitchwort.

The Stitchwort (*Stellaria Holostea*) is a smooth plant, common by hedges, with rather weak stems bearing opposite, sessile, rather stiff, lanceolate leaves, tapering to a point. The nodes are somewhat swollen: the swelling is as a rule especially marked just above the nodes, at the bases of the internodes, and it is by bending sharply at these points and not by a general curvature that the stem regains the erect position if laid prostrate.

The white flowers are borne in loose *cymose* inflorescences: the main stem ends in a flower; two axillary branches are borne in the axils of the last pair of leaves below the terminal flower, and each of these branches bears two opposite leaves and ends in a flower. In the axils of the leaves on these flowering branches other similar branches are borne, each with a terminal flower and a pair of leaves, which become smaller on the successive branches. In this way a more or less symmetrical inflorescence is formed which is called a dichasial cyme or *dichasium*. As it becomes more elaborate it becomes less regular, the opposite axillary branches developing unequally.

The flower has five small pointed sepals, five white petals, each divided to about the middle into two lobes, ten stamens in two whorls, of which the outer five on careful examination are found to be those that come not between but opposite to the petals, and a one-chambered ovary with free-central placentation, surmounted by three thread-like styles, which are stigmatic on their inner side. Nectar is secreted by glands around the bases of the stamens and can be obtained by short-tongued insects. The flower is protandrous, first the outer and then the inner whorl of stamens rising and shedding their pollen, the stigmas growing longer and spreading afterwards. The fruit ripens in a pendent position and splits at the top into six teeth.

The Chickweed (*Stellaria media*) is very similar in many of its characters, but is a smaller and weaker ephemeral plant, found commonly in waste places, by waysides and, as a weed, in cultivated ground. It is of a pale green colour, with small ovate leaves and small flowers, and is smooth, except for a line of hairs down one side of the stem and a few hairs on the leaf-stalks. The flowers are mostly self-pollinated, but receive occasional visits from flies: the petals are shorter than the sepals and divided into two lobes (cleft) nearly to the base; usually only three or five stamens are developed.

The Mouse-ear Chickweeds (genus *Cerastium*; several species are distinguished) are darker green, hairy plants, with usually five stigmas, but otherwise very similar to *Stellaria media*.

These species belong to a group within the Family, distinguished as the *Alsinoideae* from another group, the *Silenoideae*, which includes the Pinks and Campions. These have, on the whole, much larger flowers, and the sepals, instead of being separate, as in the Alsinoideae, are united, and form a tubular calyx. Along with this difference goes a difference in the form of the petals, which have, like those of the Wallflower, a narrow erect 'claw' and broad spreading 'limb.'

The White Campion (*Lychnis vespertina*) is a much more robust plant than the Stitchwort, but, like it, has opposite leaves and swollen nodes. The softly hairy erect annual shoots arise from a thin branching rhizome. The leaves are oval and more or less pointed, the lower ones stalked. The flowers are borne in loose dichasial cymes. They are usually dioecious in this species: the male flowers have ten stamens and often a small barren ovary, the female an ovary with free central placentation and five styles, and sometimes also the rudiments of stamens. They are moth-flowers, opening and emitting scent at night: the claws of the petals, held together by the tubular calyx, form a long narrow tube, so that the nectar at the bottom can best be reached by the slender tongues of moths. The petals have small scales, where the limb joins the claw, and these fringe the opening to the tube.

Thus the flowers are much more highly specialised than the more open flowers of the Alsinoideae with separate sepals. Most of the Silenoideae are pollinated by moths, butterflies, or long-tongued bees.

The Red Campion (*Lychnis dioica* or *diurna*) is very like the White Campion in structure, and, like it, has

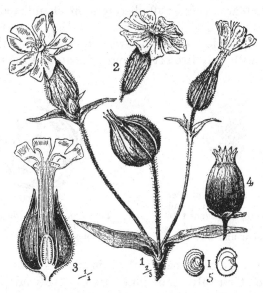

Fig. 127. White Campion. 1, inflorescence; 2, a male flower; 3, longitudinal section of a female flower; 4, fruit; 5, seed, whole and cut in half to show the curved embryo. (From Strasburger, after Wossidlo.)

dioecious flowers; but it is pollinated by day-flying insects. The female plant is much more robust than the male and has larger flowers. The fruit is a capsule, opening by ten teeth, which spread widely or roll back (Fig. 128), whereas in the capsule of the White Campion the teeth remain more or less erect. The Ragged Robin

(*Lychnis Flos-cuculi*) and the Corn Cockle (*L. Githago*) are other species that are similar in all important respects to these two Campions, except that the flowers are hermaphrodite.

The Bladder Campion (*Silene inflata*) has white moth-flowers with, as a rule, only three styles, and the calyx, which is attached some little distance below the petals on the receptacle (Fig. 130), becomes much inflated after fertilisation.   The Pinks (genus *Dianthus*) have narrow leaves of a pale blue-green, owing to a covering of wax, flowers with only two styles, and close around the base of

Fig. 128. Capsule of Red Campion with ten spreading teeth, surrounded by persistent calyx. (After Le Maout and Decaisne.)

Fig. 129.   Floral diagram of Bladder Campion; *a*, *β*, bracteoles. (After Eichler.)

the calyx a few broad scales: in the narrowness and depth of their calyx-tube Pinks are especially adapted to butterflies.   The flowers of all these plants are markedly protandrous: the stamens ripen first, the two whorls appearing in succession from the mouth of the flower, and the styles protrude and diverge later.

Among other species commonly met with are the Sandspurry (*Spergularia*), Pearlwort (*Sagina*), and Sandworts (*Arenaria*), belonging to the Alsinoideae. The Sea Purslane (*Arenaria peploides*), with short and crowded fleshy leaves, is frequent on sandy shores all round our coasts. The garden Sweet William (*Dianthus barbatus*) and the Soapwort (*Saponaria*) belong to the Silenoideae.

Considering the Family as a whole, several common features may be mentioned. In general habit the various species agree, in having opposite simple entire leaves, swollen nodes and a dichasial inflorescence. The flowers also have usually five sepals, five petals, ten stamens, and an ovary with free central placentation.

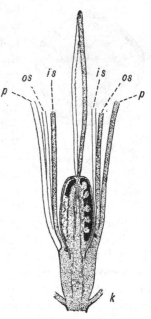

If the ovaries of a number of the species are very carefully examined variations are found in their structure from the typical one-chambered ovary, with its placenta arising from the base. In the Bladder Campion, for instance, the lower part of the ripe capsule is divided by septa, so that if cut across below the middle it presents the appearance of a syncarpous ovary with several chambers and axile placentation. In the flower the septa divide the ovary nearly to the top (Fig. 130), and still younger ovaries are completely

Fig. 130. Part of a longitudinal section of a flower of Bladder Campion. In the ovary a septum is shown breaking away at the top. Only the bases of other parts of the flower are included: the calyx (*k*) is attached some way below the rest of the flower; *p*, petals; *os*, outer, and *is*, inner stamens. The actual section is shaded with dots.

divided. They are thus at first chambered, with axile placentation, but the septa and the central column break down in the upper part, leaving the type of structure already described.

In other plants, such as the Red Campion, although no septa may be left in the fruit, traces of the septa can be found in the ovary (Fig. 131), and in young

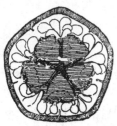

flowers they can be seen in process of breaking down. In fact, in the Caryophyllaceae as a whole, the ovary is at first divided by septa into as many chambers as there are styles, and is therefore made up of as many carpels; but during development the septa and the upper part of the central axis sooner or later cease to grow and break down during the further expansion of the ovary wall, leaving the lower part of the axis, bearing

Fig. 131. Transverse section of the ovary of Red Campion, showing the five placentae and traces of the five septa still remaining on the wall.

the placentae, rising as a free column from the base.

Free central placentation is thus in these plants derived from axile placentation, and their ovaries are syncarpous. In writing the floral formulae the number of carpels is to be indicated as for other syncarpous ovaries. As we have seen, the petals and outer stamens come opposite each other: this break in the regular alternation of parts in successive whorls is indicated in the floral formulae by a vertical line. The formula for the Stitchwort, for instance, is K5 C5 | A5+5 G(3); that of the Ragged Robin, K(5) C5 | A5+5 G(5).

In comparison with the previous Families, we may notice the following points of general interest. The parts are always definite in number, never numerous, usually in whorls of five, except in the ovary where there are in many genera only three or two carpels, as indicated by the number of styles. The highly specialised flowers in this Family, which belong to the Silenoideae, have become adapted to long-tongued insects by developing a tubular calyx. None of them are zygomorphic. In instructive contrast with these plants, most of which are perennial, are the ephemerals with small flowers, in which the stamens often do not all develop, being reduced to five or even (as in the Chickweed) sometimes fewer, and the corolla is very small and inconspicuous. (A similar contrast is to be observed among the Cruciferae—for instance, between the Wallflower and Shepherd's Purse.)

The Rosaceae.

This is another Family in which the flowers as a rule have numerous stamens, like the Ranunculaceae and many of the Papaveraceae.

The Cinquefoil, Strawberry, Blackberry, and a number of other plants have also numerous carpels, like the Buttercups. The yellow flowers of the Cinquefoil are, in fact, sometimes confused with Buttercups. When carefully examined, however, the receptacle is found to be expanded below the gynaecium into a disc, carrying the sepals, petals, and stamens outwards, so that they are no longer attached directly below the carpels. These flowers are therefore called *perigynous* (*peri.*= around) to distinguish them from flowers, like those of the Ranunculaceae and the other Families we have so far considered, called *hypogynous* (*hypo* = below), in which all the parts are attached one above another to a receptacle that is merely somewhat swollen and not specially expanded below the gynaecium.

A flower of a Blackberry, as an example, is shown in longitudinal section in Fig. 132, *B*. It has five sepals, five white or pink petals, and many one-seeded carpels. Nectar is secreted by the upper surface of the disc, and is partly concealed by the stamens and carpels, but can be reached by short-tongued insects. The mechanism of pollination is thus like that of the Buttercups, with the difference that the nectar is reached between stamens and carpels, and the anthers shed their pollen inwards.

Blackberry plants (Brambles) are shrubby hook-climbers, and their carpels ripen to fleshy drupelets (p. 235). The leaves are compound, with three or five leaflets, and narrow stipules, carried a little way up the stalk.

There are other plants that, like the Blackberries, have compound fruits made up of drupelets, and a more or less shrubby

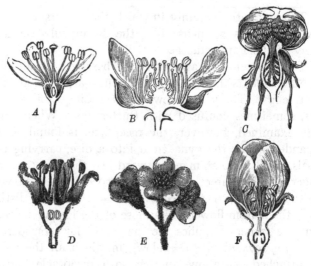

Fig. 132. Flowers of Rosaceae. *A*, Blackthorn: *B*, Blackberry: *C*, Rose; all perigynous. *D* and *E*, White Beam: *F*, Pear; epigynous. (From Ward.)

habit, and these are placed along with them in the genus *Rubus*: the best known is the Raspberry (*Rubus Idaeus*). The Blackberries themselves, though sometimes all called *Rubus fruticosus*, include a number of species and varieties, difficult to distinguish from one another, among which is the Dewberry (*Rubus caesius*), which has a waxy bloom on its fruit.

The Cinquefoil (*Potentilla reptans*) is a low herb, with creeping stems, rooting at the nodes, and palmate leaves of five leaflets, with membranous stipules. The yellow

flowers have on the outside of the calyx what appear
to be five extra sepals attached between the others.
These are small, and are sometimes divided into two or
three lobes: they are usually regarded, since the foliage
leaves have stipules, as formed by the stipules of adjacent
sepals becoming joined, and are called the *epicalyx*. The
one-seeded carpels ripen to achenes on a dry receptacle.

Belonging to the same genus (*Potentilla*) are the Silverweed,
with pinnate leaves covered by silky hairs, the Strawberry-leaved
Cinquefoil with white flowers and leaves with three leaflets, and
the Tormentil, with leaves usually of three leaflets, and flowers
with four sepals and four yellow petals. All these have an epicalyx
and a receptacle that becomes dry in the fruit.

The Strawberry (genus *Fragaria*) is very similar in
habit to the Cinquefoil, with characteristic runners. It
has leaves with stipules and three leaflets, and an epicalyx
is present in the flower; but the upper part of the
receptacle bearing the achenes becomes fleshy after pollination, and is then easily detached as a whole from the
disc that carries the calyx, etc.

The Herb Bennet (*Geum urbanum*, Fig. 115) is another
plant with many achenes, but they have long hooked
styles. The leaves are pinnate, with unequal lobes and
large leafy stipules. Here again the flowers have an
epicalyx. They are small, with yellow petals and hairy
carpels. The Water Avens (*Geum rivale*) has pendulous,
dull purplish flowers adapted to humble-bees, the sepals
standing at right angles to the disc, and so keeping the
flower half closed. The stamens are spread over the
upper surface of the disc, instead of only near its edge,
and after pollination the receptacle grows in length
between the disc and the gynaecium, bearing the cluster
of achenes with hairy hooked styles out above the
calyx.

In all the above genera there are numerous carpels on

a convex receptacle above a small disc. In the Black-thorn (Fig. 132 *A*), Cherry and Plum (genus *Prunus*), the flower has five sepals without epicalyx, five petals, and numerous stamens, but only one carpel, and this stands in a cup-shaped receptacle, so that the flower is still more obviously perigynous. The fruit is a drupe (p. 234), and the plants are shrubs or small trees.

In the Roses (p. 191), which like the Brambles are shrubby and often prickly scramblers, the receptacle grows up still farther, and becomes flask-shaped (Fig. 132 *C*), bearing the numerous stamens, five large petals and five lobed or fringed sepals above the one-seeded carpels attached within it. It afterwards becomes fleshy, enclosing the bristly achenes and forming the well-known compound fruit, the 'hip' (p. 236). In the Agrimony also (Fig. 86, II), the receptacle, which is here covered outside with hooked bristles, is deeply hollowed; it encloses two carpels. This plant is a herb, with long erect spikes of small yellow flowers.

In the genus *Pyrus*, which includes the Apple and Pear, the upgrowth of the receptacle and the enclosure of the carpel proceed a step further, for the carpels are more or less fused with it and with each other; the result is an *inferior* ovary, and a flower which is called *epi-gynous* (*epi* = above) to distinguish it from perigynous flowers, in which the carpels are still free from the recep-tacle, even though enclosed within it (as in the Rose).

The Mountain Ash (*Pyrus Aucuparia*) is an interesting intermediate form, in which the upper parts of the carpels remain separate from each other in the centre of the flower, and have separate styles. In the Pear (*Pyrus communis*) also, the styles are separate (Fig. 132 *F*). In the Apple (*Pyrus malus*), the fusion is more complete. These inferior ovaries are thus, like other complex ovaries, syncarpous. In the fruit, the receptacle grows and becomes fleshy, while an inner layer, corresponding roughly with the walls of the carpels themselves, becomes the tough 'core' that protects the seeds.

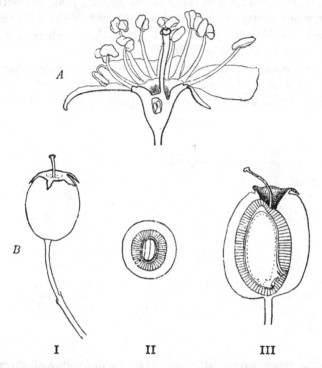

Fig. 133. Hawthorn (*Crataegus monogyna*). *A*, longitudinal section of flower, showing the single carpel largely fused with the upgrown receptacle. *B*, fruit: I, external view showing calyx and style, and the scars of two bracteoles; II and III, the same cut across and longitudinally, showing the outer fleshy and inner hard (shaded) parts into which the fruit wall is differentiated, and the single seed enclosed in it.

All the British plants of the genus *Pyrus* are small trees. The flowers, except that they are epigynous, are similar to others we have mentioned in number of sepals,

petals, and stamens, and in their general method of
pollination.

The Hawthorn (*Crataegus*) has an inferior ovary very
similar in structure to those of the genus *Pyrus*, but with
only one or two carpels and styles; but in the fruit the
part which would form in an apple the tough core becomes à hard woody shell (Fig. 133 *B*).

In addition to these genera are others which have
small flowers, lacking petals. The Lady's Mantle (*Alchemilla*) is a herb with round, palmately lobed leaves,

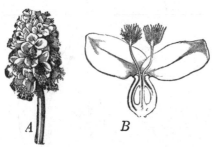

Fig. 134. Salad Burnet: *A*, inflorescence,
the male flowers below with pendulous
stamens, the female above; *B*, longitudinal section of a female flower. (After
Baillon.)

Fig. 135. Great Bur-
net; longitudinal
section of flower.
(After Baillon.)

with large green stipules, and small yellowish-green
flowers borne in clusters. These have a well-marked
disc, secreting nectar and bearing four sepals, with an
epicalyx, and at most four stamens. There are one or
two one-seeded carpels, more or less enclosed in the
hollow receptacle. Thus, notwithstanding its exceptional
features, this plant has characters connecting it with
other Rosaceae. The Salad Burnet is peculiar, in having
wind-pollinated unisexual flowers. They are borne in
globular heads, green or purplish, the lower ones male,

with four-lobed calyx and numerous pendulous stamens, the upper female, containing usually one carpel, with a long style ending in a tufted stigma that protrudes before the male flowers open. The plant is found in chalk and limestone districts, and has pinnate leaves with numerous small toothed leaflets, and broad leafy stipules. The Great Burnet is very similar in habit, but its dark purplish flowers are hermaphrodite and insect-pollinated, with usually four short erect stamens instead of many pendulous ones.

In the great majority of the Rosaceae the carpels are one-seeded. Two seeds are found in some species of *Pyrus*; but where the carpels are free, achenes are the rule. There is one genus, however, *Spiraea*, in which the carpels ripen to follicles, with two or more seeds, like so many of the Ranunculaceae. The commonest species is the Meadow-sweet, found in damp places, on the banks of ponds and ditches. It has pinnate leaves, and dense terminal inflorescences bearing many small yellowish-white sweetly scented flowers with five to eight carpels.

This large Family, the Rosaceae, thus includes plants exhibiting a wide range in their vegetative habit, from trees to herbs, and in the structure of their flowers, which range from slightly perigynous through markedly peri-gynous to epigynous. If we try to pick out the character-istic features of the group, we find the following additional points: Most of the genera have numerous free carpels, and even in some of the epigynous flowers in which the carpels have fused more or less with the receptacle, they are only partly joined with each other. Except in a few reduced (or simplified) flowers, the stamens are numerous. In both these points the Rosaceae are like the Ranunculaceae. The leaves vary very much, from simple to palmate or pinnate, but as a rule have prominent stipules; and in some genera an epicalyx is present. In none of the genera do the flowers show a very highly specialised mechanism for pollination.

As characteristic floral formulae those of the Bramble, $K_5 C_5 A\infty G\underline{\infty}$, and Apple, $K_5 C_5 A\infty G\overline{(5)}$, will serve. In *Prunus* there is only one carpel: while the formula for *Alchemilla* is $K_4 A_4 G\underline{1}$ or $\underline{2}$.

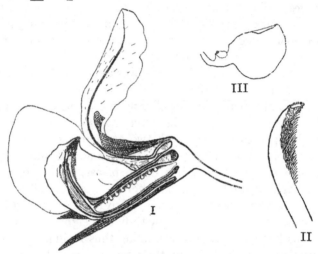

Fig. 136. Garden Pea. I, longitudinal section of the flower showing it to be distinctly perigynous: in the tip of the keel pollen is collected, ready to be brushed out by the style. II, style much enlarged, showing the brush of hairs and the terminal stigma. III, a wing-petal showing the little bulge which hooks into a hollow in the keel-petal.

The Leguminosae.

This Family is a very large one. It includes Vetches, Clovers, Gorse, Broom, and many other native British plants, as well as Peas, Beans, Lupines, and other plants of the garden, all with zygomorphic flowers very similar in type.

The large greenish-white flowers of the Garden Pea (*Pisum sativum*) show the characteristic arrangement of the petals: the large posterior petal, standing erect, called the standard; two lateral petals, the wings; and between the wings the keel, formed by the other two petals which cling together along their lower edges: in the bud the standard is folded over the other petals. Alternating with the petals on the outside are five sepals, joined into a tube below. Within the keel are ten stamens and a single carpel. The upper stamen is free, but the rest have their filaments joined together for about half their length, forming as it were a tube that is slit along the upper side where the free stamen is. Nectar collects within this tube, and can be obtained by long-tongued insects under the free stamen. The single large carpel is attached within the tube to the bottom of a hollow expanded receptacle, which bears the other parts of the flower on its edge: the flower is therefore perigynous— like the Cherry, for example. As the flower faces sideways the ovary is horizontal; but the style is erect, and is furnished on the inner side, just below the terminal stigma, with hairs pointing obliquely upwards.

The stamens also turn upwards beyond the tube, and are closely held along with the style by the keel. The anthers shed their pollen just above the hairs on the style. When a bee alights on the wings, it presses down the wings and keel, which are curiously interlocked: the tip of the style is thereby exposed, the stigma comes in contact with the under side of the bee's body, and the hairs on the style brush out some pollen from the tip of the keel. When the bee leaves the flower the keel springs back to its original position, covering the stigma and pollen once more. The advantages of this mechanism are obvious: the pollen is only exposed a little at a time, while the nectar is also protected, and can only be obtained by intelligent insects with long tongues. The garden Pea is seldom visited by insects but is self-pollinated: the mechanism can be seen in action in wild species.

The habit of the plant is well known: it is a climbing annual, with green stems bearing alternate pinnate leaves, in which the upper leaflets develop as tendrils, while the stipules are large and leafy. The characteristic fruit— the pea-pod—splits when ripe along both margins. This type of pod is characteristic of the whole family, and has received a special name, *legume*, from which the name of the Family is itself derived.

The Sweet Pea and Everlasting Pea of the garden, and other wild Peas (genus *Lathyrus*), are closely similar to the garden Pea in habit and flower. The Vetches (genus *Vicia*) differ little from them. The Broad Bean (*Vicia Faba*), belonging to the same genus, is exceptional in being an erect annual; its leaves have two pairs of large leaflets, and end in a small point instead of tendrils. In the Kidney Bean and Runner Bean (*Phaseolus*) the pollination-mechanism is somewhat similar, but the keel is long, narrow, and coiled, and the bee has to alight on the left wing, which is larger than the other: the Kidney Bean is not a climber, the Runner Bean is a twining plant (p. 423), and neither has tendrils.

Among other plants with similar flowers, several somewhat different types of pollination-mechanism are met with. In Clovers, Sainfoin, and Laburnum the stamens and style both emerge when the keel is depressed. In the Bird's-foot Trefoil, Lupines and others, the stamens push pollen before them: in the Bird's-foot Trefoil the filaments of five of the stamens are swollen

Fig. 137. Flower of Broom after it has been visited by a bee; the stamens, which are of two lengths, and the coiled style have escaped from the keel. (After Wossidlo.)

and club-shaped just below the anthers, so that they fill
the keel and act together as a piston.   In the Gorse and
Broom, the keel, which is closed on the upper as well as
the lower side, holds the stamens and style bent; the first
time it is depressed by an insect it splits, setting free the
style and stamens, which spring out and hit the insect:
the keel does not return to its original position.   In Broom
the stigma strikes the back of the first bee; the style then
curves so that later visitors meet the stigma underneath.

These explosive flowers are nectarless.

In some plants, like the Sweet Pea, mostly annuals, the flowers
are regularly self-pollinated and set seed even though unvisited.
Most of the perennials, however, cannot be automatically self-
pollinated, and some are self-sterile.   In order properly to understand the pollination-mechanisms of these Leguminosae, it is
essential very carefully to watch insects at work on them.   In
most cases the bee not only rests on the wings, but has to pull
them or the petals of the keel apart.

In classifying the numerous species into genera and the
genera in tribes, other characters are also of importance, including
especially the form of the fruit and the general habit of the
plant.   One group of genera has already been dealt with, including
the Vetches, Peas, and Beans.   Of these, the French and Runner
Beans (genus *Phaseolus*) are put in a separate tribe, owing to
their coiled keel and different habit.

The Broom, Laburnum, Gorse, and Lupines are included in
another group, with stamens all joined, and calyx two-lipped.
These are mostly shrubby plants or small trees.   The Lupines
have palmate leaves of many leaflets: but the others have
trifoliate leaves (Latin *tri* = three, *folium* = leaf) except the
spiny Gorse, and even this has trifoliate leaves in its seedling
stage (compare p. 155).   This tribe includes all the British
species with stamens all joined, except the Rest Harrows.

Another group with trifoliate leaves includes the Clovers
(*Trifolium*) and Medicks (*Medicago*).   In these the flowers are
small, and are borne many together in close inflorescences: the
Clover heads are very conspicuous.   The fruit is usually small
and indehiscent, containing very few seeds: in most species of
Clover it remains within the persistent corolla, and is short and

straight, but in the Medicks it becomes curved or spirally twisted. In the flowers of most Clovers the petals are joined to the stamen-tube. Other groups need not concern us here as they are not abundantly represented, but a few plants may be mentioned. Just as in the Cruciferae, so in the Leguminosae there are species which have exceptional fruits. Besides the Clovers and Medicks, with indehiscent fruits, the Sainfoin has one-seeded achenes, and there are several less common plants with jointed pods which break up into one-seeded pieces.

Considering together all the plants that we have mentioned, the following general characteristics of the whole group stand out:

Though the habit of the plants varies greatly, the leaves are usually compound and stipulate : in many cases

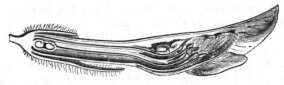

Fig. 138. Flower of Red Clover (*Trifolium pratense*) in longitudinal section : the petals are joined to the stamen-tube. (After Baillon.)

the leaves and leaflets are capable of movement, being joined to the stem or stalk by a special swollen cylindrical base or *pulvinus* (p. 160). The flowers are perigynous and zygomorphic, with very similar types of construction; the ten stamens, with filaments all joined or only the upper one free, are held along with the style in a 'keel,' formed by the two lower (anterior) petals. The gynaecium consists of a single carpel, with ovules attached along its upper margin, and ripens in most cases to a legume. A very interesting feature is the presence of characteristic nodules on the roots, inhabited by bacteria that absorb free nitrogen from the air (p. 97).

The British Leguminosae, however, only represent one

large division of the Family, the Sub-family Papilionatae, so-called from a fanciful resemblance seen in the flowers to butterflies with closed wings (Latin *papilio* = butterfly). There are two other Sub-families, one of which is especially interesting because it connects the Family very closely with the Rosaceae: this is the Mimosoideae, which includes the true Acacias[1] and the Mimosas. They are mostly trees growing in tropical or sub-tropical countries, and have small yellow flowers in little clusters, like yellow tassels. Each flower has four sepals, four petals, *numerous* stamens in the genus *Acacia*, four in *Mimosa*, and a single carpel which ripens to a legume. The flowers are, moreover, perigynous. The leaves are

Fig. 139.   Floral diagrams of Broad Bean (*Papilionatae*) and an Acacia (*Mimosoideae*).   (From Willis, after Eichler.)

delicate and bipinnate, at least in the seedling, with numerous leaflets attached by characteristic pulvini: *Mimosa pudica* is the Sensitive Plant, which folds up its leaflets and lowers its leaves if it receives a shock. There are, moreover, nodules on the roots. These plants thus resemble, on the one hand, the Papilionatae in their ovary and leaves, on the other the Rosaceae, having numerous stamens. Throughout both Families the leaves are alternate, usually have stipules, and are frequently

---

[1] The so-called Acacia, a tree often planted, with drooping racemes of white flowers, and pinnate leaves, belongs to the Papilionatae; it is *Robinia pseudacacia*, the False Acacia.

compound; while the flowers of both are perigynous (except in the epigynous Rosaceae where the modification is carried still further). We may therefore regard the Papilionatae as a group of plants with highly special-ised flowers allied to the Rosaceae[1].

The Umbelliferae.

This Family includes a very large number of species which, like the Cruciferae, exhibit a remarkable degree of uniformity in their habit and in the structure of their flowers and inflorescences. It is therefore often very difficult to distinguish the different genera and species, so many of which differ from one another only in very small and unimportant characters.

The Cow Parsnip, or Hogweed, is one of the commonest British members of the Family. Its tall, erect annual stems rise from short thick rhizomes, and end above in characteristic inflorescences. The stem is ribbed and rough, with coarse hairs, and bears large alternate pin-nate leaves, with three, five, or seven large lobed leaflets and a broad sheathing base. The internodes are hollow. In each inflorescence the flowers are small, but very numerous, and are closely aggregated together into smaller groups. In each of these groups, called *umbels*, the stalks of the flowers all rise together from the top of a common stalk, and around them at this point are usually found, at least in young inflorescences, narrow bracts, forming an involucre; but these very soon fall off. These bracts originally subtended the outer flowers, but have become displaced: many of the inner ones fail to develop. In each small umbel the youngest flowers are in the centre, the oldest on the outside. The stalks of the outer flowers are longer, bringing them at least to

[1] Compare the relationship between the Fumitories and the Poppies.

the same level as the inner ones, and the outer flowers
are slightly zygomorphic with enlarged outside petals.

In each inflorescence there are about twenty of these
small umbels, the stalks of which are attached together

Fig. 140.   Cow Parsnip, showing pinnate leaves and compound
umbels.   (After Baillon.)

at the top of the stem like the flower-stalks in the small
umbel; the whole inflorescence is thus an umbel of umbels,
or *compound umbel*.   By this arrangement all the flowers

are brought together more or less to the same level, and
the inflorescence is a very conspicuous one. The bracts
subtending the small umbels do not develop, so that there
is no involucre to the compound umbel in the Cow Parsnip;
but in many genera of the Umbelliferae there is a well
developed involucre to the main umbel and to each of
the small umbels. Lateral branches in the axils of the
upper leaves end in similar inflorescences.

Each flower has five small petals with an incurved
point, and five stamens alternating with them. The
sepals only develop as five small teeth, and in the bud the

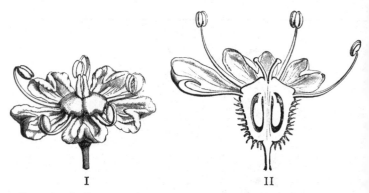

I   II

Fig. 141. I, flower of Cow Parsnip. II, longitudinal section of the
zygomorphic flower of Wild Carrot, with spiny ovary. (After Baillon.)

petals protect the stamens, which are bent inwards.
The ovary is inferior, two-chambered, with a single ovule
in each chamber: within the stamens and petals the
broad top of the ovary forms two cushion-like lobes,
which secrete nectar. The nectar is thus freely exposed
and accessible to any kind of insect. Each lobe tapers
into a short style in the centre of the flower. The two
styles are at first close together and very small, but
after the stamens have shed their pollen they grow a
little longer and diverge: the flowers are thus very

protandrous, but the stamens finally bend inwards
again so that there is a chance of self-fertilisation, if the
flower has not already been cross-fertilised.

In its main outlines this description would apply to
very many of our common Umbelliferae: erect stems,
with hollow internodes, alternate pinnate leaves, with
broad sheathing bases, and compound umbels of flowers
differing little from those of the Cow Parsnip, are all
characteristic of the great majority of the Family.

The general type of fruit is also characteristic. The
fruit of the Cow Parsnip is
flat and ribbed, and splits
when ripe into two flat one-
seeded halves, each carrying
a style; these dangle for a
time from two slender stalks
from which they are detached
by the wind. On the outside
of each piece, between the
ribs, are four dark brown
streaks, called *vittae*, which
are cavities filled with an
aromatic oily substance. On each inner face are two
similar vittae.

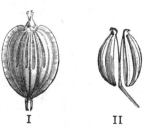

Fig. 142. I, fruit of Cow Parsnip showing ridges and vittae.
(After Baillon.) II, fruit of
Fool's Parsley splitting into
two achenes. (After Le Maout
and Decaisne.)

The characters of the ripe fruits are of great service
in distinguishing between different species and genera:
they differ in general form, being flattened or round, long
or short, sometimes beaked; in external features, being
smooth, ribbed or spiny; also in the absence or presence
of vittae, and in their number and position.

A comparison of the Cow Parsnip with the Chervil,
another very common species, will illustrate such differences in the fruit as well as in other features. The
Chervil is a rather smaller plant with a tap-root and its
stem branches more freely. The lower leaves are twice

pinnate with small leaflets pinnately divided and toothed.
The flowers are smaller than those of the Cow Parsnip
and are all slightly zygomorphic, the petals towards the
outside of the umbels being larger than the others (compare Candytuft among the Cruciferae). There are well-
developed involucres to the small umbels but not to the
compound umbel as a whole. The fruit is smooth and
shiny, narrow, rounded, and tapering at the top, and
has neither vittae nor ribs.

Fig. 143. Marsh Pennywort, *Hydrocotyle vulgaris*. (After Baillon.)

Among other plants similar in general features is the
Wild Carrot, with leaves divided into many narrow
segments and prominent involucres to both small and
compound umbels, with the bracts deeply divided into
narrow lobes. The prominent ribs on the fruit are
covered with prickles. The wild form has a slightly
thickened tap-root, which has increased greatly in thickness
in the cultivated varieties. The Parsnip is another member

of the Family, with a tap-root. In general habit it is very like the Cow Parsnip, but it has smaller, yellow flowers. Celery also belongs to this Family.

Most of the Umbelliferae have not only oil-cavities in their fruits, but also oil-canals in their stems, in the cortex. The oils are aromatic, and some of them are used as flavouring essences. Carraway seeds are the half-fruits of *Carum Carvi* ; the leaves of Parsley, used in sauces, and the green stem of Angelica, often put on iced cakes, are other examples.

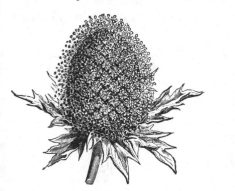 

Fig. 144.   Sea Holly: I, head of flowers; II, single flower, enlarged.
(After Baillon.)

As in the Cruciferae, there are genera which depart in some respects from the general type. One of these is the Marsh Pennywort (*Hydrocotyle vulgaris*) which has slender stems, creeping in wet mud and rooting at the nodes, bearing simple round slightly lobed leaves, with small simple umbels low down in the axils of the leaves. Another is the Sea Holly (*Eryngium maritimum*), abundant on sand along the coast, which has prickly bluish-green leaves: the blue flowers have well developed, sharply pointed sepals, and are borne in a compact rounded head, with a small spiny bract subtending

each flower, and an involucre of prickly leaves surrounding the whole head. Thus not only is the markedly
xerophytic habit of the plant distinctive, but the form
of the inflorescence is quite exceptional: but in the
structure of the flower and fruit it agrees with the rest
of the Family.

B. *Families in which the flowers have joined petals.*

The Primulaceae.

We have already examined the flower of the Primrose,
and have seen that it has a tubular corolla as well as a

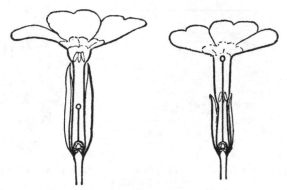

Fig. 145. Short-styled and long-styled Primroses cut in half
longitudinally.

tubular calyx. The stamens are five in number, are
epipetalous, that is, attached to the corolla-tube, and
are opposite to, instead of alternating with, the petals.
The flower is hypogynous. The ovary is one-chambered,
with free central placentation; but, unlike the ovary in
the Caryophyllaceae, it is free central from the start, and
no traces of septa are to be discovered. There is a single
style with a knob-like stigma at the top, so that in this
flower the carpels have become completely fused, including the styles. The number of carpels can, however,

be inferred from the fact that the ripe capsule splits into
five teeth (compare the Caryo-
phyllaceae). The floral formula
of the Primrose is therefore
K(5) C(5) | A5 G(5).

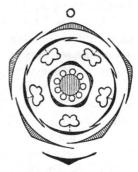

Fig. 146.   Floral diagram
of Primrose.

The plant itself has a perennial
stock with simple, more or less
hairy, wrinkled leaves. The
flowers are borne in axillary
umbels, but the common stalk
of the umbel is very short. The
Cowslip is another species of the
same genus, *Primula*, and like
the Primrose has long-styled and short-styled flowers,
borne on different plants. In general habit and in the
structure of the flowers the two species are very similar,
but in the Cowslip the umbels are borne on long stalks,
and the corolla has a much smaller concave limb, of
a deeper yellow colour, with conspicuous orange honey-
guides.

There is another British species similar to these, the
Oxlip, which has tall umbels like the Cowslip but a pale
yellow corolla with broad spreading limb more like the
Primrose. In the British Isles, the true Oxlip only
occurs in woods on boulder clay in some of the eastern
counties, though it is common on the Continent; but
where the Primrose and Cowslip grow near each other
hybrids are often found which are easily confused
with it.

There are many other species cultivated in gardens
and greenhouses under the names Primula, Auricula, and
Polyanthus. The Auriculas have smooth, fleshy leaves.
All these have flowers of two forms very similar in type.
The Water Violet (*Hottonia*) also has similar flowers: it is
a water-plant, with finely divided submerged leaves, and

inflorescences, that bear several whorls of pale purple flowers, rising above the water.

There are other plants that appear, at first sight, very different from these, which nevertheless have important features in common with them. The Creeping Jenny or Moneywort and the Yellow Loosestrife (genus *Lysimachia*), and the Scarlet Pimpernel (*Anagallis*), are the commonest species. The flowers of Lysimachia have a yellow corolla in which the petals are more or less spreading and only united at the base, forming a very short tube. The five stamens, attached opposite the petals to the base of the corolla, and the one-chambered ovary with a free central placenta are sufficient to place this genus in the same Family with the Primrose. On the other hand it is very different, not only in the shortness of the corolla-tube, but in the general habit of the plants. The Yellow Loosestrife has an erect branched stem, with leaves in whorls of three or four, bearing at the top a leafy, branching inflorescence. The Moneywort has

Fig. 147. Yellow Pimpernel (*Lysimachia nemorum*).

opposite leaves, small, rounded and smooth, borne on trailing stems that root here and there at the nodes, and the flowers occur singly in the axils of the leaves. The Scarlet Pimpernel also has opposite leaves, and solitary axillary flowers. It is a small annual weed, common in cultivated ground and waste places; the corolla is bright red, with no tube, but the petals are joined together and the five stamens to them just at the base. The capsule opens in a peculiar way, the upper half coming off as a lid.

In this Family we have, therefore, two groups of species which are very different in general habit and in the degree of specialisation of their flowers. In the *Primula* group, the flowers exhibit heterostyly and have long corolla-tubes, adapted to long-tongued insects. In the other group the flowers are open so that the nectar is accessible to short-tongued insects. In the former group the leaves are alternate and radical, in the latter opposite or whorled. On the other hand in both groups the leaves are simple and entire or slightly toothed, and the same floral formula would serve for the flowers of all the species we have mentioned.

The Cyclamen, often grown in greenhouses, belongs to this Family: its hanging flowers, with the petals turned back, have a loose-pollen mechanism (p. 216).

The Scrophulariaceae.

The Foxglove belongs to this Family. It has large hairy radical leaves and a tall erect, often branching,

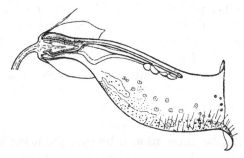

Fig. 148. Longitudinal section of flower of Foxglove.

racemose inflorescence of large purple flowers. The flower hangs obliquely, and has a long wide corolla-tube, slightly divided at the mouth into four lobes, the lower longer than the rest, and the upper one usually notched in the middle. To the base of the corolla are attached four stamens which bend upwards so that the large curiously two-lobed anthers come under the upper side of the tube and shed their pollen downwards. As the calyx has five segments, the corolla is regarded also as derived from five petals, the upper, notched lobe being formed by two which join earlier in development. The four stamens come between the petals; but between the upper, posterior two there is no stamen. The flower is hypogynous: the ovary is two-chambered, with swollen axile placentae bearing numerous ovules, and has a long style, that lies along the upper side of the corolla between the stamens, and ends in a forked stigma just beyond the anthers. The flower is pollinated by humble-bees, which crawl up the beautifully spotted 'floor' of the corolla-tube, clinging to the long hairs which cover it, to get the nectar which is secreted by honey-glands below the ovary; meanwhile they rub first the open stigma, then the anthers, with their backs. The longer stamens shed their pollen first, then the stigma expands, and the other pair of anthers dehisce at about the same time. The anthers are at first placed transversely but when ripe the two lobes turn end to end close alongside the style. The fruit is a capsule which splits at the top, the walls and septa separating from the placenta.

The Snapdragon is a smaller smooth plant with erect, sometimes rather bushy habit and small narrow alternate leaves, with racemes of flowers. The corolla here is two-lipped, the upper formed by two, the lower by three petals, and the lips are pressed tightly together closing the mouth of the flower. Within the corolla are four stamens and a style with forked stigma, arranged much as in the Foxglove. Only large humble-bees are strong enough to open the flower and reach the nectar which collects in the pouched lower side of the corolla-tube. The ovary is like that of the Foxglove, but the capsule opens by pores near the top. The Toadflax, of which the floral diagram is given in Fig. 150 B, is very similar to the

Snapdragon, but the corolla-tube grows out below into a long spur in which the nectar collects.

These plants agree in many respects, and are representative of a great part of the Family Scrophulariaceae. They have markedly zygomorphic and highly specialised flowers, with the floral formula $K(5) C(5) A4 G(2)$. In a few cases a fifth stamen develops as a staminode. In the Figwort (Fig. 149) it is a scale under the upper lip of the corolla: in other types such a scale would be in the way of the stamens and style, but in this flower first the style,

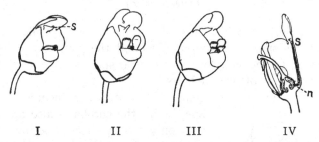

Fig. 149. Figwort (*Scrophularia nodosa*). I–III, flowers in different stages: in I, only the style protrudes; in II, the stigma is withering, two stamens have appeared and one has dehisced; in III, the other two stamens have also appeared, but one has not yet shed its pollen; IV, a flower in the second stage in longitudinal section; the stamen of the second pair is still curled up within the corolla-tube: *n*, honey-gland.

then the stamens lie against the lower lip, and the insect-visitors, chiefly wasps, reach the nectar over, not under them. The Pentstemons of the garden have a fifth barren stamen bent downwards out of the way.

In the Mulleins (*Verbascum*) the flowers differ very much from the general type. They are almost actino-morphic, with an open corolla, the yellow petals only joined at the base into a very short tube (compare the two similarly contrasted groups of Primulaceae), and there are five fertile stamens with hairy filaments. The

ovary is similar to that of the other species; the style
with its two-lobed stigma bends downwards over the

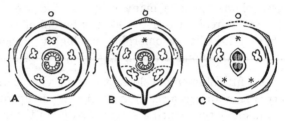

Fig. 150. Floral diagrams of *A* Mullein, *B* Toadflax, *C* Speedwell.
(After Eichler.)

lower lip on which insects usually alight, while the
stamens spread outwards. In habit the plants are very
similar to the Foxglove, but several species have a thick
covering of soft woolly hairs.

In another large genus, *Veronica*, which includes the

Speedwells, the corolla is also open,
with only a very short tube, but it
is four-lobed and there are only two
stamens. The ovary is short and
laterally flattened, but otherwise of
the usual type. Occasional flowers
with the large upper lobe divided
into two indicate that this represents
two petals: the two stamens correspond to the upper pair in other
genera. The calyx has only four
teeth, the upper sepal not developing. One of the
commonest species in the British Isles is the Germander
Speedwell (*Veronica Chamaedrys*), a slender more or less
prostrate perennial herb, with small heart-shaped hairy
opposite and nearly sessile leaves. The flowers are
borne in axillary racemes, and are of a beautiful sky-blue
colour. The style curves downwards in front of the lower

Fig. 151. Flower of
Germander Speedwell
(*Veronica Chamaedrys*),
a hover-fly flower.

petal and the stamens spread outwards on either side
(Fig. 151). The flower is pollinated by hover-flies, which
alight delicately on the lower petal, so touching the
stigma, and grasp the two stamens, pulling them towards
them so that the anthers touch the under side of their
bodies. The nectar is concealed by hairs at the mouth
of the short tube.

The shrubby Veronicas of the garden have flowers of
exactly the same type, usually white or pale blue, but
in habit are evergreens with small smooth opposite
leaves in four ranks (decussate). Among foreign members
of the Family there are a few other genera of shrubs or
trees.

A very interesting group of genera, peculiar in their
mode of life, are the Yellow Rattle, Cow-wheat, Eye-bright
and Lousewort. These are root-parasites, their roots
attaching themselves by suckers to the roots of other
plants, chiefly Grasses, and absorbing nourishment from
them (p. 179).

The flowers of all these plants have loose dry pollen
which is sprinkled on to visiting insects (p. 216). In
the Lousewort (*Pedicularis*), for instance, the three
lower lobes of the corolla form a landing stage, the two
upper ones a hood that is flattened laterally and keeps
the anthers of each pair of stamens closely in contact
with each other. The anthers dehisce towards each other,
but the pollen does not escape so long as they are close
together. The style just protrudes from the hood, and
a bee, alighting on the lower lip, first touches the stigma,
so pollinating it. Then in order to reach the nectar it
pulls apart the side of the hood, disturbs the anthers and
receives a shower of pollen on its back.

The Labiatae.

We have already studied (p. 196) the flower of the White Deadnettle (*Lamium album*), the main features of which are characteristic of most of the genera in this Family. The corolla is two-lipped, like many of the Scrophulariaceae, the lower, three-lobed lip forming a platform, the upper a hood, under which are four stamens, attached below to the corolla, and the style with a forked stigma. The ovary, however, though originally two-chambered, later becomes four-lobed by

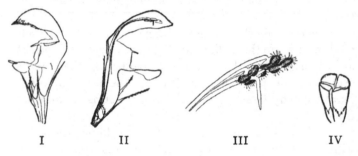

I   II   III   IV

Fig. 152. Deadnettle. I, flower of White Deadnettle. II, the same cut in half lengthways: the style is seen to rise from between the lobes of the ovary, two of which are present. III, arrangement of anthers and stigma in the similar flower of a garden Deadnettle. IV, young fruit of White Deadnettle.

the constriction of each chamber in the middle, and in each lobe a single ovule is formed: the style rises from between the four lobes. The fruit separates into four achenes or *nutlets* with hard walls, which lie at the bottom of the persistent funnel-shaped calyx until shaken out. The floral formula of the Deadnettle, $K(5)\ C(5)\ A4\ G(2)$, is exactly the same as that of many Scrophulariaceae, from which it is distinguished chiefly by the ovary and fruit.

The plant itself differs from most Scrophulariaceae

in having decussate leaves and square stems.   The leaves
are simple, hairy and without stipules.   All these features
are shared by the majority of Labiatae; though not

Fig. 153. Floral diagram
of White Deadnettle,
with the younger flowers
of the cymose cluster
indicated on either side
of the central flower.
The asterisk represents
the missing posterior
stamen. (After Eichler.)

Fig. 154.   Marsh Stachys.
(After Praeger.)

every plant with square stems and decussate leaves
belongs to this Family.

Another feature common to most of the genera is the

inflorescences. The flowers appear to be clustered around each node, as if whorled, but there is really a condensed inflorescence (that is, one in which the stalks do not develop) in the axil of each leaf: this inflorescence is cymose in construction, the oldest flower being in the centre, the next two on either side of it and the younger flowers being grouped as shown in Figure 153, each ending a lateral branch from the axis of the last.

Other species of *Lamium* are the Red Deadnettle, with smaller reddish purple flowers, a smaller plant and an annual, and the Yellow Archangel which is common in some Oak and Ash woods. Other very common plants with similar flowers are the Hedge Stachys or Woundwort (*Stachys sylvatica*), the Betony (*S. Betonica*) and the Marsh Stachys (*S. palustris*), with their clusters of purple flowers borne in the axils of smaller leaves at the top of the plant, in a compound inflorescence (Fig. 154); the Ground Ivy (*Nepeta Glechoma*), a low plant, with heart-shaped crenate leaves and blue flowers, female on some plants, hermaphrodite on others; and the Self-heal (*Prunella vulgaris*) with creeping stems ending in close erect compound spikes of purple flowers, the leaves in the spike broad and bract-like, the calyx two-lipped.

In several genera the stamens and style protrude from the mouth of the flower. In the blue flowers of the Bugle (*Ajuga reptans*), for example (Fig. 155, I), the upper lip is very short and erect, but the lower lip forms the usual three-lobed platform, while in the small pale yellow flowers of the Wood Sage (*Teucrium Scorodonia*) there is no upper lip, the posterior petals forming part of the platform. The flowers of both these plants are protandrous, the latter markedly so.

The Mints and Thyme have small flowers aggregated in dense clusters or terminal heads, with a short corolla-tube and almost spreading upper lip, and protruding

stamens and style. The Wild Thyme is a small-leaved
more or less prostrate plant, with hard stems that last
through the winter; the flowers are borne in numerous
terminal 'spikes' with about three flowers in the axil of
each leaf. They are very protandrous: the style is
at first short but grows beyond the anthers after they
have dehisced and the stigma lobes then diverge. On
some plants the flowers are all without stamens. The
Water-Mint (*Mentha aquatica*), a very common plant by
the side of streams and in marshy places, has an erect
habit, of the type usual in the Family, and is softly hairy,

Fig. 155. I, flower of Bugle with upper lip undeveloped. II, flowers
of a Mint (*Mentha sativa*) in two stages: in the one the style is
short and the stigma closed, in the other the style has lengthened
and the lobes of the stigma have separated. The left-hand flower
has the posterior lobe of the corolla notched: this lobe corresponds
to the upper lip of other Labiatae.

with the characteristic Mint smell. The flowers are borne
in dense terminal heads, and are almost actinomorphic,
with a bell-shaped corolla and four stamens, erect and
equal in size.

The pollination-mechanism in the genus *Salvia*,
including Sage, is unique. Only the two lower stamens
are present, and these are peculiar in form. The filament
is short, and the two lobes of the anther are separated by
a long connective: only the upper lobe is fertile, the other
is barren and usually modified in form. The long connective carries the fertile half-anther under the upper lip;

the barren half is in the way of bees probing for nectar and the connective is hinged a little above it to the filament. When a bee enters the flower it pushes against the barren end of the anther, and so brings the fertile end down on its back. Later the style lengthens and brings the stigma down over the lower lip. In some species the upper pair of stamens are represented by small staminodes.

Mint, Thyme, Sage and several other species, such as Marjoram, Rosemary and Lavender, are aromatic herbs, used in flavouring and for their scent, due to volatile oils, which are distilled from some of them for making perfumes. Several of these are interesting xerophytic plants. Lavender has narrow leaves covered thickly with hairs, Rosemary has narrow leaves rolled back, with hairs crowding the groove on the lower side where the stomata occur.

The Compositae.

This Family is a very remarkable one for it includes more than ten per cent. of all the species of flowering plants, and is abundantly represented all over the world. It is thus an extremely successful Family. Many of our commonest and hardiest weeds belong to it, such as the Daisy, Dandelion, Groundsel, Coltsfoot and Thistles.

Although the species are estimated at over thirteen thousand, they have certain features so characteristic that there is seldom any difficulty in distinguishing members of this Family from those of any other. On the other hand, species so numerous yet similar are bound to be very often difficult to distinguish from one another (compare the Cruciferae and Umbelliferae).

The most obvious feature is the inflorescence: what is usually called the flower is really a head of very small flowers (often called florets) borne on a broad receptacle and surrounded by an involucre of bracts. As the oldest flowers are outside and the youngest in the middle this inflorescence is racemose; it may be regarded as a condensed and modified spike, in which the axis bearing the sessile flowers has expanded laterally instead of elongating. The bracts subtending the flower may be present as scales or (modified) as hairs, or they may not develop at all (as in some Umbelliferae). This form of inflorescence is called a *capitulum* (Latin, = little head).

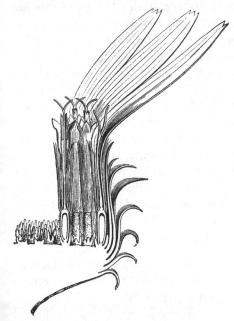

Fig. 156. Part of a longitudinal section of a capitulum of Aster. (After Baillon.)

Fig. 157. A disc floret of the Tansy; the expanded stigmas end in tufts of hairs; the calyx is reduced to a mere fringe. (After Baillon.)

Capitula are found in some genera outside this Family; but the structure of the individual florets is also characteristic. The capitulum of the Ox-eye Daisy has a yellow centre, called the 'disc,' of small yellow flowers surrounded by a white fringe of narrow petal-like organs, each belonging to a flower of a different type, called a 'ray-flower.' If the disc-flowers be examined they are found to vary in appearance according to age. In the centre they may still be closed. The youngest open ones show the five-toothed tubular corolla, from which projects a column with a mass of pollen at the top. In older flowers a forked stigma projects from the column, which is therefore tubular. Each lobe of the stigma ends in a tuft of hairs: when it first protrudes the two lobes are together and the hairs form a terminal brush, which at an earlier stage pushes pollen before it up the tube, acting like a piston in a cylinder. When a flower is dissected out and its corolla split down one side, the projecting tube can be seen to consist of five long anthers, joined together edge to edge, the filaments of which are separate and attached to the corolla-tube, alternating with its teeth. Below the corolla is an inferior ovary: as is indicated by the forked stigma, it is derived from two carpels, but only a single ovule develops. The calyx forms a hardly visible rim, faintly toothed, round the top of the ovary. The floral formula of a disc floret is thus $K_5 C_{(5)} A_5 \overline{G_{(2)}}$:

Fig. 158. Floral diagram of a Composite with a pappus, indicated by dots. (After Eichler.)

this is the characteristic formula of hermaphrodite flowers throughout the Family.

A ray-flower is very different in structure. The corolla is very zygomorphic: it has a short tube but it is expanded on the outer side above the tube into a long strap-shaped lip. Such a flower is called a *ligulate* flower. At the tip of this long lip are indications of three teeth: it is formed, therefore, mainly from the three anterior petals, and there are no teeth corresponding to the other two petals. No stamens are present, only a

style with a forked stigma, the branches of which are
without the terminal brush of hairs possessed by those in
the disc-flowers and are more slender. The ray-flowers
greatly increase the conspicuousness of the capitulum as
a whole.

The ovaries ripen to small ribbed achenes, which in
this genus gradually become dislodged from the receptacle;
but have no elaborate means for dispersal.

Other members of the Family differ from this in
various details, but the main features of the pollination-
mechanism are the same in all. The anthers form a cylinder
round the style, which acts like a piston. As the style
grows the pollen is pushed before it and is thus exposed
at the top little by little. Finally the stigma protrudes
and exposes its stigmatic surfaces, which till then were
closed together out of contact with the pollen.

A few other examples will illustrate the kind of
differences of detail. In the
Thistles there are no special ray-
flowers, the flowers are all tubular
and hermaphrodite. The corolla-
tube is much longer, so that the
nectar cannot be reached by
short-tongued insects, and the
purple heads are visited chiefly
by bees and butterflies. The
style bears a ring of hairs just
below the fork instead of a
terminal brush. Other smaller
differences are the presence of

Fig. 159. Capitulum of Spear
Thistle. (After Praeger.)

bristles on the receptacle between the flowers, perhaps
representing bracts, and the prickly habit, which usually
extends to the involucre. A very important difference
is the structure of the calyx which develops as a ring of
long hairs, called a *pappus*: in most Thistles, including

the commonest species, the hairs are branched and feathery. This pappus persists as a plume at the top of the fruit, which is thereby rendered very light and buoyant and can be carried even by gentle breezes. In the Ox-eye Daisy and a few other plants, including the Daisy and the Sunflower, the calyx, not being needed for protection, hardly develops at all: but in the Thistles and the majority of the Compositae it forms a pappus and serves

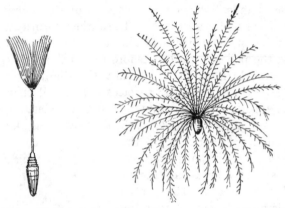

Fig. 160. Plumed fruits of Dandelion and Spear Thistle. (After Praeger.)

the new function of aiding dispersal of the achenes by wind.

The Knapweeds and the Blue Cornflower (genus *Centaurea*) have ray-flowers with an enlarged corolla but these are tubular like the disc-flowers, not ligulate, and they are quite barren, having neither stamens nor style and producing no seed in their ovary. The method of pollination is peculiarly interesting for pollen is pushed out just when an insect alights. The anther-tube is long and projects far out of the corolla-tube. The style has a ring of hairs, as in the Thistles, just below the fork. When

an insect touches the tip of the anthers, the filaments
suddenly shorten and draw the
tube a little way down over the
style so that some pollen is pushed
out at the top. This can be seen
happening if an anther-tube is
touched and then watched for a
moment. Afterwards the filaments
gradually recover their former
length, and push the anther-tube
up again: meanwhile the style
grows a little longer so that it will
push more pollen out when the
stamens next contract. This mechanism, depending on stamens that
are sensitive to touch, has the advantage that pollen is exposed as a
rule only when required. In other
genera pollen is gradually exposed
independently of the visits of insects, and is thus more liable to
be spoilt by rain. In many cases,
however, the involucre or the ray-
flowers close over the disc and afford

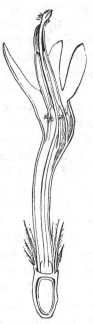

Fig. 161. Longitudinal
section of flower of
Centaurea. (After Le
Maout and Decaisne.)

the pollen protection, and in some the whole capitulum
bends down: these movements take place regularly at
night, as well as in dull weather.

The Cornflower shows very well another feature which
is common to most Compositae. The lobes of the stigma
curl eventually right back till their tips come in contact
with the hairs that formed the pollen-brush on the style,
so that self-pollination is likely to occur, and self-fertilisation if cross-fertilisation fails.

The Dandelion and the Hawkweeds are examples of
another group of genera in which the flowers are all

ligulate and differ from the ligulate ray-flowers of other
genera in having five distinct teeth
at the tip of the corolla, showing
that all five petals share equally
in the formation of the strap-
shaped expansion.   The plants also
exude white milky *latex* when
wounded.   These two characters
mark them off from all the others,
and they are placed in a Sub-family,
Liguliflorae, other genera forming
the Sub-family *Tubuliflorae*.

In the Dandelion the tubular
base of the corolla is of medium
depth and the flower is visited
chiefly by insects with tongues of
some length, especially by bees.
The style is covered for some

Fig. 162. Ligulate flower
of a Hawkweed. (After
Baillon.)

distance on the outside with upwardly directed hairs, and
its two arms eventually bend back and roll up so that
the stigmatic inner surfaces touch the hairs.   Between
the pappus and the ovary there is a constriction which
lengthens as the fruit ripens forming the characteristic
stalk that bears the plume (Fig. 160).

The Goat's Beard, another member of the Liguliflorae, has
a similar stalked plume: this species is commonly called
'John-go-to-bed-at-noon' because the capitula close about mid-
day.   Other common Liguliflorae are the Sow-thistle, Hawk's
Beard, and Hawkbit.

Among common Tubuliflorae are, like the Thistles with all
the flowers tubular and hermaphrodite, the Burdock, with an
involucre of hooked bracts (p. 233); and with ligulate ray-flowers,
the Colt's-foot, Golden-rod, Ragwort, Daisy, Yarrow and numerous
others, most of them with short-tubed disc-flowers, visited
largely by flies. The flowers are usually yellow or white, the ray-
flowers of the same or a different colour from those of the disc.
The Groundsel is a very common annual weed, belonging to the

same genus as the Ragwort, but it is regularly self-pollinated:
no ray-florets develop and the heads are small and inconspicuous
so that they receive few visits from insects although nectar is produced and is easily accessible.   The fruit of the Bur-marigold is
crowned by toothed bristles which catch in the fur of animals.

The Garden Marigold has all the flowers unisexual, the disc-
flowers male, the ray-flowers female: in the former the style is

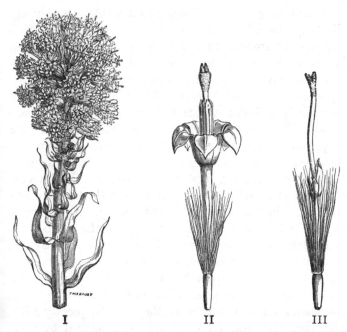

I   II   III

Fig. 163. Butterbur: I, male inflorescence; II, male flower;
III, female flower, with reduced corolla. (After Baillon.)

covered with hairs and acts as a piston, but there are no
stigmatic papillae.   The Butterbur is another plant with unisexual flowers.   It is usually dioecious.   The inflorescences
appear before the leaves in spring, at the ends of shoots which
bore leaves the previous year: often large patches of plants
are all of one sex, having been derived vegetatively by means

of the vigorous rhizomes from a single stock. The male inflorescences are easily distinguished because they have fewer,
larger flower-heads of a brighter pink, from which protrude
styles with conspicuous white swollen tips. In the female
inflorescences the florets are smaller, the pappus often partly
obscures the corolla even at a very early stage, and the style
ends in a forked stigma. Though it forms no ray-flowers this
species is closely allied to the Colt's-foot.

Considering now the Family as a whole, we may say
that capitula, piston mechanisms, of the type described,
and achenes are found throughout the group. The leaves
have no stipules and are simple, but are often pinnately
lobed or divided, sometimes into many· very narrow
segments, as in the Yarrow. The plants are generally
very hardy and vigorous, and the perennial herbs, which
include the great majority of the Family, as a rule
propagate themselves very readily by quickly growing
rhizomes (e.g. Michaelmas Daisy), by tubers (Artichoke,
Dahlia), by creeping branches (Mouse-ear Hawkweed),
or by short ascending branches from a thickened 'root-
stock' (as in the Dandelion and Daisy). The success
of the Family as a whole, as tested by the abundance
and very wide distribution of its members, may be
attributed to their possessing a unique combination of
features. (1) The arrangements for ensuring pollination
are simple and effective. The flower-heads are very
conspicuous, still more so as they are frequently very
numerous (e.g. Ragwort). Nectar is produced by each
flower in the head and is thus far more abundant
than in large single flowers. The capitulum usually
affords a comfortable resting place for insects regaling
themselves with nectar and they often remain for some
time on a single capitulum, meanwhile pollinating
thoroughly all the open stigmas and getting thoroughly
dusted with fresh pollen. As the corolla-tubes are
usually not long the flowers do not depend for their

pollination upon a limited class of insects and receive
numerous visitors. The automatic self-pollination which
takes place finally in the majority of cases ensures fertilisation, even if cross-fertilisation fails. The piston
mechanism presents pollen little by little, so minimising
waste, and the nectar is also protected from rain and
from short-tongued marauders in the narrow corolla-
tubes, closed by the stamens, which sometimes have hairs
on the filaments.

(2) The small plumed achenes possessed by most
species are carried immense distances by wind. This
accounts for their wide distribution, and moreover
increases the likelihood of some seeds falling in congenial
situations.

(3) It is often said that the massing together of flowers
in a close inflorescence means economy of material, but
it is questionable if this is really true, or at any rate
important. Of course much less food-material is expended
in developing a small floret than a large elaborate flower;
but large flowers may have as many seeds in their ovary
as are present in a whole flower-head of some of the
Compositae. Each floret has only a single ovule and
each ovule is thus provided with a separate stigma and
corolla. A similar reduction in the number of ovules is
seen in other cases in which numerous small flowers are
aggregated together, as in the Umbelliferae. It must be
partly a necessary consequence of their aggregation; for
instance, there are limits to the amount of food which
can be supplied through a stalk of a certain size. But it
is possible that the rigorous selection of pollen which
must take place, when only one of the grains deposited
on each stigma can fertilise an ovule, is of importance
and tends towards increased vigour in the offspring.

C. *Families with the whole perianth petaloid, all parts of the flower usually in whorls of three, and leaves with parallel veins.*

The flowers of most of the plants described collectively as 'bulbous plants' by gardeners,—for example, the Snowdrop, Crocus, Daffodil and Tulip,—have flowers in which the perianth shows no differentiation into green protective calyx and conspicuous corolla but is all petaloid. Although they have not all true bulbs, a more or less thickened perennial storage-organ of some kind, a rhizome, corm, or bulb, is characteristic of them: indeed only with a large store of food-material already available could they produce their large flowers so early in the season. Another feature in which they resemble one another is the parallel venation of their leaves.

These plants fall along with other similar plants into Families which are distinguished from each other by the structure of their flowers.

The Liliaceae.

This Family, called after the Garden Lilies, includes, for example, the common Bluebell and the Tulip. The Lilies grow from bulbs with numerous overlapping scales. These send up erect shoots bearing many crowded parallel-veined leaves and several large flowers at the top. Each flower has a perianth of six brightly coloured leaves, arranged in two alternating whorls, but all alike. There are two whorls of three stamens, and a three-chambered superior ovary with a long style ending in a three-rayed stigma. The floral formula is thus $P_{3+3} A_{3+3} \underline{G}_{(3)}$, the whorls alternating regularly all through. The flowers are adapted to large butterflies or moths; they secrete nectar in long grooves, covered over by hairy ridges, down the middle of the perianth-leaves near their bases, and the anthers are so attached to the tapering filaments that they turn readily in any direction (*versatile* anthers).

The Tulip has a bulb of concentric sheathing scales, which sends up a stem, bearing several broad sheathing leaves with parallel venation and ending in a single flower, which differs only in details from that of a Lily. The perianth only opens widely in warm bright weather,

Fig. 164. Flower of an Orange Lily in longitudinal section: *g*, groove in which nectar accumulates, covered below by a thick mat of over-arching hairs; *b*, bracteole; the style is hollow.

closing up if it is dull or cold and also at night. There is no nectar but the large anthers provide abundant pollen.

The common Bluebell has long and narrow radical leaves and a long stalk bearing a raceme of flowers, each in the axil of a bract. The perianth leaves form a hanging bell though they are at most slightly joined at the base, and the stamens are just attached to them there. These blue pendulous flowers are pollinated by bees. Nectar is secreted by the ovary itself; glandular tissue is present in the ovary-wall along the margins of the septa.

In the garden Hyacinth, which in most respects closely resembles the Bluebell, the perianth is tubular below, and the stamens are attached about half-way up this tube. In the Lily-of-the-Valley also the perianth is tubular with only short teeth: the two whorls are nevertheless still distinct, the edges of the outer teeth overlapping the inner. The floral formula of these flowers is thus $P(3+3) A3+3 G(3)$.

The fruit of most of the Liliaceae is a capsule, splitting as a rule at the top and forming a censer mechanism: in the Bluebell the flower-stalk turns up so as to carry the fruit erect. The Lily-of-the-Valley differs in that it forms berries. It also has not a bulb but a rhizome, the tip of which turns up and forms broad leaves with sheathing bases, and the raceme of flowers. Other plants with rhizomes and berries are Solomon's Seal and Asparagus. The thick rhizome of Solomon's Seal (p. 167) turns up each year and forms an arching aërial shoot, bearing alternate broad leaves each with a few greenish-white flowers in its axil. The aërial shoots of Asparagus are very much branched and bear small scale-leaves with axillary tufts of narrow green leaf-like short-shoots (p. 173) that carry on the work of photosynthesis. It is the young aërial shoots that are eaten, soon after they appear above the ground. The flowers are mostly unisexual and dioecious, each with vestiges of organs of the other sex.

The Onion and Garlic belong to this Family. The inflorescence is a cymose umbel, which in the Onion, where the flowers are numerous and small, forms a spherical head. The fruit is a small capsule with few seeds. The Onion is a bulbous biennial (p. 172).

The Herb Paris is exceptional in several respects. The aërial stem bears a whorl of four broad leaves and a single terminal yellowish-green flower, in which the parts are in whorls of four. The perianth leaves of the inner whorl are narrower and rather yellower than the outer. The ovary and stigma are dark purple; their colour is said to deceive carrion-flies into visiting the flowers, which are markedly protandrous. The fruit is a blue-black berry. Occasionally plants are found with five leaves and whorls of five in the flower.

The shrubby Butcher's Broom (*Ruscus aculeatus*, p. 172), with phylloclades, also belongs to this Family. The small flowers are unisexual and dioecious and the perianth is greenish and inconspicuous. In the male flowers only the outer three stamens develop.

A few tropical members of the Liliaceae are trees, with an exceptional type of secondary growth: the Dragon Tree of the Canary Islands, etc., is the best known example. Others are the Yuccas, with a thick erect stem bearing a rosette of many large narrow closely arranged leaves at the top and forming large erect inflorescences of numerous white flowers, and the Aloes, with a similar rosette of large thick sword-like leaves. The method of pollination of the flowers of Yucca is very remarkable: it depends upon a single species of moth, the female of which gathers pollen from the stamens before leaving a flower and makes it into a large ball, which it carries off to another flower; there it first deposits eggs in the ovary with its sharp ovipositor and then runs to the top of the style and presses the ball of pollen on to the stigma. The result is that more ovules are fertilised than are required to feed the larvae.

We have seen that a considerable variety of habit is

exhibited in the Family. There are differences, too, in the form and size of the flowers and in the degree of union of the perianth-leaves; but the general structure of the flower is otherwise very uniform throughout the Family.

The Amaryllidaceae.

Plants belonging to this Family are readily distinguished from Liliaceae because they have an inferior ovary. The Snowdrop, Daffodil and Narcissi are the best known examples.

The Snowdrop (*Galanthus nivalis*) has a solitary pendulous flower, adapted to bees. When young it is enfolded by a large membranous sheathing bract, called a *spathe*. The two whorls of the perianth differ in shape and colour, the outer white and spreading, the inner erect, with green grooves on their inner side, secreting nectar. The six anthers lie close around the style, and dehisce by apical slits. Each stamen has an outwardly directed appendage; when these are disturbed by bees visiting the flower the anthers are shaken and a shower of dry, dusty pollen falls on the visitor. As the style is longer than the stamens the stigma is touched first.

In the Narcissi and Daffodil (genus *Narcissus*) the flower has a perianth-tube, to which the stamens are attached, and a *corona*, looking like an extra perianth, forming a sort of prolongation of the otherwise short perianth-tube of the Daffodil, but in the Pheasant-eye Narcissus, for instance, merely a fringe around the mouth of the long narrow tube.

The Daffodil (*Narcissus Pseudonarcissus*) has a humble-bee flower, with a large golden yellow corona. The bud, like that of the Snowdrop, is enclosed when young by a membranous spathe, and is at first erect, but the stalk bends over so that the open flower is horizontal or slightly drooping. The stamens surround the style very closely, and the nectar, which is secreted by glands in the top of the septa of the inferior ovary, can only be reached through narrow spaces between the bases of the filaments, by comparatively long-tongued bees. A humble-bee, creeping up the flower, touches first the projecting stigma, then the stamens, with its back.

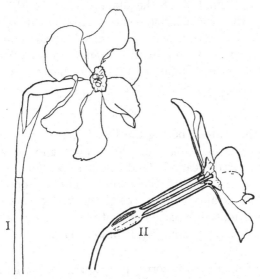

Fig. 165. Pheasant-eye Narcissus (*N. poëticus*). I, flower, with sheathing membranous bract, showing the two whorls of stamens and the three-lobed stigma at the mouth of the perianth-tube, surrounded by the corona. II, a flower in longitudinal section.

The fragrant white flower of the Pheasant-eye Narcissus (Fig. 165) is adapted to moths. It has a long narrow tube, the mouth of which is almost blocked by two whorls of sessile anthers, one a little above the other, and the round stigma in the midst of them. The greenish yellow corona, with its crimped bright red margin, marks out the centre of the flower very conspicuously. It requires a long and slender proboscis to reach the nectar, which is secreted as in the Daffodil by the top of the ovary, and moths in probing for it cannot help touching both the stigma and the anthers. Owing to the close proximity of these, self-pollination may occur, apart from the visits of insects.

Some species of *Narcissus* have a cymose umbel of flowers at the top of the stalk; in these the spathe encloses the whole inflorescence. All these plants have bulbs, and the fruit is an erect capsule. There are some members of the Family with rhizomes, and the American Aloe (or Century Plant, p. 181), like the true Aloes, has a stout erect stem and a rosette of thick fleshy leaves.

The Iridaceae.

To this Family belong the Iris, Crocus, and Gladiolus. The flowers have an inferior ovary and only three stamens: as the carpels (and the chambers of the ovary) do not alternate with the stamens but are opposite to them, the arrangement of the parts corresponds to that in the Amaryllidaceae but with the inner whorl of stamens missing. The floral formula is therefore written $P_{3+3} A_{3+0} G_{\overline{(3)}}$.

The Crocus has a corm (p. 170, Fig. 54) with several buds on the top. Each bud gives rise to a tuft of radical leaves, surrounded below by sheathing scale-leaves, with the flowers amongst them each at first enclosed in a papery sheath: one of the flowers is terminal on the main axis. The perianth forms a very long and narrow tube, and the ovary remains below the ground. The stamens are attached round the mouth of the perianth-tube, and dehisce outwards. The stigma has three lobes, which do not separate until a day or two after the anthers have begun to shed their pollen. Honey-glands are present in the top of the septa of the ovary; but the nectar rises

high in the very narrow tube so that bees can reach it
as well as butterflies and moths.

The genus *Iris* includes the Yellow Flag, which is
common in marshy places, and the Gladdon or Roastbeef-
plant (*Iris foetidissima*, so-called because of its smell
when bruised) as well as the garden Irises. They are
plants with thick rhizomes (except the Spanish Iris,

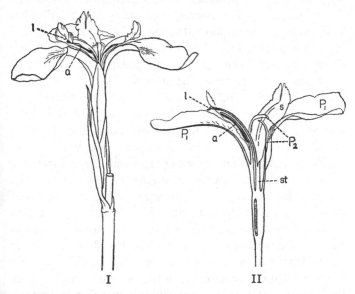

Fig. 166. Yellow Flag (*Iris Pseudacorus*). I, flower; II, the same in
 longitudinal section. In this species the inner erect leaves of the
 perianth, P_2, are small. P_1, outer leaves of perianth; *a*, anther,
 under lobe of stigma (*s*); *l*, lip of stigma, with long papillae on upper
 (receptive) surface; *st*, style, rising from the inferior ovary.

which has a bulb), the branches of which turn up at the
end and form flattened shoots bearing two ranks of
leaves and an inflorescence of few flowers. The leaves
have folded sheathing bases and a vertical sword-like
blade. The upper leaves are modified as green sheathing

bracts or spathes. The first flower is terminal, a second is borne in the axil of the spathe that enclosed the terminal flower, and has a thin sheathing bract (or bracteole) of its own hidden within the spathe. Other flowers may also be produced in the axils of lower spathes. The flower has a highly specialised mechanism adapted to humble-bees (Fig. 166). The stigmas are modified into three large petaloid lobes, each of which arches over one of the outer leaves of the perianth, then turns up and divides. The inner perianth-leaves stand erect between the spreading outer ones. The latter afford a convenient alighting place for bees: they are conspicuously marked and in most species bear, under the stigma-lobes and extending a little beyond them, long close hairs. The stamens are completely hidden under the lobes of the stigma, and the stigmatic surface is situated a little beyond them on a transverse flap, just where the stigma turns up.

Bees alight on an outer perianth-leaf, force their way between it and the arching stigma to reach the nectar, secreted by the top of the ovary, and in so doing first deposit pollen on the stigmatic side of the flap, and then rub their backs against the anther, which dehisces downwards. On withdrawing they close the flap so that no pollen from the same flower is deposited on the stigmatic surface. Thus, although the flower is actinomorphic, bees have to approach it in a particular way; it acts, in fact, like three zygomorphic closed flowers.

The Gladiolus has a corm from which arise shoots like those of the Iris but smaller and bearing a number of sessile flowers in a long spike. The flowers all turn and face towards the same side of the inflorescence: they are distinctly zygomorphic, the lower perianth-leaf forming the alighting place distinguished by special honey-guides, the stamens and style arching upwards, and the anthers all dehiscing downwards. An insect visiting the flower touches first the projecting three-lobed stigma, then the anthers, with its back.

These three Families have flowers constructed on a similar plan, and in vegetative habit have features which distinguish them from other Families we have dealt with. They are a good example of Families which can be brought together into one larger group, called an Order (or a Cohort) of Families. The name of this Order is the *Liliiflorae*.

The Orchidaceae.

One other Family remains to be considered along with them; this is one in which the flowers are highly specialised to a remarkable degree.

In habit the Orchids are usually very similar to plants of the other three Families, being all perennial herbs, often bulbous, with parallel-veined leaves. The flowers, too, have a petaloid perianth of six leaves in two whorls. Some species of the genus *Orchis* are fairly common, flowering in spring or early summer—for instance, the Early Purple Orchis (*O. mascula*), Spotted Orchis (*O. maculata*), and Marsh Orchis (*O. latifolia*); any of these will serve as an example.

The flower of the Spotted Orchis is represented in Fig. 167. It is zygomorphic. One of the perianth-leaves, belonging to the inner whorl, is expanded into a conspicuous alighting place for insects (*l*) and also forms a spur (*sp*): this leaf is called the *labellum*. Insects have to bore into the tissues of the spur to extract sweet sap. The inferior ovary is twisted so that the flower is turned through 180 degrees and the labellum, which is really posterior, is brought underneath into an apparently anterior position. The ovary, though it differs from those of the Liliiflorae in being one-chambered, is also of three carpels, since it has three parietal placentae.

The rest of the flower is at first sight unrecognisable, forming a stout column, projecting from the centre of the flower, just above the entrance to the spur; two inner perianth-leaves arch over it. When a pointed pencil is

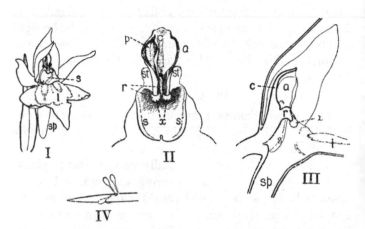

Fig. 167. Spotted Orchis (*Orchis maculata*). I, flower, with bract: *l*, labellum, with spur (*sp*); *s*, stigma, above which projects the *rostellum*. II, central column enlarged: *c*, connective of the anther (*a*) in each half of which the pollen forms a pollen-mass (*p*) which is connected by a thin stalk with a sticky gland inside one of the lobes (*x*) of the projecting part of the rostellum; *s*, *s'*, two lobes of the stigma, of which the third lobe forms the rostellum (*r*); *st*, staminodes. III, column and part of the perianth in longitudinal section, showing how the stamen and stigma are joined; the stigma has developed disproportionately on the upper side and its upper lobe become much modified, forming the rostellum (*r*, *x*). IV, the two stalked pollen-masses (*pollinia*) withdrawn on a needle by pressing the two lobes (*x*) of the rostellum downwards, so disclosing and bringing the needle into contact with the sticky glands; the dotted outlines show the positions assumed after a short exposure to the air.

pressed into the mouth of the spur of a newly opened flower, just as the head and proboscis of a bee would be, and withdrawn again, it carries with it two club-shaped objects, attached like a pair of antennae (Fig. 167, IV). Examined under the microscope these are found each to consist of a mass of pollen, attached by a slender thread to a sticky lump of tissue by which it adheres.

If a second flower is treated in the same way, and the withdrawal of the stalked pollen-masses (or pollinia) carefully observed, the pollen is found to come from two recesses, each curtained by flaps of tissue on either side (Fig. 167, II): these two, with the thick connective that joins them, form the anther of a single fertile stamen. The sticky masses, at the bottom of the pollinia, and the lower part of each stalk come from a structure called the *rostellum* (*r*), which projects below the anther and over-hangs a large two-lobed stigma (*s, s'*). Soon after the pollinia are withdrawn they bend down, in drying, so that if the pencil is inserted again the pollen-masses come against the stigmatic surface. There are two other small prominences on the column, on either side of the anther (*st*).

This curious structure can be explained by reference to a type of flower, such as is found in the Amaryllidaceae, in which there are two whorls of stamens and a three-lobed stigma. Two lobes of the stigma are fer-tile, the other is modified as the rostellum. The fertile anther is the anterior one, and therefore belongs to the outer whorl, and the two prominences close on either side of it are staminodes, repre-senting two of the inner whorl of stamens. The other three stamens are absent.

Fig. 168. Floral diagram of *Orchis*, showing the parts in their original positions in relation to the axis, before the ovary twisted: *LAB*, labellum; *STD*, stami-node. (After Eichler, modified.)

The large majority of Orchids have flowers constructed

on a similar plan, with a single fertile stamen: they form
the Sub-Family *Monandrae*. The
Lady's Slipper Orchid (*Cypripedium*) is different, and represents the other Sub-Family, the
Diandrae, so called because two
fertile stamens are present. The
labellum is pouch-like, with very
smooth curved edges. Insects
alighting on it easily find their
way right into the pouch, but
have to creep out on one side
or the other of the column.
On either side they must brush
first a lobe of the stigma and
afterwards an anther. They
leave the flower, smeared on
one side with pollen which they
will sooner or later deposit on
the stigma of another flower.

Fig. 169. Flower of a Lady's
Slipper Orchid. (From
Veitch.)

It will be observed that
neither in *Orchis* nor in *Cypripedium* can self-pollination be brought about.

The only British *Cypripedium* is rare: all the other British
Orchids belong to the *Monandrae*. Outside the genus *Orchis*
there are few common species. The Butterfly Orchids (genus
Habenaria, Fig. 170) have white waxy moth-flowers with long
spurs. The Twayblade has a raceme of greenish flowers the stalk
of which bears two broad opposite leaves. The flowers are visited
chiefly by ichneumon flies, which crawl up the forked labellum,
licking nectar from a groove which runs along the middle of it.
When an insect reaches the top it comes into contact with the
rostellum: in response to the touch the rostellum immediately
exudes a sticky fluid which cements the pollinia to the insect's
head.

The Fly Orchid is common in some parts of England, and is

interesting because it is very seldom visited by insects and so usually remains unpollinated; while the Bee Orchid has never been observed to receive a visit, but is regularly self-pollinated by the pollinia falling out of the anthers on to the stigma: these two Orchids (genus *Ophrys*) have been called after insects because a resemblance to them has been seen in the labellum. The Fly Orchid is apparently adapted, by its brownish colour and the presence of shiny spots on the labellum, to carrion-flies.

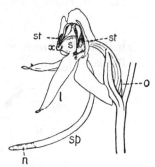

Fig. 170. Butterfly Orchid (*Habenaria chlorantha*). Flower, with long slender spur, *sp*, containing nectar, *n*; *s*, stigma; *x*, one of the lobes of the rostellum in which are the glands connected with the pollen-masses; *e*, entrance to spur; *st*, staminodes; *o*, ovary, twisted at the base. In this Orchid the rostellum and the connective of the anther are broad, so that the glands especially are wide apart.

The fruits of Orchids are capsules, with numerous very minute seeds, little bigger than pollen-grains, which are readily carried like dust by the wind. This feature is of great advantage to the many tropical Orchids that live as *epiphytes* on the upper branches of trees. Many of these show special adaptations for epiphytic life. They have *clinging* roots, by which they are held fast, and around which humus collects, *absorbing* roots which penetrate the humus, and *aërial* roots that hang in the air and absorb any water trickling down them and even water vapour by means of a spongy dead tissue with which they are covered. The aërial roots contain chlorophyll and help in photosynthesis. Many of these epiphytes have swollen fleshy internodes that store water and food during the dry season.

A few Orchids, for instance, the Bird's-nest Orchid (*Neottia*), found in Beech woods (p. 460), are *saprophytes*, devoid of chlorophyll and living on organic matter found in humus.

D. *A Family with simple unisexual flowers.*

The Salicaceae.

This Family includes the Willows (genus *Salix*) and the Poplars (genus *Populus*). They are all trees or

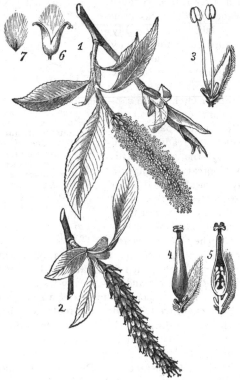

Fig. 171. Crack Willow (*Salix fragilis*). 1, male, and 2, female flowering shoot; 3, male flower; 4, female flower, and 5 longitudinal section of the same; 6, ripe capsule; 7, seed. (After Wossidlo.)

shrubs, having simple alternate leaves, with small stipules. The male and female flowers are borne on separate plants

in *catkins*: these are spike-like, often pendulous inflorescences consisting of an axis bearing bracts with the
flowers in their axils.

In the Willows the flowers are of extremely simple
construction. A male flower consists usually of two
yellow stamens, in some species of three or five, and one
or two small fleshy scales, which are honey-glands;
these parts are borne, without any trace of a perianth, in
the axil of a bract of the catkin. A female flower consists
of an ovary on a very short stalk, together with one or
two honey-glands as in the male flower. The ovary is
one-chambered, with a number of ovules on two parietal
placentae; it tapers above and bears a two-lobed stigma.
The flowers thus include only the organs essential for the
production of seed, and honey-glands.

The catkins appear early in spring, usually before the
leaves, and their sweet scent attracts numerous insects to
visit them for nectar, including bees and moths; these
carry pollen from the male to the female catkins. The
fertilised ovaries ripen to small capsules, that split into
two valves. Each seed is enveloped in a tuft of silky
hairs, that acts as a parachute and catches the wind.

There are a number of species of Willows, some of
which are easy to distinguish; but they hybridise very
freely, and hybrids are frequently found, as well as
possible varieties, which are sometimes confusing. The
chief characters by which the species are distinguished
are the form and hairiness or smoothness of the leaves, the
shape and position of the catkins, and the details of the
flowers. The White Willow (*Salix alba*), for instance, one
of our two commonest species, has grey foliage, owing to
the silky hairs with which at least the young leaves are
covered; the leaves are narrow, lanceolate and pointed,
with small teeth. The catkins are cylindrical, borne at
the ends of leafy short shoots, and have greenish-yellow

silky bracts: in the male flowers there are two stamens
and two glands, in the female a pear-shaped ovary and
two or one gland. This species is a tree that grows by
the water-side and in some parts of the country is
frequently pollarded (p. 411). Another species similar to
this is represented in Fig. 171. The other commonest
species is the Sallow or Goat Willow (*Salix Capraea*),
found in hedges as a shrub or small tree: it has broad
greyish-green wrinkled leaves, whitish underneath, with
a short cottony, not silky, down, and the stipules are

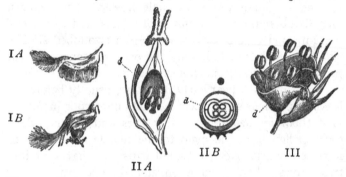

Fig. 172. I, Aspen; *A*, male, and *B*, female flower in the natural
position, with subtending bract. (After Willkomm.) II, *A*, longitudinal section of female flower of another Poplar; *B*, floral
diagram of the same. III, male flower of a Poplar. *d* = disc.
(After Engler and Prantl.)

conspicuous and broad. The catkins·are sessile on the
twigs, not at the ends of leafy short shoots, and are short
and thick, especially the male, which owing to the length
of the stamens are ovate or oblong in outline. The young
catkins are very silky owing to fringes of long silky hairs
on the bracts; twigs bearing them are often worn as
'palm' on Palm Sunday. In the male flower are two
stamens with long filaments, in the female a conical
ovary on a distinct stalk, and in each is a gland next
the axis of the catkin.

The Poplars have broad leaves on long stalks. The cat-kins are pendulous; the bracts are brownish, deeply toothed and fringed with long hairs, and soon fall off. The male flower consists of a broad disc shortly stalked and more or less cup-shaped bearing a number of stamens. In the female flower there is also a cup-shaped disc, around the base of the ovary, which is very similar in structure to those of Willows, but bears a much larger stigma. There are no honey-glands; the pollen is dry and powdery and is carried by wind.

This Family is clearly very far removed from any others that we have studied. The flowers are extremely simple, and are very simply yet efficiently adapted for pollination, in the one genus by insects, in the other by wind. (For other details see pages 410-12.)

CHAPTER XX

Of the Families with which we have dealt, the Salicaceae appear to stand apart from the rest on account of their very simple flowers; while the Liliiflorae and Orchidaceae also form a separate group distinguished by the construction of their flowers and the parallel venation of their leaves.

There are other features which separate the latter group from all the others: their embryos have but one cotyledon (p. 249) and their internal structure is of the Maize type (p. 144), the vascular bundles in the stem being scattered and having no cambium; no secondary growth in thickness takes place, except in a few peculiar cases. The same features are shared by the Grasses, Sedges, Rushes, Palms and other plants. All these are therefore placed in a large class and called, because of their monocotyledonous embryos, the *Monocotyledons*.

The rest, including the Salicaceae, have dicotyledonous embryos, net-veined leaves, and a type of structure similar to that found in the stem of a Sunflower, with the primary bundles in a ring: cambium is present and secondary growth in thickness occurs. These are called *Dicotyledons*.

Within the Dicotyledons we may note the presence of simple-flowered Families such as the Salicaceae; and among the rest the important and well-characterised Families with joined petals which are conveniently grouped together as the *Sympetalae* (Greek *syn* = together).

Dicotyledons and Monocotyledons alike have their ovules enclosed, in ovaries, and are therefore called *Angiosperms* (Greek *angeion* = case, *sperma* = seed) to distinguish them from the Pines, Firs and other plants, with seeds borne on open scales, which are called Gymnosperms (Greek *gymnos* = naked).

There are other plants, including Ferns, Mosses and Seaweeds, which do not bear seeds like Angiosperms and Gymnosperms; the latter are therefore distinguished from them as the *Seed Plants*, or more often as the *Flowering Plants*.

So we have been dealing with plants representing only one part of the Vegetable Kingdom, chiefly the Angiosperms.

In classifying the Angiosperms in Families, the reproductive organs, as we have seen, prove to be most important. The vegetative organs in one and the same Family often have a wide range of form, varying much with the conditions of life, and their external features cannot always be relied upon as a guide to relationship. Nevertheless in some Families a certain general resemblance is shown by most members even in their vegetative features (e.g. Umbelliferae and Labiatae).

The form of the embryo, and the type of arrangement of the vascular system in stem and leaves, being characters which are common to large groups of plants, must have remained little altered during the evolution of great changes in other characters.

We have already tried to trace a few of these changes in the Ranunculaceae. In that Family highly specialised flowers have been evolved from simpler ones. We have now seen that species with flowers highly specialised in a variety of ways are found in many Families; in all these Families the evolution of special adaptations must therefore have been going on. Specialisation of flowers to the longer-tongued and more intelligent insects has involved the concealment of the nectar and its protection at the bottom of a tubular structure of some kind. We have remarked in passing on the different ways in which this has been brought about in the specialised genera of Ranunculaceae (Aconite, etc.), Cruciferae (e.g. Wallflower) and Caryophyllaceae (Silenoideae) and in the Sympetalae. The most successful departure appears to have been the fusion of petals to form a tubular corolla, for among the Sympetalae are whole Families (e.g. Labiatae) with flowers of a highly specialised type. But the Families which are sympetalous have not all been derived from a single stock—the Primulaceae, for instance, show many more points of resemblance to the Caryophyllaceae than to the Labiatae or Compositae. Sympetaly has therefore been evolved independently along perhaps many different lines, and sympetalous Families are classed together as the Sympetalae not because they are allied to one another, but, mainly for convenience, because they have all evolved the same very important modification in the structure of the flower.

Zygomorphy is another feature which appears over and over again in different Families and must therefore have been evolved many times independently.

Another direction in which evolution has very generally proceeded is towards fusion of carpels. Examples of this are found even in the Ranunculaceae and Rosaceae, where in most genera the carpels are free. In all the other Families we have studied, the ovaries are syncarpous; but the carpels arise separately, indicating that they were separate in the ancestors. Free carpels are therefore regarded as primitive.

In most of the genera in which the carpels still remain separate they are numerous, and are spirally arranged, like the leaves of most plants and the cone-scales of Pines and Firs. Where they join they are few in number and arranged in a whorl. These facts suggest that a large number and the spiral arrangement of carpels are also primitive characters, and that the small number of carpels found in the great majority of Families is due to reduction. In the Ranunculaceae, Rosaceae and Leguminosae we have instances of reduction in the number of carpels, even to one, without any accompanying fusion, so that this is another direction which has very generally been followed in evolution.

Reduction in number of the stamens is another very general direction. How far this reduction may proceed is illustrated by the Orchidaceae where, in the Monandrae, only one fertile stamen remains. In the Ranunculaceae the numerous stamens are arranged spirally, but in Families with few stamens these are arranged in whorls. Even the perianth of some species in the Ranunculaceae has its numerous parts spirally arranged like the foliage leaves and bracts, but in other species the perianth has become arranged in definite whorls.

All these changes, from many spirally arranged parts to few in whorls, have formed part of the widespread evolution of specialised flowers.

Another general direction is towards the upgrowth of the receptacle, seen in various degrees in the Rosaceae, leading finally to the formation of an epigynous flower with an inferior ovary. In all the Umbelliferae, Compositae, Amaryllidaceae, etc., the flower is epigynous, so that along this particular line of development these Families are far advanced as compared with Families with perigynous or hypogynous flowers, although some of these other Families may be further advanced in other directions. The Labiatae for instance are sympetalous but hypogynous, the Umbelliferae are epigynous but have separate petals.

One other common direction may be mentioned. In many genera, and in some whole Families, flowers have become aggregated together into dense inflorescences. If the Clovers, Labiatae, Umbelliferae and Compositae are compared, it is seen that, in becoming closely massed together, the flowers have in most cases become small and the number of ovules has been reduced to very few, even to one in each flower (compare p. 355). In comparison with these it is interesting to see that among the Ranunculaceae and Rosaceae the species with the largest number of carpels in a flower form only one ovule in each carpel.

The Compositae have advanced far along each of these general lines of development. The carpels are joined and are only two in number, the stamens five; the corolla is tubular; the ovary is inferior; the flowers are massed together, small, with a single ovule to each. This Family is therefore on the whole further advanced (that is more modified and less primitive) than other Families which have advanced along one or more but not as many of these lines at the same time. That evolution has occurred so generally in these directions, and that very successful Families have resulted from each of the new departures, must mean that they involved distinct advantages, so that there is little wonder that the Compositae, which have obtained all these advantages and added others more or less peculiar to themselves, should be the most successful of all the Families.

Notes on the morphology of fruits.

The classification of fruits according to their form and structure is usually very easy, but there are a few which often give difficulty because they do not fall properly into any one class. In order to make these cases clear it will be well to recapitulate the main classes into which fruits may be divided.

These classes depend in the first instance on the kind of gynaecium from which the fruits are derived. The gynaecium may consist of one or more separate carpels, when it is called *apocarpous*, or it may be *syncarpous*; again it may be either *superior* or *inferior*.

If we include in the term fruit all the parts of a single flower that persist after fertilisation[1], apocarpous fruits are frequently compound or *aggregate* fruits, consisting of several separate *monocarpellary* fruits (Greek *mono* = one); or they may be simple fruits, as in the Pea or Cherry where only one carpel is present.

Monocarpellary fruits, whether single or aggregated, may be
(1) succulent—drupes and drupelets (Plum, Blackberry);
(2) dry and indehiscent (a) achenes (Buttercup),
(b) schizocarps (some Leguminosae);
(3) dehiscent—legumes and follicles (Leguminosae; many Ranunculaceae).

Superior syncarpous fruits, as well as monocarpellary fruits, are recognisable as superior by the presence of the style or its scar at one end and the scar of attachment or the persistent calyx along with the stalk at the other end. These again may be
(1) succulent (berry of Tomato);
(2) dry and indehiscent (a) one-seeded—achenes or nuts,
(b) schizocarps (Geranium, Mallow);
(3) dehiscent—capsules of various kinds, including the silique and the silicule.

Inferior fruits have the remains of the calyx or any other persistent parts of the flower at the end *opposite* to the stalk. These also may be
(1) succulent (berry of Gooseberry, etc.);

[1] Aggregate fruits formed from inflorescences and including parts from more than one flower may be distinguished as *infructescences* (e.g. a fig).

(2) dry and indehiscent (a) achenes or nuts (Compositae, etc.),
(b) schizocarps (Umbelliferae);

(3) dehiscent—capsules.

But it is necessary to be on guard against calling an aggregate fruit like that of the Rose inferior because it has the remains of the calyx at the top, for the achenes are not fused with the receptacle. It is here that the difficulties arise, for there are cases, especially among the succulent fruits of the Rosaceae, where the fruit is not fully inferior nor completely syncarpous. In the so-called 'berry' of the Mountain Ash (Fig. 173) the

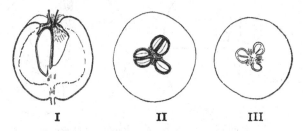

I   II   III

Fig. 173. 'Berry' of Mountain Ash (*Pyrus Aucuparia*). I, fruit cut in half lengthways; one whole carpel and another in section are shown. The carpels are only joined with each other at the base and at their tips they are also free from the upgrown receptacle. Some of the vascular bundles supplying the different parts of flower and fruit are shown, and at the top the remains of stamens, as well as two styles. II and III, transverse sections: II across the middle showing the carpels separate from one another in the centre, but joined with the fleshy receptacle outside; III at a lower level, where the carpels are joined.

carpels are free from one another except just at the base, and at the tip are also free from the receptacle. If they were wholly free from one another and from the receptacle, the fruit would be an aggregate fruit like the hip of the Rose. If on the other hand the carpels were joined together and to the receptacle right to the top the fruit would be inferior: this is the case in the Apple (Fig. 174).

In the Apple, Mountain Ash 'berry,' and other similar fruits the fleshy part is clearly derived chiefly from the receptacle, but it is a mistake to think that any sharp dividing line exists between

receptacle and carpel-wall where these have become joined[1]:
all we can say is that the combined receptacle and carpel-wall,
forming the wall of the inferior ovary, becomes differentiated
into (1) a fleshy mass of tissue bounded by an outer skin and
(2) the horny lining that protects the chambers within.

The Apple and Pear, Mountain Ash 'berry,' haw of the Hawthorn and other similar fruits have been called 'false' or
'spurious' fruits because the receptacle as well as the ovary
proper enters into their formation; but, if the term be used, *all*

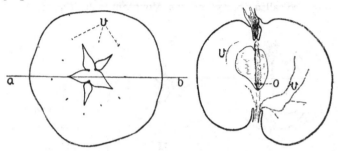

Fig. 174. An Apple in transverse and longitudinal sections. In this
    specimen the central axis of the syncarpous ovary had been split
    during the swelling of the fruit so that three of the five chambers
    are open to the centre, two still remaining closed. The longitudinal
    section shows that the carpels were joined with each other and
    with the receptacle right up to the base of the styles. The ovules,
    o, in this fruit had not grown; nevertheless the fruit had swelled,
    after pollination. At the top are the remains of the sepals, stamens
    and styles. *ab*, direction in which the longitudinal section was
    made; *v*, parts of vascular bundles.

inferior fruits should be called false fruits. Nowadays it is falling
into disuse: so-called 'false fruits' are either inferior fruits, or
aggregate fruits, or else they fall between these two classes.

    [1] This point can perhaps be best realised if we think, for instance,
of the carpels in the 'berry' of the Mountain Ash as *sessile* carpels,
of which the broad bases have become still further broadened by the
growth of that part of the receptacle to which they are attached. In
the aggregate fruit of the Rose the bases of the carpels remain narrow
and the receptacle expands between them. In neither case is it possible
to say exactly where carpel ends and receptacle begins in the region
of attachment, just as no sharp line can be drawn at the base of a leaf
between leaf and stem.

The fruit wall, or *pericarp*, whether of a superior or an inferior fruit, may become dry and leathery or brittle, or it may all become woody, or all succulent except for an outer skin; or it may form an inner protective layer as well as a fleshy or spongy part. Many of the names applied to fruits depend upon these characters. *Nuts* are one-seeded fruits with a hard woody pericarp—other one-seeded indehiscent fruits are *achenes*. Succulent fruits in which the bulk of the pericarp becomes succulent, so that the seeds are embedded in pulp, are called *berries*: the seed-coats may be woody but no hard protective layer is formed by the pericarp itself. Succulent fruits in which an inner layer becomes woody, forming a 'stone,' are *stone-fruits* or *drupes*. Typical drupes (Plum, Cherry) are monocarpellary fruits: when aggregated, as in the Blackberry, they are called drupelets. The Haw, Medlar and Cotoneaster are inferior or semi-inferior stone-fruits, with from one to five chambers, each of which forms a stone easily separated from the rest. In the Apple, Pear, Quince, and Mountain Ash, a thin inner layer of the pericarp becomes horny, and also the septa: these fruits, which are all inferior or semi-inferior, are called *pomes*.

A few peculiar examples will further illustrate the use of these terms. The Date is not a stone-fruit, but a one-seeded berry, because the 'stone' is the seed, and the whole of the pericarp is fleshy. The Coconut is a drupe formed from a three-chambered ovary, in which only one seed develops: the shell is the inner part of the pericarp; the thick outer part, which is removed before export, is not juicy but spongy and fibrous, and acts as a float. The fruit of the Walnut is an inferior drupe; the green fleshy, but not edible, outer part splits, and sets free the 'nut' containing the seed. The fruit of the Almond is a drupe formed from a single carpel, like a plum, but the outer layer, which at first encloses the 'nut,' is spongy, not juicy. In the 'berries' of the Holly and Elder, each chamber of the syncarpous inferior ovary forms a separate 'stone,' surrounded by its own shell, formed by the pericarp: these are, therefore, not berries, but stone-fruits containing more than one stone.

The Rose hip and the Strawberry are aggregate fruits in which the receptacle becomes succulent, in the former enclosing the achenes, in the latter bearing them exposed on its surface. The Fig, Mulberry, and Pineapple are fleshy infructescences (p. 236).

SECTION V

PLANTS IN RELATION TO THEIR ENVIRONMENT

CHAPTER XXI

'FITNESS'

It is clear on a little consideration that the continued existence of any species of plants has depended in the past and still depends upon its success in at least two respects. Each plant must be so adapted to its environment that from its seedling days onwards it can obtain and use air, light, and soil, and continue to live under conditions which are ever changing; but sooner or later every individual plant dies, and to preserve the species it must produce offspring and provide for them well.

The mode of life and the method of reproduction must, moreover, be suited to each other.

The balance of characters
Annuals, for instance, and especially ephemerals (p. 181) are not adapted to endure hardship. By starting early to flower they ensure the production of seed as soon as possible. They seldom have highly specialised flowers; fertilisation is frequently secured in the surest and most economical way—by self-pollination.

Perennials are provided with means of persisting through seasons unfavourable to active life. Many of these, it appears, have risked the occasional failure of the seed-crop, which may result when pollination depends upon a limited number of insects[1], for the economy and other probable advantages which greater specialisation of the flower makes possible. They do not, as a rule, bloom until the plant is well established; and in proportion to their size they often produce far less seed in any one season than annual plants.

In considering the fitness of plants for their environment, we have, therefore, to take into account the whole of their characters.

We must also recognise that different plants may be **Different** fitted for different environments. Plants **environ-** which cannot make headway in an open **ments.** meadow may flourish in the shade and shelter of a wood, and vice versa. Success or failure thus depends upon environment. It is obvious, for instance, that wind-pollination and the dispersal of seeds by wind are not suited to positions which are protected from the wind, nor insect-pollination where insects are few. Besides, a plant can keep a proper balance between its vegetative functions, especially between the absorption and transpiration of water, only within certain limits, which are different for different species (compare Chapter x).

We have already seen how plants are not all equally adapted to withstand drought: we can, in fact, distinguish markedly *xerophytic* plants, capable of living in places where the water supply is scanty and precarious, and markedly *hygrophytic* plants, which can only live where there is plenty of moisture. Between these extremes are plants showing intermediate degrees of adaptation to dry or moist conditions, so that no hard and fast lines can be drawn; but plants which are neither distinctly hygrophytic nor distinctly xerophytic are sometimes

---

[1] Some plants seem, so to speak, to have over-reached themselves in this respect, and although very highly specialised, are seldom pollinated, or have to rely on self-pollination almost entirely.

called *mesophytes*. In countries where the conditions during one season of the year (the cold winter, or the dry season) make moisture difficult to obtain, plants are common which are xerophytic during that season, and at other times mesophytic or hygrophytic; such plants are distinguished as *tropophytes*. The Tulip, for instance, in the summer exposes its green leaves to the sun and air, while in the winter its succulent bulb is protected beneath the ground. Similarly, trees which shed their leaves in autumn are tropophytes.

Another very important variable factor is temperature. Any given kind of plant lives most healthily within a certain small range of temperature. The more the temperature rises or falls from this most favourable *optimum* temperature, the less successfully does the plant live and grow, while growth ceases altogether beyond certain maximum and minimum temperatures. These temperature limits vary for different plants, and it is largely owing to this fact that the character of the vegetation differs so greatly in countries of different latitudes and on mountains at different altitudes.

On temperature and moisture together depends the general character of the vegetation in different parts of the globe. The luxuriance of tropical rain-forests, like those of Brazil, in the basin of the Amazon, is only possible where the temperature is high and the air always moist; whereas in districts where drought and high temperatures go together, deserts are found where only extreme xerophytes, like the Prickly Pear and other Cacti, can exist.

Study of plants in relation to environment. In this Section we shall be concerned with the special relation of plants to the environment in which they are found growing in nature— their natural habitat. Our object will be to discover as far as we can in what special ways the plants around us are adapted to their environment, and why they grow where they do and not somewhere else.

There are, in the first place, certain special habits of growth and modes of life which are so well characterised that the plants which have them are grouped together as special *forms* of plants. Some such forms we have already studied: annual herbs, for instance, may be grouped as a special form of plants. Other well-marked forms which we shall study are trees, water plants, and climbing plants.

The study of forms of plants, however, represents only one part of the subject, or rather one method of attacking it—by concentrating attention on individual plants. There is another valuable method of attack which opens up many new points of view; this is to study how the *distribution* of plants is connected with special features of the environment. This has proved so promising a method and has opened so wide a field for research that a special department of Botany has come into existence called *Plant Ecology* (Gr. *oikos* = dwelling-place), the study of plants in their homes.

Although Plant Ecology is concerned primarily with the distribution of plants, in seeking to explain their distribution it must clearly take into account the special features of the plants themselves; it therefore comes to include the whole subject matter of this section, supplementing the study of distribution by the careful study of individual plants.

CHAPTER XXII

TREES

Trees are giants among plants, and often outlive many generations of men. Beeches, Oaks, and Yews occur in this country with huge trunks many feet in diameter, whose age must be measured in centuries; while some of the giant Wellingtonias of California, or Californian Pines, must have lived more than a thousand years[1].

Trees are, therefore, fitly called perennials. They differ much in habit, however, from most other perennials. These spread sideways, generally underground, and sooner or later break up, usually by the decay of their older parts, into portions which continue their life as separate plants. The stem of a tree, on the contrary, grows first upwards, and though it branches repeatedly, its parts remain connected in one colossal shoot-system. In very many perennials the parts which appear above ground die down completely at the approach of winter; only the underground parts remain alive till the spring, when they send up new annual shoots. Trees do not die down in this way, but brave exposure to winter weather. Year by year they grow in size and strength, the main

[1] A section of the trunk of one of these trees, in the British Museum, shows 1335 annual rings (see p. 393); it was cut down in 1890, so that it dates back to about 555 A.D. Allowing for the fact that the number of annual rings is not a sure guide to the age of a tree, we may, at any rate, say that this tree dates back to Saxon times, before William the Conqueror came to England.

stem forming a pillar-like trunk capable of supporting the weight of the branches with their heavy masses of foliage. Below the ground the root-system grows in proportion thicker and stronger, spreads farther and puts out new rootlets, so obtaining access to fresh supplies of mineral salts. With an aërial shoot-system, often fully exposed to wind in winter, when the ground is cold and absorption difficult, it must clearly be of very great importance for trees to be able to reduce to a minimum the loss of water by transpiration or evaporation. Most of our common trees are *deciduous*, shedding their leaves at the approach of winter; while others which are *evergreen* have tough leathery leaves which can resist drying up because of their thick cuticle, and so can safely remain on the tree all the year round. The trunk and branches of all trees are protected from loss of moisture by their bark.

In order to get a more complete view of the special features exhibited by trees let us study carefully some examples.

The Horse-chestnut.

In the first chapter we took a brief glance at the twigs of the Horse-chestnut in summer, and noticed the large opposite palmate leaves (Fig. 183, p. 401) with their axillary buds, its terminal buds and terminal inflorescences, and the scars left by fallen leaves of previous years. If we examine a twig in its leafless winter condition we shall find at its tip either a terminal bud, now large, brown and sticky, or a pair of similar buds with a saddle-shaped scar between them left by the inflorescence. In the latter case a shield-shaped scar, left by the pair of leaves in the axils of which the buds were borne, is present below each bud: these are thus lateral buds.

Below, at intervals, are several pairs of leaf-scars, each pair alternating with the next; each scar has a brown bud just above it, though some of the buds are very small. Farther down the twig is found a close belt of crowded scars, which are narrow vertically, but are still arranged in alternating pairs. Below this girdle of scars, large leaf-scars are found again, farther apart, followed by another girdle of narrow scars. It is often possible to trace several such series, but the surface gets rougher where the branch is thicker and the marks are more difficult to decipher, till farther down they are quite obliterated.

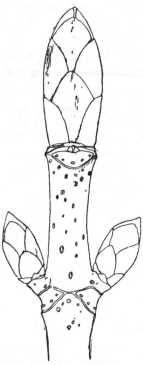

Fig. 175. Horse-chestnut twig. (From Ward after Dawson.)

A girdle of narrow scars is also found at the base of each lateral branch; to understand the meaning of these scars we must study the buds from which such twigs grow. Choosing a large bud for convenience we find it covered outside by overlapping brown scales which are sticky with a resinous substance that glues them together into a close waterproof covering. When these outer scales are removed thinner pale green scales are discovered. The arrangement throughout is decussate (p. 160), like that of the foliage-leaves. Within the scales is a woolly mass in which can be distinguished tiny compound leaves, each with several leaflets folded along their midribs. In some of the largest terminal buds little woolly inflorescences are also found within the foliage-leaves. The delicate parts are thus well protected from being robbed of their moisture by dry winds and also from excessive damp which would make them specially liable to damage by frost and disease[1].

When the bud opens in spring the outer scales separate, the inner ones grow out and finally the foliage-leaves appear, with the inflorescence, if there be one. Often one or more of the inner scales bear at the top several teeth which when closely examined are seen to be rudimentary leaflets. When such a scale is compared with a foliage-leaf on the one hand and an ordinary scale on the other, it is seen that a bud-scale is a leaf in which the leaflets have not developed, while the stalk has remained short but has broadened.

Fig. 176. Expanded bud-scale of Horse-chestnut, bearing rudiments of leaflets.

Soon the scale-leaves fall, leaving clean narrow scars close together, the scars of the girdle. The stem above them grows in length, bearing the foliage-leaves farther apart. At the apex is either the terminal bud or an inflorescence, while in the axils of the leaves minute lateral buds appear. As the year passes, the terminal bud and the lateral buds grow larger and become new winter buds.

Clearly, each girdle of scale-leaf scars marks the base of one season's growth, so that it is possible by means of these girdles to tell the age of different parts of a twig and

---

[1] It is probable that the woolly hairs protect the young leaves from excessive loss of water *while they are expanding*. They do not prevent the bud from being frozen.

to follow its rate of growth from year to year. If the length added yearly during a series of years is measured for a number of twigs on the tree it is often found that in certain years growth was uniformly slow, in other years fast, depending partly on the weather in spring while the twigs were expanding, partly on the weather during the previous summer and other conditions that may have affected the amount of available food-materials.

Two smaller kinds of markings found on the twig have not yet been mentioned. On the leaf-scars are dots arranged in a crescent, like the nails in a horse-shoe. If a leaf-stalk is cut across, the vascular bundles are seen arranged in a similar curve. These marks are therefore the scars of the vascular bundles that passed from the stem into the leaf-stalk and so to the veins of the leaf.

Irregularly scattered over the surface of the twig are other dots or tiny oval spots, slightly raised. These are called *lenticels*. Their nature is shown by the following very simple experiment. The cut end of a twig is con-
Exp. 59.   nected by a piece of stout rubber tubing to a bicycle pump. When the twig is immersed in water and air pumped into it, bubbles of air emerge from the lenticels. Lenticels are, therefore, air-passages through the bark of the twig. Through them oxygen can enter for the respiration of the living cells of the stem, and the carbon dioxide produced can pass out. Some moisture also must escape through them; in the winter, when loss of water must be checked, and when on the other hand the tree is dormant and respires far less than when actively growing, the lenticels are found to allow air to pass less readily if at all. The experiment proves also that air can pass through the tissues of the twig and therefore that air-spaces exist by which gases can diffuse to and from the lenticels.

Lenticels are less obvious in the older rougher bark of

the Horse-chestnut, although they are larger there, but they can be seen on close inspection. They are also present in the bark of the roots.

Growth in thickness, and internal structure. When a twig first grows out from the winter bud its stem is soft and green, but later it turns brown and becomes tough and woody, growing thicker and rougher year by year. If such a twig be cut through at the end of its first year and the cut surface examined with a lens, around the large white pith is found a continuous band of wood, surrounded by a darker juicy layer of phloëm. This structure is very similar to that of other dicotyledonous stems that have grown in thickness. As we have seen, such growth is due to the activity of a cylindrical layer of tissue, the cambium, from which cells are cut off on the outside to form phloëm and on the inside to form xylem. The outer tissues of the twig are readily peeled as a rind from the wood, because the cambium is delicate and easily torn.

The skin (or *periderm*) can be peeled off separately: it is much thicker than the epidermis of an old Sunflower stem, besides being brown, and is found under the microscope to consist mainly of *cork* (p. 141). Cork is formed in great quantity by the Cork Oak, from which it is stripped to make bottle corks. The use of these to prevent the escape of liquids and gases enclosed in bottles (as when ginger beer is corked in stone bottles) shows that cork is practically impermeable to water and air. It is, therefore, an excellent protection for stems against loss of moisture: but its imperviousness makes lenticels necessary to enable the tissues of the twig to respire. Lenticels may be seen in bottle corks as dark brown streaks passing through from side to side: large bungs, which can only be cut out in such a way that the lenticels pass straight through them, have to be soaked in

wax to make them air-tight. Cork is formed by a cambium, like that which forms the secondary wood and
phloëm, and in the Horse-chestnut first arises just under
the hairy epidermis. The dry black remains of the epidermis are visible in the section outside the brown cork.

If the cut surface of a twig be wetted with a solution
of iodine and more closely examined it appears as represented in Fig. 177. Round the edge of the pith a blackish

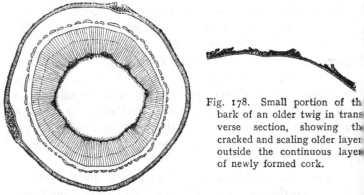

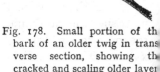

Fig. 178. Small portion of the
bark of an older twig in transverse section, showing the
cracked and scaling older layers
outside the continuous layer
of newly formed cork.

Fig. 177. Transverse section of a two-year-old twig of Horse-chestnut, viewed
through a lens, after staining with iodine. The tissues, from the surface
inwards, are: cork (shaded), with lenticels, bounded outside by dead
epidermis; cortex (unshaded); pericycle fibres in groups; phloëm; broad
zone of wood comprising two annual rings, bounded outside by the
cambium, with groups of primary wood next the pith. (See text.)

colour shows that starch is present in the cells. Black
lines extend from this starchy pith through the wood—
lines of cells full of starch, called medullary rays, running
between the masses of wood, which the iodine stains
brown. These medullary rays really extend through the
phloëm also, but there they contain no starch[1] and
cannot be clearly distinguished except in a thin section
under the microscope. They not only store food, but

[1] In the Horse-chestnut, at least in spring.

conduct food and water to the growing cambium. On the inner edge of the wood the xylem of some of the primary vascular bundles can be seen. Outside the phloëm are many bright brown groups of fibres—the pericycle fibres.

In the two-year-old twig, shown in Fig. 177, an inner and an outer zone are distinguishable in the thick woody cylinder. If a thin slice be held up to the light, it is seen that just within the boundary between the two zones there are very few of the tiny holes which are numerous outside it. These holes are the vessels, which are numerous also near the inner edge of the wood, but diminish in size and number towards the outside of each zone, the wood there becoming closer and harder in texture. When the growth of the wood is followed by examining twigs at different times through the year it is found that the wood containing many large vessels is formed in spring, but as the season advances far fewer, smaller vessels and a larger proportion of hard fibres are formed, till at the approach of winter growth ceases. When growth begins the following spring, the new wood again contains many large vessels; there is thus a sharp contrast between the *spring wood* and the *autumn wood* of the previous season. As each zone which becomes marked off in this way is a year's growth of wood it is called an *annual ring*, and from the number of these rings it is possible to tell the age of a branch, or of a tree.

Occasionally, however, the rings are less clearly marked; at other times, a fresh burst of activity may occur in the summer, and another lot of 'spring wood' be formed, so that two rings appear during one year. This may happen, for instance, if the first crop of leaves is ravaged by caterpillars. The number of annual rings, therefore, gives only the approximate age of a tree. It is instructive in this connexion to compare in several twigs the number of annual rings with the number of girdles of bud-scale scars. In some trees with thin slowly growing twigs the two numbers may often fail to agree, owing partly to the difficulty of distinguishing the thin annual rings.

The great growth in bulk of the wood stretches the phloëm and cortex enormously, and they tend also to be compressed between the taut resistant periderm and the expanding wood. For this reason the phloëm does not remain long in action, and must continually be replaced by new phloëm formed by the cambium.

The dead cork cannot grow, nor can it stretch much, so it soon shows numerous cracks. New cork formed underneath prevents these cracks from exposing the living tissues below. The old cork appears as a rough scaly covering outside (Fig. 178).

In old bark large thick scales are found which are separating from the trunk, beginning at their edges. These consist of more than cork; new layers of cork are formed, in patches, dipping into the tissues of the rind, and so cutting off scale-like portions of the outer rind, including the old and disused phloëm; these scales of tissue, deprived of water, die, and eventually split off; but meanwhile, underneath them, other scales of tissue have been cut off by still deeper layers of cork, so that the bark, consisting of dead tissues as well as cork, is always very thick.

The Beech and the Lime.

Trees like the Beech and the Lime differ in many points from the Horse-chestnut. They have simple alternate leaves borne on either side of their zig-zag twigs. The branches bear the foliage in more or less horizontal layers, tier above tier, instead of in large rounded masses. Further differences appear when the trees are examined in detail.

External features. The buds of the *Beech* are long and narrow, tapering to a point, and covered by many

overlapping light brown scales.   The twigs, like those of
the Horse-chestnut, show girdles of thin scars at intervals.
The small leaf-scars are raised on small humps: when a

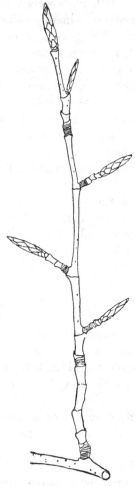

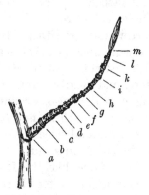

Fig. 180.   Dwarf-shoot of
Beech.   The girdles of scars
of fallen bud-scales, *a–m*,
show that the twig is eleven
years old.   Between the
girdles are crowded leaf-
scars.   (After Ward.)

Fig. 179.   Beech twig, showing
three years' growth.   (From
Ward after Dawson.)

leaf falls it is cut off just above the point of attachment, and a part of the leaf-base is thus left.

As in the Horse-chestnut some of the lateral buds form branches, others remain dormant. Others, however, become *dwarf-shoots* (Fig. 180), the stems remaining short and bearing leaves close together. The girdles of bud-scale scars show that such dwarf twigs, even though only a centimetre or two in length, may be many years old.

When the bud opens the green leaves appear from *between* the inner bud-scales, and these scales are found to be attached in pairs, each pair with a leaf between them. These scales are clearly *stipules*. They soon fall, leaving narrow scars on either side of the leaf-base. Below them are a number of other scales, without foliage-leaves but also distinctly paired, each pair together encircling the stem and nearly alternating with the pair next to it. These scales also we infer to be stipules, the leaves belonging to them not having developed. Thus in the Beech the bud-scales are not modified leaves as they are in the Horse-chestnut.

The *Lime*, like the Beech, has slender zig-zag twigs; they bear in winter short round reddish buds, in which only two scales are visible—a small outer one on one side, and an inner one which enwraps completely the rest of the bud (Fig. 181). When the bud opens, other scales appear amongst the young foliage-leaves (Fig. 181, II): these scales are in pairs as in the Beech, on either side of foliage-leaves, that is, are stipules. The two outermost scales are borne singly, and therefore do not each represent a stipule, but perhaps a modified leaf or a pair of stipules joined.

On one side of the single bud found at the end of a Lime twig is a leaf-scar, flanked by narrow scars on either side of it left by its stipules: this leaf-scar shows a characteristic crescent of dots representing the vascular bundles. On the opposite side of the bud is another scar, oval, with a complete ring of vascular bundles. This is the scar of a bud; it has no leaf-scar below it, and must therefore have been the *terminal* bud protecting the growing point of the twig. The bud which now ends the twig is situated between the terminal bud-scar and the leaf-scar: it is therefore axillary. When it grows out it will add a further length to the twig. Such a twig

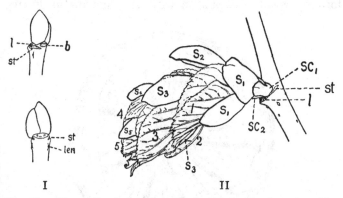

I   II

Fig. 181. Lime. I, a bud at the end of a twig as seen from two sides: in the upper are seen the scar of the terminal bud, *b*, opposite it a leaf-scar, *l*, with the scar of a stipule, *st*; in the lower the leaf-scar and parts of the scars of its two stipules are seen: *len*, lenticel. II, an opening lateral bud: *l* and *st*, scars of the subtending leaf and one of its stipules; sc_1 and sc_2, first and second bud-scales; 1, 2, 3, etc., successive foliage-leaves emerging from between their stipules, s_1, s_2, s_3, etc., which have served as additional bud-scales.

is therefore really built up of a series of lateral branches, each borne in the axil of a leaf near the end of the previous branch. In fact, each growing point of the Lime only forms one year's growth.

In the Beech also the end bud is usually a lateral bud with the scar of its subtending leaf beside it; but the remains of the terminal growing point can scarcely be detected.

Internal structure. In their internal structure the twigs of the Beech and Lime are like the Horse-chestnut, though they are thinner and differ in details.

In a Beech twig, broad medullary rays divide the wood into wedges, while at the point of each wedge a group of primary xylem can be distinguished. Thus the cambium continues the primary medullary ray, by forming a broad band of parenchymatous cells, instead of forming wood and phloëm with very narrow medullary

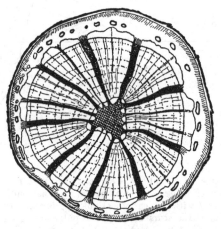

Fig. 182. Transverse section of eight-year-old twig of Beech, treated with iodine, and seen under a lens.

rays, as in the Horse-chestnut. Iodine stains the pith and the rays both in the xylem and in the phloëm dusky or black, showing the presence of starch, and finer medullary rays can also be seen as black lines which do not pass right through to the pith. The annual rings are very thin; growth in thickness is much slower than in the Horse-chestnut.

Just below the periderm is a layer of other tissues containing starch and staining dusky black with iodine.

Before staining, this is deep green in colour; it is therefore assimilating tissue. By stripping off the thin cork layer and holding it up to the light it may be seen to allow some light of a reddish colour to penetrate, which makes some photosynthesis possible. Even in the old bark of the trunk of large Beeches, which long remains smooth, the corky covering is very thin and under it green tissue is found. This tissue has been formed towards the inside by the same cambium that on its outside forms the cork.

If a small piece of rind is cut out and stripped from the wood of a Beech branch, the medullary rays can be plainly seen as longitudinal lines and lozenge-shaped streaks, both on the exposed surface of the wood and on the inner surface of the rind. They are still of two sizes, as in the young twig, though most even of the large ones are secondary. They do not extend far in the vertical direction.

In the Lime the phloëm is strengthened by fibres. It is therefore able to resist stretching, so that as the woody cylinder enlarges the medullary rays yield most, and in a cross-section appear fan-like in the phloëm. The bark of older Limes consists largely of disused phloëm cut off by new and deeper layers of cork; and it is owing to the fibrous nature of this phloëm that the bark shows the numerous narrow longitudinal fissures characteristic of it.

Evergreen trees: Holly, Firs, etc.

The Horse-chestnut, Beech and Lime are all deciduous; they shed all their leaves in autumn. The Holly is an example of an evergreen tree, the leaves of which are able to withstand the severity of winter and remain on the tree for several years. Holly leaves are shiny and leathery, owing to the thickness of their cuticle, which protects them from dry cold winds. The growing points are protected

in winter buds, covered by a few scale-leaves which are
very similar in form to the young leaves. Each year,
winter buds grow out into new shoots, bearing the young
leaves. Each year, too, some of the old leaves fall,
especially during the winter: their scars are visible on
the older parts of the twigs.

The twigs of the Holly long remain shiny and deep
green, covered only by their epidermis, which has a
very thick cuticle. But gradually dark spots and streaks
appear here and there where the epidermis has cracked
and cork has developed. Small patches of deep green,
where there is still no cork, can be seen in stout twigs
several years old. Old Holly bark is smooth and greyish
brown, with dark lenticels.

Most of our evergreen trees, the Pines, Firs, and
Spruces, and the Yew, have small narrow 'needle-leaves,'
though their leaves are usually numerous. They are
covered by a thick cuticle, and the stomata are sunk below
the surface, so retarding somewhat the escape of water
vapour through them even when open.

Shrubs and other woody perennials.

There are many plants with woody perennial stems
above ground which do not grow large enough to be called
trees, but which otherwise have many features both in
their structure and mode of life in common with them.
No sharp line can, in fact, be drawn between trees on
the one hand and shrubs on the other, all kinds of intermediate forms existing, down to low shrubby plants like
Ling and Heather. Among the Willows are some species
which become tall trees, others shrubs, and still others,
the 'Alpine' Willows, on high mountains, with woody
but prostrate stems. There are also scrambling and
climbing plants with woody perennial stems, like the
Briars and Brambles, Honeysuckle and Ivy. The stems

of all such plants grow in thickness, and are protected against evaporation of water in winter; some are evergreen, with tough hardy leaves, others are deciduous.

Some general features of trees.

When a leaf is shed, whether of an evergreen or of a **Leaf-fall.** deciduous tree, in nearly all cases a smooth scar is left, which is covered by a layer of cork continuous with that covering the rest of the twig.

Fig. 183. Leaf-fall in Horse-chestnut showing detachment of petioles from the twig and the scars that are left, also the fall of the leaflets from the petiole. (After Kerner.)

Usually, some time before the leaf would naturally fall, it will come off readily and cleanly at the same point, but if broken at any other point a ragged tear results. This shows that the place where severance occurs is prepared beforehand, a layer of tissue across the base of the leaf becoming weakened; this layer is called the *absciss layer*.

Just below it, cork is formed to protect the tissues lying underneath the surface that will be exposed. The vessels and other woody parts of the vascular bundles, being dead, cannot take part in this preparation of the absciss layer and remain intact, even when the living cells of the absciss layer have completely broken down; so that the leaf remains attached merely by parts of the vascular bundles, until it is broken off by wind or frost. In the Horse-chestnut, Ash and Virginian Creeper an absciss layer is formed at the base of each leaflet, as well as at the base of the leaf-stalk itself, and the leaflets often fall separately.

Twigs and small branches may sometimes be found in great numbers under trees, especially after a strong wind. When the broken ends are examined they are found in many cases to be quite smooth, showing that an absciss layer had been prepared just as for leaf-fall, the wind only finishing what the tree had already begun. Many trees regularly prune themselves in this way: were it not for the removal of large numbers of twigs, in addition to those broken off by strong winds or under the weight of snow, many more branches would be found than actually remain: this is clearly seen if the much branched twigs of a Beech, for instance, be compared with the older branches.

Branch-casting and tree form.

Tree form may also be affected by strong and dry prevailing winds, the buds of the windward side being dried up and development on that side therefore interfered with. Similarly in exposed places trees are unable to grow tall because above a certain level the buds are killed.

Healing of wounds.

When a branch is sawn off a tree, the cambium grows out where it is exposed forming a ring of tissue, called callus, all round near the edge of the circular scar. In this tissue xylem and phloëm and an outer covering of cork develop. This ring of tissue extends inwards over the wound until it meets in the middle and completely covers over the exposed surface of the scar. The cambium unites and becomes continuous, so that later layers of wood cover the wound completely.

Cambium is capable of uniting under other circumstances. Trees of the same kind growing very close together may in expanding come to press against each

Fig. 184. A wind-blown tree.

other; the whole rind may in consequence be split by the pressure, and their cambiums be brought into contact. The new phloëm and xylem formed by the two trees is then continuous around both and in this way the two trunks grow into a single trunk. Branches which cross each other may unite in a similar way.

The successful grafting of one plant upon another depends upon bringing the cambium of the graft in contact with that of the stock. One of the simplest and best known cases is the budding of roses. A bud of the rose which it is desired to propagate is removed, along with a three-cornered patch of rind separated at the cambium from the wood. A T-shaped slit is made in the rind of the briar stem on which the bud is to be grafted, the rind being lifted so that the bud can be inserted with its cambium in contact with that of the briar. When this has been done the whole is bound up tightly until fusion and healing are complete.

It is usually easy to separate the phloëm from the **Ringing** xylem in trees, where much secondary **experi-** growth has occurred. For this reason it is **ments.** possible to obtain evidence by experiments with trees that the phloëm conducts the food-material from the leaves, a fact of which we have so far had no proof.

If a complete ring of rind be removed from the trunk of a tree, rings of callus begin to grow out over the exposed wood from the cambium at the two cut edges of the rind; at the lower edge this tissue soon stops growing, but from the upper edge growth continues as long as the tree lives. The food-material coming from the leaves thus reaches the upper cut edge of the rind, but cannot cross through the wood to the lower edge; so that there the growth of tissue to cover the wound ceases as soon as the reserve food in the lower part of the tree is exhausted. Moreover, unless the upper callus ring soon covers the exposed wood and so connects the phloëm across it again, the tree dies because its roots are starved. On the other hand, if only the outer tissues are removed, and the phloëm is left, the tree does not die. It is, therefore, the phloëm which is essential for the translocation of the food supplies.

This fact may be illustrated by a simpler and less wasteful experiment. A ring of rind is removed from near **Exp. 60.** the base of a cutting of Willow or Rose, which is then planted so that the soil covers the exposed

surface. The roots which it puts out are found to come
from callus which grows from the *upper* edge of the ring.
This is as far as the food supplies from above can reach,
because here the phloëm that conducts it is interrupted.

**Heart
and sap
wood.**
If a large leafy branch be placed with its cut end in
water coloured with eosin (or red ink) the
water is found to ascend chiefly in the *outer*
layers of the wood. The cut surface of
the stump of an Oak, a Birch or other tree yielding

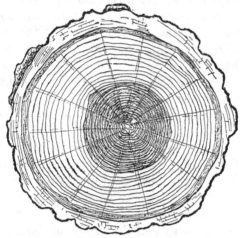

Fig. 185. Cross-section of the trunk of an Oak, showing the dark
heart-wood in the centre, and the lighter sap-wood; also the
annual rings, phloëm, and thick bark. (After Gregson.)

useful timber, shows that the inner, old wood becomes
harder and darker in colour than the outer layers. This
heart-wood no longer conducts water; the vessels have
been blocked up and the whole mass hardened, and it
now serves merely a mechanical function. It is only the
heart-wood which makes durable timber. The outer
wood, which still conducts water, is called the *sap-wood*:
it is softer and moister, and of no value as timber. In

some trees (e.g. Horse-chestnut) the heart-wood remains
soft.   In the Willow it rots away leaving the tree hollow.

Planks cut from the trunk of a tree show various
markings, the character of which varies much
with the kind of tree.   As a rule the principal
kinds of 'graining' are due to annual rings
and medullary rays.   The appearance which
they give to the cut surface differs according to the
direction in which the wood is cut.   This is readily
understood if we consider their distribution in the trunk.

The markings of timber.

If the trunk is sawn longitudinally through the middle
the cut passes in the same direction as the medullary
rays but across the rings; the smooth surface will therefore be likely to show the medullary rays as streaks
(usually lighter in colour, often silvery) running transversely (across the grain); while the annual rings will
form markings running longitudinally (with the grain)
and close together.

If on the other hand the longitudinal cut does not pass
through the centre of the trunk but some distance away
from it, it will pass obliquely through the annual rings,
which will therefore appear as much wider markings;
while the medullary rays will be cut at right angles to
their greater length down the middle of the section, but
obliquely towards its edges, and their appearance will
vary in consequence from short narrow vertical lines to
broader streaks approaching in appearance those seen in
the section through the middle of the trunk.

Smaller markings or mottlings depend on the distribution of fibres and wood parenchyma in groups amongst
the water conducting elements; while in Oak the vessels
of the spring wood are so large that they are readily seen
with the naked eye and appear as dark streaks or dots.

Here and there in a plank (of deal, for instance)
knots usually occur.   These are portions of small cylinders

of wood that run through the main mass across the annual rings and have annual rings of their own around a central pith. Such knots are, in fact, the wood of branches: the cambium of the branch is continuous with that of the main trunk and the new wood is laid down in branch and trunk together; the annual rings in the trunk tend to bulge out round the knot. Tiny knots occur, which represent the wood going out to supply dormant

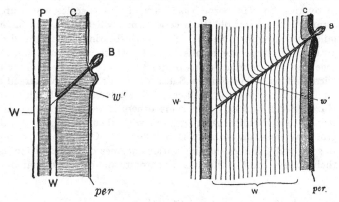

Fig. 186. Diagrams of longitudinal sections through a branch and its dormant bud, in the first and twentieth years respectively. *B*, the bud; *W*, the wood; *w'*, connexion of bud with wood; *per*, bark; *C*, phloëm and cortex; *P*, pith. (After Ward.)

buds (Fig. 186). Where branching is frequent the grain becomes very irregular, and very beautiful markings result, such as in 'briar' wood, from a shrubby Heather of the south of France, which is used for making tobacco pipes.

Notes on Common Trees.

In studying trees, all the points already mentioned in connexion with the Horse-chestnut and other examples should be examined, to see in what respects each treé agrees with or differs from others. In addition to the form and arrangement of leaves and buds, nature of the bud-scales, twig characters, bark, etc., the general form of the tree should be observed. This differs under different conditions, but is often characteristic of the species. On it depends largely the effective exposure of the foliage to the light; it will be found that the size and arrangement of the leaves on the twigs, the disposition of branches and the general form of the tree are connected with and suited to each other.

The trees which are common in the British Isles belong to a number of different Families. The needle-leaved evergreen trees, the Pines, etc., do not bear flowers like those of the Dicotyledons and Monocotyledons: their ovules are not enclosed in ovaries, but are borne *naked*, usually upon scales which are arranged in *cones*: old Pine cones from which the seeds have fallen are familiar objects. The pollen-bearing scales are also arranged in separate smaller cones. These trees are classed together as Coniferae (cone-bearers). Our other trees are all Dicotyledons. The Family Salicaceae, which we have already studied, comprises the Willows and Poplars. The Birch and Hazel, Beech and Oak and some others have points of resemblance to the Salicaceae and form with them a large natural group of trees, all with unisexual flowers. Other trees belong to scattered Families showing various degrees of specialisation in their flowers, most of them hermaphrodite and insect-pollinated.

Conifers.

The Scots Pine (*Pinus sylvestris*) has its long needle-leaves borne in pairs on short shoots, sheathed at their base by several membranous scales. In the terminal bud young short shoots can be found in the axils of the numerous bud-scales. When the bud expands the scales fall, leaving only a woody decurrent base. The outer scales have no axillary short shoots, while in the axils of those just around the new terminal bud a circle of similar lateral buds are formed, which later grow out into branches. The Scots Pine has usually a tall bare trunk capped by a spreading mass of foliage at the top; but in the open its lower branches often persist for a long time. The bark scales off, at least from the branches and upper part of the trunk, leaving a characteristic orange-red surface.

The male flowers or cones are borne in clusters on shoots which later grow on beyond them and bear foliage. A cone consists of a number of stamens packed together spirally on a

central axis, each bearing two large pollen-sacs on its under side.
These split open longitudinally in early summer, and shed large
quantities of light dusty pollen: each grain, seen under the
microscope, has two wings, which increase its surface and make
it still more readily buoyed up by the air and carried by the
wind.   The female cones, which are borne laterally near the
ends of the twigs, appear from the outside to consist of thick
woody scales spirally arranged.   On dissecting out one of these,
another thin short scale is found beneath it, and two ovules above

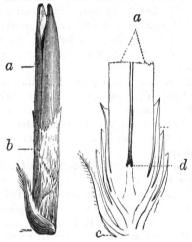

Fig. 187. Dwarf-shoot of Pine, just expanding, to the right in longitudinal section and more highly magnified: *a*, the two needles
face to face; *b*, the sheathing scale-leaves; *c*, the scale-leaf in the
axil of which the dwarf-shoot stands; *d*, the stem-apex of the
dwarf-shoot. (After Willkomm.)

it near the axis. The Scots Pine takes three years to produce ripe
seeds. In the first year the scales separate so that wind can blow
pollen between them to the micropyle of the ovules where they are
caught by a drop of sticky fluid. The scales then close together
again, opening two years later to set free the winged seeds.

A seedling Pine has several cotyledons and the stem does
not at first bear membranous scale-leaves but green needle-leaves
with no short shoots in their axils. This is like the mature
structure in the Silver Fir and other Conifers.

The Silver Fir (*Abies pectinata*) and the Spruce (*Picea excelsa*) are similar to each other in many respects, having numerous needle-leaves borne directly on the twigs and a very regular pyramidal form with branches in whorls. They have no short shoots like those of the Pine. They can be distinguished by their twigs and their female cones. The Silver Fir has flattened leaves, with short stalks which leave clean round scars on the smooth stem. The leaves of the Spruce are not flattened, spread more uniformly on all sides of the twig and have decurrent bases.

The female cones are also different: that of the Spruce grows pendant, and is like that of the Scots Pine in structure, but with thinner scales. The cone of the Silver Fir grows erect, and the narrow lower scales protrude from between the broader scales that bear the ovules: the scales fall from the ripe cone, leaving the central axis.

The original terminal growing point is still found at the top of many a well-grown Fir or Spruce: hence their tall straight trunks, used for masts and poles. The terminal winter-bud is closely surrounded by a cluster of other large buds; the branches which arise from these therefore appear whorled, and the trees are conical in form.

The Larch (*Larix europaea*) is a deciduous Conifer. Its small needle-leaves are borne in clusters on short shoots. The female cones are small, with thin ovule-bearing scales, red when young.

The Yew (*Taxus baccata*) has dense dark foliage and usually a rounded form. The leaves are flattened, and turn over to either side of the twig so that they are all nearly in one plane. It has small male cones but the ovules are borne singly, each in a tiny axillary bud, covered only by a few scales. After fertilisation a fleshy cup grows up around the seed-coat, forming the red or yellow juicy part of the Yew 'berry.' The seedling, unlike those of the Pine, Larch, Fir, and Spruce, has only two cotyledons.

Dicotyledons.

The Willows and Poplars are trees or shrubs with simple leaves, and bearing male and female catkins on different plants: each bract (p. 371) has a simple flower in its axil. The foliage is not dense and the trees are usually found in open situations on deep moist soil, commonly on the banks of rivers.

Sali-caceae.

Most *Willows* have pale green lanceolate or ovate leaves on short stalks, with small stipules that soon fall. The buds are alternate, small and narrow, covered by a single outer scale; they are usually pressed close to the stem (Fig. 188) and grow either into long twigs or into short shoots ending in catkins. There are numerous species: the common White Willow, *S. alba*, has narrow leaves and catkins. *S. Capraea*, the Goat Willow, or Sallow, has ovate leaves, white with short close down below and wrinkled above, and stout, nearly sessile catkins.

Willows are pollarded in some parts of the country, i.e. cut back periodically about ten feet above the ground; pollard willows have a characteristic form, a large number of straight branches arising near together from the top of the trunk.

The *Poplars* have broad, usually ovate and slightly lobed, leaves on long thin stalks which are mostly flattened at right angles to the blade so that the leaves flutter in the wind. This fluttering movement is especially characteristic of the Aspen (*Populus tremula*), the rather small round leaves of which, borne on very thin stalks, are seldom still. The twigs of Poplars tend upwards at the end and bear the leaves loosely all round them. The buds are covered by many overlapping scale-leaves.

The White Poplar (*P. alba*) is a tall tree with broad ovate leaves with

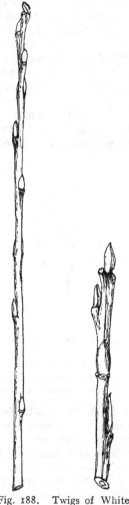

Fig. 188. Twigs of White Willow. (From Ward, after Dawson.)

a close white cottony covering of hairs underneath. The Black Poplar is the best known as it is most frequently planted. Its leaves are roughly triangular, green below as well as above. The Lombardy Poplar is possibly a variety of it with upturned branches, giving the whole tree a tall narrow 'pyramidal' form.

The Birch, Hazel, Alder, Hornbeam, Beech, Oak, Walnut and

Other catkin-bearing trees.
Chestnut all have at least the male inflorescence more or less catkin-like. Both male and female catkins are borne on the same plant. The leaves, which are simple except in the Walnut, are alternate, and the bud-scales are stipules.

The *Birch* has small, toothed leaves loosely arranged on thin more or less drooping twigs. The light foliage and thin straight trunk covered with white or pinkish papery bark give it a delicate

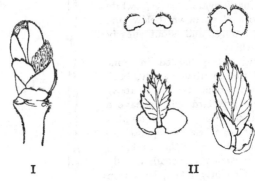

I   II

Fig. 189. Hazel. I, opening bud, with leaves emerging from between the stipular bud-scales; below the bud are shown, on the left, the scar of the leaf that subtended it, with the scar of a stipule; and, on the right, the scar of the terminal bud. II, parts dissected from four successive nodes of the bud.

and graceful aspect. It requires a great deal of light and is not found among trees with dense foliage. The male catkins are formed in the autumn; the female catkins appear in early spring from winter buds, and the male catkins open at the same time and shed their pollen. After fertilisation the scales become tough and the ovaries develop into small winged achenes (p. 232). There are two common species of Birch: *Betula alba*, with smooth, thin twigs, many prominent lenticels and pointed doubly serrate leaves, and *B. pubescens*, with downy twigs and leaves. *B. nana* is a low shrub, found on Scotch mountains.

The *Hazel*, commonly found in hedges or in the undergrowth

of open woods such as Oak woods, is a shrub or small bushy tree, with obovate pointed, doubly serrate, hairy leaves, rather like large Elm leaves, on pendulous twigs with glandular hairs. The bark of the branches is shining, red brown, with numerous transverse lenticels: the older bark becomes scaly. The bud is covered by a few pairs of stipular bud-scales (Fig. 189) within which are other stipular scales with young leaves. The male catkins are similar to those of the Birch, the female inflorescence is like an ordinary winter bud (p. 200). The ovary develops after fertilisation into a nut—the hazel-nut—surrounded by a green lobed *cupule* formed by fused bracts.

The *Alder* is a tree which grows chiefly in moist places, like

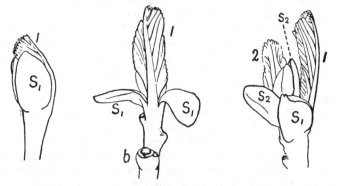

Fig. 190. Opening buds of Alder: 1, 2, first and second leaves; s_1, s_1, stipules of first leaf; s_2, s_2, stipules of second leaf; b, leaf-base and leaf-scar, with the scar of a stipule visible on one side.

the margins of streams. The leaves are broad and blunt, tapering a little at the base, and nearly glabrous. The young twigs and leaves are sticky. The bud is shortly stalked and covered by the stipules of the lowest young foliage-leaf. The whole bud is thus enclosed in the stipules of its oldest leaf and each leaf within is protected by its own stipules. The bracts of the female catkin form after fertilisation thick woody scales, and the whole catkin then looks very like a small black cone. Clusters of the old female catkins usually remain attached to the trees in winter and are a ready means of identifying the tree. The male and female catkins are both formed in the autumn ready to open in the early spring.

The *Hornbeam* is found in hedges and woods. It is rather like the Beech in habit; but it seldom reaches a great size. Its buds are shorter than those of the Beech and its leaves doubly serrate, not shiny, and nearly glabrous, never fringed with hairs like Beech leaves. The catkins are not formed till the spring. The female catkins appear with the leaves and are thin and loose; the bracts develop into large leafy appendages, each clasping in its base a small nut, for which it serves as a wing.

We have already studied the general features of the *Beech*, its smooth grey bark, long tapering buds, and characteristic dense foliage. The leaves are ovate and entire, fringed with hairs when young, their upper surface shiny. The inflorescences appear with the leaves. The male flowers are borne in silky clusters at the ends of long stalks; the female flowers are borne in pairs on short thick stalks. In each female flower is a three-angled ovary which develops into a nut; two flowers together are surrounded by a cupule which grows woody and spiny around the ripe nuts, and splits into four valves to set them free.

The wavy outline of an *Oak* leaf is well known. The gnarled and bent branches are equally characteristic, also its spreading habit; the buds are round, showing numerous scales, and are clustered round the tips of the twigs. The bark is rough and deeply furrowed. The inflorescences appear with the leaves. The loose male catkins are slender and drooping and arise in clusters from lateral buds. The female flowers contain only an ovary, the base of which is covered by a number of scales which fuse later and become the woody cup of the acorn. There are two British Oaks. *Quercus Robur*, the pedunculate or Common Oak, usually bears two or three acorns on a long stalk or peduncle: its leaves are nearly sessile, with the base of the blade turned back on either side of the midrib. *Q. sessiliflora* has usually almost sessile acorns; its leaves on the other hand have distinct stalks, and white stellate hairs on the under side in the angles between veins and midrib.

The *Sweet Chestnut* and *Walnut* have been planted in some parts of the country. The former has large lanceolate or oval sharply toothed leaves, and spikes bearing clusters of flowers at intervals, the lowest female, the upper male. The cupule that surrounds each female cluster becomes spiny and woody, like that of the Beech, though larger; and splits similarly into four valves to allow the three nuts to escape. The Walnut has pinnate leaves, dense drooping male catkins arising from lateral buds, and sessile female flowers at the ends of the twigs. The ovary wall in ripening forms an outer green fleshy layer, and an inner woody layer, the shell of the nut, which encloses the single seed.

Of our other commoner trees, the Elms and the Lime have alternate leaves, the Sycamore and Maple, the Ash, and the Horse-chestnut, have opposite leaves. The Sycamore and Horse-chestnut are not native, nor is the Lime except perhaps in the South.

Elms have rough ovate pointed irregularly toothed leaves, with one side larger than the other at the base (see Fig. 192). The bark is furrowed. The buds are small, covered by several

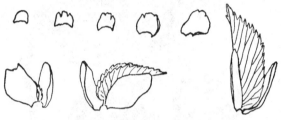

Fig. 191.   Elm: parts dissected from successive nodes of an expanding bud; some of the lower scales have been split at the top in the expansion of the bud.

bud-scales, the outer ones single, the inner ones in pairs on each side of the young leaves (Fig. 191): occasionally, between the single scales and the paired stipules, a two-lobed scale is found, suggesting that each of the lower scales corresponds to a pair of joined stipules.

The dense clusters of reddish hermaphrodite flowers appear before the leaves early in spring. The fruit is an achene surrounded by a thin wing.

There are two chief kinds of Elm in the British Isles, the English Elm and the Wych Elm. The Wych Elm is commoner in the north, the English Elm in the south. The former has usually a more spreading habit and larger leaves than the English Elm, and its fruits are larger, the wing slightly notched at the top. The fruit of the English Elm is smaller and deeply notched, nearly to the seed-chamber. It usually grows tall rather than spreading, and often shows a characteristic distribution of the foliage in two masses, the upper large and rounded, separated by a gap from the more spreading lower mass. Its fruits are said not to germinate in Britain and it is probably not a native; but it puts up numerous suckers from its roots and so propagates itself. It occurs most commonly in hedgerows, and unlike the Wych Elm is seldom found in woods.

Fig. 192. Twigs of the English Elm, *Ulmus campestris*: 1, flowering twig; 2, twig of preceding year, with tuft of winged fruits and a dwarf-shoot bearing leaves; 3, a twig in its winter condition. (After Willkomm.)

The leaves of the *Lime* are thin and broad, heart-shaped at the base, with toothed edges and a tapering drip-tip. The old bark is blackish with narrow fissures. (For twig characters, etc., see p. 396). The pale yellow flowers hang in small clusters from long stalks, joined for part of their length to narrow yellowish bracts, which later serve as wings to the clusters of globular fruits. The fragrance of the flowers and their abundant nectar are very attractive to bees.

Another tree with alternate leaves is the *Plane*, which is often planted, especially in towns. Its bark is smooth and from it scales are cast annually, leaving light-coloured patches which gradually darken to an olive grey. The leaves are palmately veined with pointed lobes, on long stalks, with bases which completely envelop the buds, fitting over them like extinguishers: when quite young the leaf-base grows up around the bud and encloses it except for a small hole on the upper side. The fruits are in globular almost prickly heads.

The *Sycamore*, or more correctly, Sycamore Maple, is one of

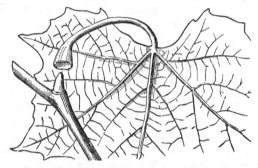

Fig. 193. Leaf of Plane, *Platanus*, detached to show the bud which was hidden in the base of the petiole, and the circular leaf-scar left round the base of the bud (from Ward).

our most commonly planted trees, though it is not native. It is readily distinguished by its rather large opposite oval buds on stout, smooth twigs, some of which have a pair of lateral buds at the end, where a terminal inflorescence was produced; by its large palmately lobed opposite leaves, with toothed pointed lobes; and by its fairly smooth, lavender-grey bark, which becomes darker and scaly when very old. The greenish flowers are borne in spring in drooping racemes on long stalks. Each flower has usually five sepals, five petals, and about eight stamens on a thick honey-secreting disc below a two-chambered ovary. The two wings of the fruit are inclined to one another at about a right angle.

The *Hedge Maple* is smaller, a native of Britain, with small dark green leaves with rounded lobes, not toothed. The old twigs often bear prominent ridges of cork. The flowers are borne in small erect branched inflorescences. The wings of the fruit are in a line with each other. The *Norway Maple*, often planted, has similar but larger inflorescences, and large fruits with the wings relatively short and inclined to each other; and its leaves have few large and pointed teeth.

The *Ash* (*Fraxinus excelsior*) has long opposite pinnate leaves with toothed pointed leaflets, on stout grey-green upturned twigs flattened below the nodes. The squat winter buds are sooty black, and bluntly pyramidal. The branched inflorescences appear from lateral buds before the leaves, and are at first reddish, later dark purple. The flowers are male, female, or hermaphrodite. The hermaphrodite flowers have a two-chambered ovary with two spreading stigmas and two stamens. The male flowers consist simply of two stamens: the female flowers have a minute perianth round the single ovary. The fruits are one-seeded and winged, hanging in bunches and known as Ash 'keys.'

Other species of Ash, e.g. *Fraxinus Ornus*, the Manna Ash of South Europe, have flowers with calyx and corolla. These show that the Ashes belong to the same Family (Oleaceae) as the Privet and Lilac, which have flowers with a tubular corolla, two epipetalous stamens and a two-chambered ovary. The *Privet* is an evergreen shrub, native in southern England, with white flowers and blue-black berries. The *Lilac* is a deciduous shrub, not a native, but largely cultivated. It has opposite entire heart-shaped leaves, tapering to a point, pale green, thin, and smooth. Its buds resemble those of the Sycamore but the bud-scales are only slightly modified leaves. The terminal bud regularly dies, a pair of lateral buds on either side of it taking its place.

The *Horse-chestnut* (p. 387) is readily distinguished by its very stout, long and strongly upturned twigs, with large shield-shaped scars and, in winter, large sticky buds,—in summer, large palmate leaves: when young the leaflets hang downwards, in this way protecting the lower surface from the wind. The brightly coloured flowers borne in the pyramidal inflorescence are zygomorphic, highly-specialised bee-flowers (Fig. 70, p. 206).

In addition to the trees already mentioned are a number of smaller trees and shrubs, most of them not so common or so widely distributed, including a number belonging to the Rosaceae, with succulent fruits. Most of our orchard trees are cultivated varieties of these. They can be best distinguished by their flowers and fruits (see p. 318).

The Mountain Ash (*Pyrus Aucuparia*), with pinnate leaves and clusters of red berries (p. 379, Fig. 173), belongs to the same genus as the Apple and Pear. So also do the Bcam-tree (*Pyrus Aria*), and the wild Service-tree (*P. torminalis*), as well as the

Fig. 194. Blackthorn.

Crab-Apple, the wild species of which our cultivated apples are varieties. Nearly allied to this genus is the common Hawthorn. The Cherry and Plum belong to the genus *Prunus*. *Prunus spinosa* is the Blackthorn or Sloe of our hedges; its clusters of white flowers appear on the thorny twigs before the leaves unfold (Fig. 194). *Prunus Padus*, the Bird Cherry, has long drooping racemes of white flowers, and black bitter fruits.

The Elder is a common shrub or small tree with opposite pinnate leaves and brittle twigs with a large pith. The buds survive the winter with no special covering, and the lenticels remain fully open. The large inflorescences bear numerous small white flowers at one level, so that they are very conspicuous. The corolla has a short tube with five teeth, there are five stamens, and the 3–5-celled inferior ovary ripens to a black berry. The Elder belongs to the same Family as the Honeysuckle, which is a woody climber, and the Guelder-rose, *Viburnum Opulus*, with simple lobed leaves, and Wayfaring-tree, *V. Lantana*, both of which have inflorescences of white flowers resembling those of the Elder. In the Guelder-rose the outer flowers are large and barren.

The *Holly* (p. 399) is an evergreen with spiny leaves. Its white flowers appear in the summer in axillary tufts, and the fruit ripens to the familiar red berry, in which the seeds are enclosed in separate woody shells.

CHAPTER XXIII

CLIMBING PLANTS

Another group of plants which exhibit a special habit of growth are climbing plants. These make use of neighbouring plants or other supports by which to raise themselves from amongst the surrounding vegetation. They are especially characteristic of tropical forests, where they develop strong rope-like or woody stems which cling to the trees and intertwine, forming almost impenetrable thickets: these *lianes*, as they are called, are able to reach the light in thick forests, into which so little light penetrates that the ground is almost bare of plants. In our own country, climbers are met with in woods and hedgerows and in other places where plants grow tall and thickly together. They vary greatly in their methods of climbing. Some twine round their support, others attach themselves by climbing organs of various kinds. But in comparison with plants which reach an equal height without help they all have thin rapidly growing stems with long internodes, and so spend far less building material in raising their leaves, flowers and fruits to the light.

The Hop is a perennial which puts out vigorous **Twining** annual twining shoots with long internodes. **plants.** After growing upright for a short distance these shoots bend over and begin to describe a circle in the air, pointing successively east, south, west and north;

they thus follow the same direction as the sun[1], but
complete a circle in two or three hours. Vigorous shoots
may sweep a circle some feet in diameter in search of
support. This circling movement is called *circumnutation*.

Careful observation and measurement (by the method

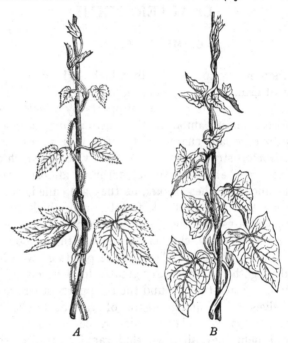

Fig. 195. *A*, Hop twining with the sun; *B*, Convolvulus
twining against the sun. (After Payer.)

of Exp. 47, p. 265) show that it is the rapidly growing
part of the stem which performs this exploring movement; the bent region is always about the same distance

[1] This is the direction in which the hands of a watch, placed face
upwards, move, and is called the clockwise direction; the opposite
direction is called counter-clockwise.

from the tip of the stem as it grows in length. When the stem comes against a support only the part beyond the support is free to continue the movement, with the result that it coils round the support. Its roughness, due to small prickles, helps it to hold on.

The Scarlet Runner Bean (*Phaseolus multiflorus*) is another twining plant; it is not a native of the British Isles, but is commonly cultivated, and it can easily be grown in pots for experimental study.

The direction of circumnutation can be observed and its rate
Exp. 61   measured for a plant in which the stem has begun to bend over, by standing the pot on a piece of paper. The base of the stem is fastened to a stick stuck in the soil of the pot, and the direction of the free upper end recorded at intervals by lines drawn immediately underneath it, the time being noted against each. The records may be made more accurately with the aid of a plumb-line. It will be found that the Runner Bean, unlike the Hop, moves and twines against the sun, in the counter-clock-wise direction. By repeating the experiment under different conditions it may be shown how temperature, moisture, etc., which affect the rate of ordinary growth, affect this growth movement.

If a plant be slowly revolved in a horizontal position (on a klinostat) the shoot no longer twines, but even loosens the younger coils already formed round its support. This shows that the twining movement is a special kind of response to gravity. Another experiment which shows this is very simple. A plant which has made a few coils round a stick is turned upside down;
Exp. 62   the change of position reverses the coiling move-ment, for the youngest two or three coils loosen themselves from the stick; the stem then begins to twine upwards towards the pot. The bending of the young stem of a twining plant into the horizontal position when it begins to twine is brought about by the more rapid growth of the convex side. In the circling move-ment that follows, different sides become convex in succession: that is, rapid growth takes place on different sides in succession, passing round the stem in the direction in which it twines. In whatever position the stem may be, it is stimulated by the force of gravity to change its direction of growth, whereas for the

stems of other plants there is a position of rest in which they are not stimulated.

Common British twining plants. The Honeysuckle (*Lonicera Periclymenum*), found in woods and hedges, has perennial twining stems which become woody: its leaves are opposite and its flowers (p. 214) are borne in cymose heads. The fruit is a red berry.

The Woody Nightshade (*Solanum Dulcamara*) is also woody at the base, sending out each year long twining shoots which die back a long way in the winter. The flowers are like those of the Potato: the blue corolla has a short tube the mouth of which is closed by the five anthers which project as a yellow cone.

All the other British twiners are annuals or have only an underground perennial part, sending up annual shoots.

The two Bindweeds—the large White Convolvulus of our hedges (*Calystegia sepium*) and the Field Bindweed (*Convolvulus arvensis*) with smaller pink or white sweetly scented flowers—have creeping rhizomes. The annual shoots of the Field Bindweed are as often found creeping on the ground as climbing on other plants; unlike most twining stems they can coil round a horizontal support.

The Dodder (*Cuscuta europaea* and other species), allied to these Bindweeds (Family Convolvulaceae), not only twines round other plants but puts suckers into their tissues and absorbs from them the whole of its nourishment. It has no green leaves, only small scales, and the whole plant is whitish in colour. The seedling stem like those of other twiners sweeps round in search of suitable support, but it is the actual contact which stimulates it to twine round its host, as well as to put out suckers. The stem of the Dodder is thus, like tendrils, sensitive to contact.

The Black Bindweed (*Polygonum Convolvulus*) is an annual, resembling *Convolvulus arvensis* in appearance, but distinguished from it by its wind-pollinated flowers and the peculiar membranous stipule which sheathes the stem at each node.

Among the Monocotyledons the only British climbing plant is the Black Bryony (*Tamus communis*). From an underground tuber long slender green twining stems grow up, bearing shiny heart-shaped leaves which, unlike so many Monocotyledons, are net-veined. It has small greenish unisexual flowers in separate drooping racemes: there is a six-lobed perianth, with six stamens in the male flowers and an inferior three-chambered ovary in the female flowers (cf. Amaryllidaceae).

Other climbing plants bear organs sensitive to con-
Tendril   tact by which they fasten themselves to
climbers.   suitable supports.

The *leaf-stalks* of the Traveller's Joy (*Clematis Vitalba*)
curl round twigs and branches with which they come in
contact, and then grow thick and strong.   The climbing
'Nasturtiums' (*Tropaeolum*) of the garden and greenhouse
also climb by means of their leaf-stalks, which curl round
the support and then bring the leaf-blade face to the
light.   In the Rampant Fumitory (*F. capreolata*) the
blades of the leaves are sensitive.   They are divided into
many narrow segments any of which may coil if stimulated by contact with a support.

In most cases, however, the sensitive organs are thin
highly specialised *tendrils.*

In the Sweet Pea (*Lathyrus odorata*) branched tendrils
form the upper part of compound leaves, which bear
below the tendrils a single pair of leaflets and have two
small ear-like stipules at the base.   The way in which the
branches of the tendril occur in pairs suggests that these
correspond to other leaflets.   The stem is expanded into
green wings, whereby the surface for photosynthesis is
increased.   The young tendrils are nearly straight;
older ones have their branches coiled and twisted around
each other or round a support.

The sensitiveness of the young tendrils may be shown
     Exp. 63   experimentally by stroking them very care-
fully with a pencil or a piece of stick.   A rough twig is
most effective, whereas stroking with a glass rod covered
with thin gelatine, which is perfectly smooth, does not
stimulate them at all; nor has dropping water (e.g., rain)
any effect.

In nature, plants are seldom still, and when a tendril
meets a twig the continual friction soon results in the
curvature of the tendril into a hook which holds on to the

twig; in this way, too, the tendril is brought into closer contact whereby the friction and therefore the stimulation are increased.   Careful rubbing of a tendril in one spot will

Exp. 64   show that it curves only in the immediate neighbourhood of the stimulated part.   It follows from this that when the tendril has hooked on to the support it curves where it is in contact and so brings successive portions of the tendril against the twig to be stimulated and continue the coiling.   In this way the tendril coils *closely* round the stem; a little consideration will make clear that if it went on coiling from the beginning along its whole length it could not readily draw the coil tight afterwards round a rough stick.

Like the Sweet Pea are the Everlasting Pea of our gardens (*Lathyrus latifolius*) and the very similar wild Everlasting Pea, the annual green winged stems of which arise from a creeping rhizome; also the Meadow Pea (*L. pratensis*), another perennial, with racemes of yellow flowers.   All these have one pair of leaflets below the branched tendril.   The leaves of the common Tufted Vetch (*Vicia Cracca*), found in hedges, the less common Wood Vetch (*Vicia sylvatica*) and the Garden Pea (*Pisum sativum*) have several pairs of leaflets below the tendril.

In the Yellow Vetchling (*Lathyrus Aphaca*), a small weak annual with yellow, mostly solitary flowers, the whole leaf develops as a slender simple or slightly branched tendril, and the work of photosynthesis is performed by large broad leafy stipules which look at first sight like opposite leaves.

In the Passion-flower (often grown in greenhouses) the tendrils occur in the axils of the leaves and are therefore lateral branches: at their base they bear a flower-bud which often appears to spring from the same leaf axil.   These tendrils are extremely sensitive, responding to the slightest friction, so long as they are quite turgid[1].

The branched tendrils of the Vine occur opposite the leaves. It would appear at first sight that each tendril might be one of a

---

[1] This caution applies, of course, to all experiments with tendrils: on cut branches they often appear quite insensitive because their growth is stopped by lack of turgor.

pair of opposite leaves, but about every third leaf has no tendril opposite to it; besides, this interpretation would not fit the inflorescences, which are borne in similar positions to the tendrils. An inflorescence must be a stem structure; and as no trace of a leaf can be found below it, it is terminal not axillary. There-

Fig. 196   Shoot of Garden Pea, showing compound leaf with tendrils.
(After Baillon.)

fore the stem arising between it and the leaf must be an axillary branch, although it is apparently a continuation of the main stem. This interpretation will fit the tendrils equally well. The main stem ends in a tendril or an inflorescence, and the bud in the axil of the last leaf grows out, following the direction of the internode below; in this way the main axis is made up of a number of successive branches (compare the structure of cymose

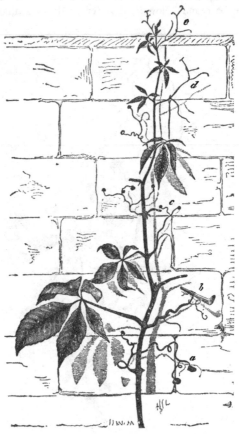

Fig. 197. Shoot of Virginian Creeper, with branch tendrils, some developing adhesive discs. (After Sachs.)

inflorescences, and the mode of growth of the branches of trees in which the terminal bud dies).

The tendrils of the Virginian Creeper (*Ampelopsis Veitchii*), which is a near ally of the Vine, arise in the same way. They

differ from other tendrils as they turn away from the light and
fix their tips into cracks and roughnesses on walls or the bark of
trees. Their sensitiveness to contact is shown by the develop·
ment of sucker-like swellings at the tip by which they make
themselves fast.

The White Bryony (*Bryonia dioica*) which grows wild in most
parts of England has long tendrils which arise beside the leaves,
and the nature of which it is difficult to determine. The lower
part of the tendril is stout and stiff, the upper part more slender
and sensitive to contact. Similar tendrils occur in other members
of the same Family (Cucurbitaceae) such as the Cucumber, Vegetable Marrow and Pumpkin. In the latter, abnormal forms often
occur intermediate between the simple slender tendril and a long
leafy structure; it is probable therefore that the stiff base is a
stem, really axillary though it has become displaced, bearing a
leaf tendril.

The tendrils of the White Bryony are long, simple and
rather stouter than in most other plants: they are very suitable
for experiment, not requiring as skilful handling as the smaller
and more delicate tendrils of Peas and Vetches. As in the
other Cucurbitaceae just mentioned, the tendrils are peculiar
in being closely coiled up underneath when they first appear.
They then gradually uncoil and, when they have straightened,
move round in search of a support: this movement is quite
spontaneous, and not dependent upon gravitation like the circumnutation of twining stems. Curvature follows only when the
tendril is stroked underneath and is confined to the near
neighbourhood of the part which is stroked.

Old tendrils which have clasped a support are coiled into a
spring between the stem and the support, and have become
tough and strong. By coiling in this way the tendrils draw the
plant closer to the supporting twigs and fix it more firmly; they
also form springs which yield when strong winds might otherwise
break the plant from its hold or damage it. It will be observed
that there is a point or sometimes three points at which the
direction of coiling is reversed—this is because both ends were
fixed before coiling began.

The Ivy is a root climber. On the side of the stem

Root climbers. next the support, whether wall or tree-trunk, are found numerous small adventitious roots. These arise only on the side away from the light and make their way into cracks and crevices. Thus they are negatively heliotropic and are not geotropic like ordinary roots. That they *are* roots can be seen by cutting the stems across, especially at levels where young roots are forming: they clearly arise from within the tissues, i.e., endogenously.

The Ivy is an evergreen; its stem becomes strong and woody. When it attaches itself to trees the dense shade cast by its leaves sooner or later interferes with the growth of the tree; for, as shaded buds remain dormant, few branches are formed and the tree comes to look thin and gaunt.

Another method of climbing is used by plants like the

Scramblers with hooks. Bramble, the stems of which scramble through hedges and thickets, hooking themselves by recurved prickles. It is easy to pull a cut bramble out of a hedge tip foremost but very difficult backwards against the prickles. Thus the prickles do not interfere with the forward growth of the stem, especially as they are still young and flexible on the rapidly growing part of the stem, but they prevent it from slipping back.

The prickles are merely surface structures, having no vascular supply, so that it is not difficult to break them from the stem; they are therefore classed with hairs and glands as *emergences*. Many Roses have similar prickles. A scrambling hook-climber with a very different appearance is Goose-grass or Cleavers (*Galium Aparine*), an ally of the Bedstraws (*Galium verum* is the common Yellow Bedstraw). It is an annual with thin green stems and whorls of small leaves which may be found in spring growing straight up into a hedge. The whole plant is very rough to the touch, particularly when drawn through the hand backwards.

On examining it with a lens small stiff hairs curved downwards can be seen. These act like the prickles of the Bramble, catching on to roughnesses on the stems of the hedge plants: they readily cling to the clothes (compare the fruit, p. 226).

All these climbing plants have weak stems with long internodes, and if loosened from their support fall to the ground unable to bear their own weight. In this they resemble plants which creep and rest their weight on the ground; some creeping stems, like the runners of the Strawberry, have rather long internodes, though this is not so marked or so general a feature as it is among climbers.

The conditions of life of climbing plants, moreover, differ greatly from those of creeping plants in two respects which affect their internal structure. In the first place, while creeping plants put out roots at frequent intervals along their stems, climbing plants are rooted at the base only, and all supplies of water and mineral salts must travel long distances to reach the leaves, while the food required by the root or to be stored in underground organs must pass as far downwards. It is interesting to find therefore that climbing plants have unusually wide vessels and sieve-tubes, through which more rapid conduction is possible. It is for this reason that the structure of sieve-tubes can best be made out in the stems of large and vigorous climbers like the Vegetable Marrow.

In the second place the stems of climbers have frequently to resist being pulled upon: the weight of leaves, flowers, or fruit always hangs upon the part of the stem above them, and when they are swayed by the wind they pull upon the stem, tending to pull it away from its hold. Tendril climbers, which attach themselves to many different twigs, may frequently be pulled in opposite directions at different points. To strengthen them against these pulls they form strands of mechanical tissue in which the fibres are long and firmly interlocked. The fibrous nature of old stems of the Honeysuckle is very evident. These woody stems are very tough, and do not as readily break across as the twigs of trees with shorter fibres that do not interlock as much. Ash twigs, for instance, are very brittle, having short fibres. Tough-wooded trees like the Oak have longer fibres, but even the twigs of such trees break more easily than the stringy stems of woody climbers.

It is interesting that climbing plants belong to widely different Families, so that the climbing habit must have been evolved many times quite independently. In some cases it is a comparatively recent development, for one or two species of climbers (e.g. the Black Bindweed, and the genus *Clematis*) are closely allied to a number of other species which do not climb.

CHAPTER XXIV

WATER-PLANTS

Trees and climbers are forms of plants adapted in special ways to the ordinary environment of soil and air. Plants which live wholly or in part submerged in water are adapted, on the other hand, to a special environment, in which the conditions of life are very different from those of ordinary land-plants.

Some of the best known water-plants are the Water-lilies. The Yellow Water-lily (*Nuphar lutea*) and the White Water-lily (*Nymphaea alba*) grow wild in the British Isles in the margins of streams and lakes, and are grown along with others of their kind in the waters of parks and gardens. Their large round heart-shaped floating leaves are borne on stout stalks arising from a thick rhizome in the mud at the bottom of the water. It is this rhizome which lasts through severe winters when the old leaves are killed off. The young leaves are at first rolled up to the midrib; so they readily push their way up between the floating blades of older leaves even when these are crowded, and then open out above them or spread themselves upon the surface of the water. Should the level of the water rise the stalks of the leaves grow longer and bring the leaf-blade to the surface again.

These floating leaves are peculiar in several respects. When taken out of the water they hang limp and flabby, being quite unable in air to support their own weight; in their natural medium they are, however, supported by the water. They are thick and fleshy, quite a pale green below, deep green and shiny above. If pushed below the water they come up again and the water collects into drops and runs off them. Thus the upper surface is not wetted by water. This is because of a coating of wax which can be removed by wiping with alcohol or warm water, after which water spreads over the wiped surface as it does over the under side.

When examined under the microscope the leaves are found to have numerous stomata, on their *upper* surface only, in contact with the air. They have palisade tissue above, like ordinary air-leaves, with numerous chloroplasts; but below this is a thick layer of spongy tissue with very large air-spaces, and the cuticle of the lower epidermis is so thin that it can only be seen after careful staining.

In the leaf-stalk the air-spaces are very large and readily visible to the naked eye when the stalk is cut across (Fig. 198), the cells themselves forming a network of partitions bounding the air-spaces. Small bundles occur here and there at the corners of the meshes. A length of the stalk cut at both ends offers very little resistance to the passage of air. If blown through, with one end below water, large bubbles of air rapidly emerge. The air-

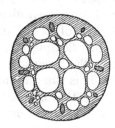

Fig. 198. Section across a flower-stalk of a White Water-lily, showing the large air-channels. The principal vascular bundles are also indicated.

spaces, therefore, are not only large but form continuous passages. Transverse partitions, or *diaphragms*, helping to strengthen the stalk, occur at intervals, but there are numerous air-spaces between the cells that compose them, which connect the large air-chambers above and below.

In the rhizome, and in the roots which grow from it into the mud, large air-spaces are also found. In the stalk and floating leaf they give greater buoyancy, but this effect is of no benefit in the case of roots and rhizome. When, however, the peculiar conditions under which the plant has to carry on its functions are considered, the advantage of a large air-circulating system will be clear. All parts of plants must respire, and for this they need a supply of oxygen. The roots of most land-plants cannot grow in water-logged soil where the circulation of air in the soil is impeded. Now the mud at the bottom of a pond or stream is always water-logged; oxygen is only slightly soluble in water[1], and so but a very small supply penetrates slowly by diffusion into the mud. Hence the advantage of large air-spaces in the roots and rhizome, communicating through those in the leaf-stalk and leaf with the atmosphere and so providing a path for the more rapid diffusion of oxygen. In the daytime some of the oxygen set free by the leaf in photosynthesis will find its way down: carbon dioxide formed in respiration below ground will also pass upwards to the leaves as an additional supply for photosynthesis, though, as carbon dioxide is soluble in water (at 15° C. water will dissolve an equal volume of carbon dioxide) some of it passes through the mud into the water by diffusion.

Water vapour is transpired through the stomata of the leaf less rapidly than from the leaves of land-plants, for the air just above the water is usually moist. As the lower epidermis has so thin a cuticle, part of the moisture needed by the leaf can be absorbed directly

---

[1] At 15° C. water in contact with ordinary air dissolves only about 0·007 of its volume of gaseous oxygen.

from the water. The amount which need be absorbed by the roots is therefore far less than in the case of land-plants. Two facts fit in with this: (1) the plant forms little xylem, and (2) the root-system is not a large one. It may indeed be that the roots are more important as holdfasts than as water-absorbing organs; but they probably absorb mineral salts from the mud.

The flower-buds are carried up on long stalks to the surface of the water. There they open and are pollinated by insects.

The flowers have many spirally arranged petals and stamens, and a many-celled ovary, with a flat top on which the stigmas radiate as in the Poppy. In the White Water-lily there is a gradual transition from petals to stamens through intermediate structures like narrow petals bearing anthers on their edges.

The fruit of the Yellow Water-lily splits up into separate carpels which float, and are so dispersed over the surface of the water; later the wall of the carpel decays, setting free the seeds, which sink to the bottom. The fruit of the White Water-lily ripens and opens under water, but the seeds float to the surface, each buoyed by a spongy swelling on the seed-coat. This eventually decays, and the seed sinks, but meanwhile it has probably travelled far, like the carpels of the Yellow Water-lily.

The Water-lilies, then, like land-plants, have their roots in the soil, and their leaves obtain carbon dioxide through their stomata from the air. Like many land-plants, they have a perennial rhizome, and their flowers are pollinated by insects. Their chief peculiarities depend upon their less vigorous transpiration, upon their being rooted in water-logged mud, and upon the mechanical requirements of floating structures: their root-system, mechanical tissues, and water-conducting tissue are not strongly developed, large air-spaces penetrate their underground tissues and communicate with similar air-channels in leaf-stalk and leaf; these are thereby made lighter, and the blade, in addition, is broad, to resist submergence. Water readily runs off the waxy surface, and so the stomata are not blocked up. The seeds or fruits are dispersed by water.

Occasionally, Water-lilies are found growing in deep water. The leaves do not reach the surface, but remain completely submerged, and appear very different from the ordinary floating leaves: they are of a paler green, more irregular in outline and more delicate, yielding

Plants growing wholly submerged.

readily to the currents which flow by them. The upper side, as well as the lower, has but a feebly developed cuticle and is permeable to water. Their method of nutrition must, of course, be very different from that of the floating leaves. Their supply of carbon dioxide must come from the water around them, and enter by diffusion through the whole of their surface.

Elodea. In studying photosynthesis we had occasion to use a water-plant, the Canadian Water-weed (*Elodea canadensis*), which grows always submerged and so depends entirely on the carbon dioxide dissolved in the water (p. 26). This plant grows in shallow water, feebly rooted in the mud, or sometimes floating quite freely. Its weak branching stem bears whorls of small narrow delicate leaves, and the whole plant readily yields to the smallest movement of the water. The leaves are dark green and transparent. Examined under the microscope they are found to consist of but a few layers of cells, with the intervening air-spaces running as narrow black lines along the leaf. In the midrib are elongated conducting cells, but no xylem can be found. There are no stomata; and chloroplasts are present in the epidermis even more abundantly than in the internal tissues, indicating that the epidermis takes the most active part in photosynthesis. The carbon

dioxide diffuses directly into the epidermal cells, whereas
in leaves which obtain their carbon dioxide from the air it
enters first those cells that are in contact with the internal
air-spaces, being hardly able to penetrate directly through
the thick cuticle into the epidermis.

Being feebly rooted or even quite free, this plant
must be able to obtain not only carbon dioxide but
everything it requires directly from the water. Mineral
salts present in the water diffuse in through the whole
surface of the plant; oxygen for respiration is obtained
in the same way—by the leaves and young stem at night
when photosynthesis is not supplying them with it, and
always by roots that are not in the mud. The water
needed for growth can also be absorbed directly wherever
it is wanted. As no transpiration is possible from the
leaves of a completely submerged plant, there would
seem to be no need for any flow of water from roots
to leaves. We find, indeed, no vessels at all in the
internodes of the stem; there is only a small canal
running up the middle, surrounded by phloëm, the whole
of the conducting tissues forming but a very small central
strand.

Elodea is a plant with male and female flowers on
different individuals. It is a native of North America,
and was unknown in Europe before 1847 when it was
recorded in the North Eastern counties of England.
Only the female plant had been brought over, but although
it has therefore never been able to produce seeds it has
since spread all over Europe by the breaking off of branches.
In winter the plant sinks to the bottom of the water,
the younger leaves at the tip merely becoming a little
more closely packed than when growth is active.

The flowers are very small. The female flowers have
an inferior ovary, and a perianth of six segments which
is carried up with the stigma to the surface by the growth

of the fused perianth-tube and style. Where the male
flower also occurs, pollination takes place at the surface of
the water: the male flower-bud breaks off and floats to
the surface before it opens; if it is wafted near enough
to a female flower it is drawn to it by capillary attraction,
just as two floating corks placed within a short distance
of one another are drawn together.

Elodea is thus adapted in a high degree to life in the
water. It remains submerged throughout life, and even
though its flowers come to the surface they are pollinated
by water. Its structure differs greatly from that of
land-plants, corresponding to the different conditions of
nutrition and growth. On the whole these conditions
seem favourable, for its success in spreading entirely by
vegetative propagation is very striking. In fact, submerged water-plants, as a rule, grow rapidly and
luxuriantly. This is probably due to several causes.

(1) In the first place, water, which is necessary for
growth, is always abundant: on land, growth is often
retarded or even stopped during hot sunny days when
transpiration carries away more water than the roots can
supply.

(2) Water supports submerged and floating plants,
and they can grow with very small expenditure of building
material; in fact, water-plants seldom contain more
than 5 per cent. of dry matter, often as little as 2 per cent.,
as compared with 10 to 40 per cent. in land-plants. Thus
a little food-material goes a long way.

(3) Carbon dioxide is more abundant in the water
than in the air: we have seen that the leaves of *Elodea*
are thin, and so for their bulk present a large surface to
the water for the absorption of carbon dioxide, as well as
light.

(4) Finally, water keeps plants at a more even temperature than land-plants. The latter are often exposed

to sharp frosts; or the heat of the sun may raise the temperature of their leaves many degrees higher than that of the air around them, and respiration even become so intense that food is used up as fast as it is manufactured. Water, on the other hand, gets warmer or colder very slowly because it takes a great deal of heat to raise its temperature (in other words, it has a high specific heat, compared with metals, or soil, or air). On hot days, water only gets slowly warmer, and so never gets hot; at night it loses its heat slowly, and so does not get very cold. In winter it may gradually cool down to a low temperature; but, except in small shallow ponds and ditches, the water at the bottom does not freeze owing to a peculiar property which water possesses. As water cools, like most substances, it contracts in volume, and so gets heavier, until the temperature reaches 4° C.; the colder water therefore sinks and the warmer water rises, so that there is a continual circulation of water, which keeps the temperature fairly uniform. When the temperature has reached 4° C., however, the water no longer contracts, but expands again as it is further cooled, and becomes lighter. The colder water, therefore, now remains on the surface, while the water below, instead of coming to the surface in convection currents, only very slowly loses its heat by conduction through the cold upper layers. Thus, water freezes on the surface, the ice thickens but slowly downwards, and the bottom water remains at 4° C., the temperature at which water is heaviest.

On the other hand, in hot weather, the warmer, and therefore lighter, water remains on the surface, and the temperature of the deeper water fluctuates very little indeed.

All these facts can be readily observed by making records of the temperature of the water in a pond or even a tub at the

surface and a foot below, at different times of day, in sun and
shade, in various weather, at different seasons of the year, noting
at the same time the shade temperature of the air. Parallel
records should also be made of the temperature of the soil near
by. Maximum and minimum thermometers are most useful,
if they can be left in the water with their bulbs at the proper
level, as they give the highest and lowest temperatures reached
since they were last set, and show at once the range of temperature.

Other common British water-plants.

Most of the Pondweeds (*Potamogeton crispus, densus, gram-
ineus,* etc.), like *Elodea,* are entirely submerged except for the

Fig. 199. A Pondweed, with broad floating and narrow submerged
leaves. (Copied from Reichenbach, with modification.)

inflorescence, a spike of inconspicuous, wind-pollinated flowers,
which rises above the water. The leaves, usually ribbon-like,
though they vary very much in shape and size, are thin, flexible,
and transparent, like those of *Elodea,* with parallel veins; they
are borne on stems which rise from a creeping rhizome in the mud.
These stems, like the stem of *Elodea,* have no vessels in their
internodes; water is conducted through a few canals which run
between the cells of the small central strand of conducting tissue;

ordinary xylem only occurs at the nodes. Some species have floating leaves as well (Fig. 199), while the common *P. natans* has oval floating leaves and under water only reduced leaves.

There are other water-plants which live partly submerged like *Elodea*, partly like the Water-lilies obtaining carbon dioxide from the air, by floating or even aërial leaves.

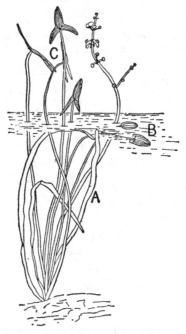

Fig. 200.   Arrowhead (*Sagittaria sagittifolia*) with *A* submerged, *B* floating, and *C* aërial leaves.   (M. G. T.)

A very interesting and well-known plant of this kind is the Arrowhead, so called from its arrow-shaped leaves, which stand up out of the water.   In addition to these aërial leaves, when growing in deep water the plant has narrow, ribbon-like submerged leaves, which are very flexible and yield to the moving water, and floating leaves with oval blades, all springing from a short rhizome (Fig. 200). The arrow-shaped leaves are, however, usually most numerous.

The inflorescence also stands erect, well out of the water: it bears several whorls of rather large white insect-pollinated flowers, the upper ones male, the lower female; the perianth consists of three green sepals and three petals, but unlike most other Monocotyledons the stamens in the male flower are numerous and there are many *free* carpels in the female, suggesting comparison with the Ranunculaceae among the Dicotyledons. From the axils of the leaves grow shoots, which bury themselves in the mud, and at their ends form buds with swollen stems full of reserve materials. These tubers last through the winter, and in the spring grow into new plants.

The Water Crowfoots are another interesting group of plants, most of them rather small, some of which have both floating and submerged leaves, others leaves of one kind only. The floating leaves are round, and more or less lobed, with a smooth and shiny upper surface, having stomata, like Water-lily leaves. The submerged leaves are cut up into many narrow segments: leaves of this form, like the thin ribbon-shaped leaves of the Arrowhead and the similar leaves of *Elodea*, present little resistance to the moving water, and expose a large surface for the absorption of carbon dioxide compared with their volume. The flowers, like white Buttercups, we have studied already (p. 282). They are usually borne just above the surface of the water.

An interesting feature of the Water Crowfoots is that many of them appear very different when grown under different conditions. For instance, a plant grown in deep water may produce only submerged leaves, in shallow water floating leaves as well; in still shallower water or mud it may only produce floating leaves. It is instructive to grow seeds or portions of the same plant under different conditions, in soil, mud, and water, to see what form the plant takes.

There are other species of Water Crowfoots, distinguished by the character of their flowers as well as by their vegetative habit, which are found only with one kind of leaf. Some grow habitually in the deeper waters of rivers (e.g. *R. fluitans*) and have only

much divided submerged leaves; others (like the Ivy-leaved
Crowfoot, *R. hederaceus*, with very small flowers) are always
found in shallow water or wet mud, and have only floating leaves.
If the mud becomes drier such plants may even lift leaves above
the mud.

There are many plants, like the Water Plantain, the
Bulrush, and the Water Dock, which grow in the edges

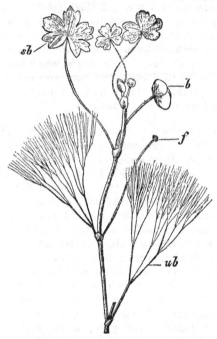

Fig. 201.　A Water Crowfoot;　*sb*, floating and *ub*, submerged leaves;
b, flower; *f*, fruit.　(After Strasburger.)

of streams and lakes, and in ditches, which have no submerged or floating leaves.　They are rooted in the mud,
and their lower parts are submerged, but all their leaves
are aërial.　Such plants show few or no striking differences
from land-plants, and are connected with them by many

other plants growing in mud or marshy places. In common with water-plants like the Water-lilies, air-spaces are well developed in the underground parts, and the leaf-stalks and ascending stems are often hollow; for in marshes, as well as under water, the soil is water-logged and very deficient in oxygen. These features become less prominent where the soil is less water-logged or less permanently so. In fact, it is well to distinguish between permanently wet places, and places which dry up at times, not only because occasional drying improves the aëration of the soil, but also because it makes greater precautions necessary against too rapid evaporation. Curiously enough, many marsh plants show distinctly xeromorphic characters even though they grow in permanently wet places; and it is an unsolved puzzle in many cases why this should be so.

CHAPTER XXV

THE DISTRIBUTION OF PLANTS AND THE FACTORS WHICH GOVERN IT

It is an obvious fact, that vegetation varies from
place to place. Meadow, moorland, heath,
wood and wayside have each their characteristic flora. When the plants which compose
this flora are examined a striking degree of uniformity is
often discovered. In a particular meadow, for instance,
a few species of Grass may be found abundant in all parts,
together with Buttercups, Plantains, etc., and it is usual
to find other meadows in the same neighbourhood with a
flora very similar in composition. In a wood near by, few
or none of these plants may occur: if an Oak wood, it will
consist of large Oaks, with perhaps an undergrowth of
Hazel, and Primroses or Bluebells on the ground. It is
clear that the shade cast by the trees is very important in
determining the kind of plants that grow beneath them,
for in the open rides and glades Grasses may be found
again in abundance. These examples will serve to illustrate the fundamental fact, from which Plant Ecology
starts, that plants are distributed, as a rule, not singly
and haphazard but in groups, which may be called *plant
communities*.

Within a plant community it is often possible to distinguish smaller groups of plants. In a meadow some
plants may be found on the tops of hummocks or ridges,

Plant
commu-
nities.

where the soil is drier, which are not found on the moister level ground. In such cases it is convenient to have different names by which to distinguish between the community as a whole and the smaller groups; the vegetation, of the same general character, covering the whole area is called a *plant association*, and the small groups within it *plant societies*.

Careful study of the distribution of the societies in one or two associations will suggest at once some of the factors on which the distribution of different plant communities depends.

Differences of level, leading to differences in drainage
Conditions affecting communities.
and water supply, have just been mentioned: in a low-lying meadow, moisture-loving plants, like the Marsh Marigold, may occur only in the wetter places.   The north and south sides of ridges, banks, and hedges often show marked differences, which may be attributed to differences in the amount of light and warmth which they receive from the sun.   These are all differences due to physical factors.

Again, in a hayfield, erect plants like Buttercups flourish, and the Grasses flower, and set abundant seed; but in a meadow where animals graze only Grasses and rosette-plants, on the one hand, which keep close to the ground and can propagate themselves vegetatively, and, on the other, plants like Thistles and Ragwort, which the animals avoid, continue to flourish.   The presence of animals thus profoundly affects both the presence or relative abundance of the different species of plants and their growth and manner of life.

On a well-kept lawn, regular mowing is even more effective, and, in addition to the close cutting, Daisies, Plantains, and other rosette-plants are weeded out.   Man thus selects and fosters the plants which he wishes to grow, and keeps others back.   Even in pasture he brings his selective influence to bear, cutting down Thistles and other plants that are avoided by animals and would spread at the expense of the Grasses.

Indeed, in a thickly populated country like ours, man **Man's** controls the greater part of its vegetation. **influence.** History tells us that a large part of this country was once covered with dense forests, in which roamed wild boars and wolves, deer and other animals that have long since disappeared. These forests have gradually been cut down, and more and more land cleared for grazing or cultivation. Only in the New Forest and a few other small areas of natural or semi-natural woodland are their relics to be seen, more or less altered from their original state by the removal of trees for timber. Man has thus completely altered the face of the country, and he continues to influence its vegetation, consciously and unconsciously. He tills the soil, and grows corn and other crops for his own use; on the cultivated land characteristic weeds, mostly annuals, are able to flourish. The cattle and sheep which he breeds influence the vegetation on their grazing grounds; they are largely instrumental in preventing trees from springing up and converting the meadow-land into forest again. In other places he plants trees. Hedgerows, waysides, and ditches also owe their existence to his activity, though most of the plants are self-sown; many of them belonged to the primeval vegetation, and have taken refuge there.

Thus even in the midst of cultivated land are many plant communities, all well worth study; but, in seeking to understand the reasons for their distribution and why they are composed of certain plants rather than others, man's influence has to be considered as a very important factor in the environment. Other factors also exercise, however, a determining influence, even the cultivated plants themselves differing in different districts, for reasons, such as climate and kind of soil, which the local farmers would be able to suggest.

Still more interesting are wild uncultivated areas where the influence of natural conditions on the distribution of plants is not complicated by man's interference to any great extent. We may mention at once moors and heaths, rocky mountains, sea-shores and cliffs, rivers and lakes. They are, in most cases, areas which, owing to poorness of soil, exposure, drought, excessive moisture, or other conditions, are unsuitable for cultivation.

Before we attempt to classify the various factors, such as rainfall, temperature, soil, etc., on which the character and distribution of the plant communities of these areas depend, there is one point of great interest to be borne in mind, in studying their composition. We have seen that in a wood the shade of the trees favours some plants and excludes others. This is a simple illustration of how the plants which grow together in the same community may influence each other. The species of plants which are most abundant in a community must always influence greatly any other plants which grow with them. If these *dominant* species are tall, and not too rank and crowded, small plants requiring shade and shelter may grow beneath them, whereas such plants would have no chance in a community dominated by low-growing species. Where several species are abundant in the same community their requirements are usually different, often complementary, so that the struggle for existence between them is not severe, but rather they help each other in the struggle with plants from other communities.

In the fens we have an excellent illustration of the interdependence of plants in an association (Fig. 202). Both above and below ground the plants occupy various levels, so that we may speak of a *stratification* of the vegetation. Below ground the rhizomes of some plants

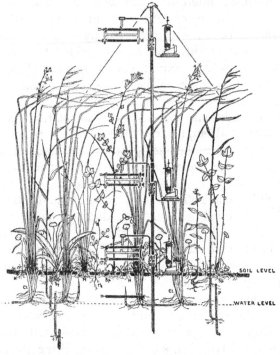

Fig. 202.   Diagram of a section through the vegetation in a fen, showing stratification of both shoots and roots. Note that only in the upper parts of the plants is the wind directly felt, as shown by the leaves being blown to one side. (After Yapp. The instruments were used by Prof. Yapp for comparing the conditions that affect transpiration at different levels.)

The plants are as follows: *Cl* (*Cladium*) Fen Sedge; *P* (*Phragmites*) Reed; *Sp* (*Spiraea*) Meadow-sweet; *M* (*Molinia*) Purple Bent-grass; *L* (*Lythrum*) Purple Loosestrife; *H* (*Hydrocotyle*) Marsh Pennywort; *Sy* (*Symphytum*) Comfrey; *Th* (*Thalictrum* Meadow-rue, young plants.

are deeply buried, those of others are not so deep, while others again have roots which are confined to the surface layers of soil. Above ground some plants creep in the shade of others which are tall, and the plants vary in character according to their position and degree of exposure to sun and wind.

Another example of well-marked stratification occurs in certain Oak woods, where Soft-grass (*Holcus mollis*), Bracken, and Bluebells form a society in which the roots of the grass occupy the surface soil, the Bracken rhizomes are below them and the bulbs of the Bluebells are found deeply buried. Here, moreover, we find the Bluebells coming up long before the Bracken, a simple illustration of another way in which plants in the same community may fit in with one another, by vegetating, flowering, or distributing their seed at different times.

The co-existence of plants in a community may therefore depend not only upon their being adapted to the physical conditions of their common habitat but also upon their being suited to each other.

Summary of the various Conditions that affect Vegetation.

The various factors which may influence vegetation fall conveniently into four groups:

(1) *Climatic factors* determine the general character of the vegetation over wide areas. They are *temperature*, *rainfall*, amount of *sunshine*, and direction and force of the prevailing *winds*; they depend upon latitude, altitude, and geographical situation.

In a small country like ours rainfall is most conspicuous in its effects. It is greatest in the west among the mountains, and diminishes eastwards; wet moors are characteristic of the west, dry heaths of the east; ferns are most abundant in the moister woods of the west.

The difference of temperature between the south of England and the north of Scotland is sufficient to affect very distinctly the farm crops: north of Aberdeen even oats often fail to ripen.

In winter, heavy snow often falls in the north of England when none falls farther south, and there is always a difference of several weeks in the time of opening of buds and flowers in the spring.

On the higher mountains the lower temperature and greater force of the wind prevent the growth of trees above a certain level; it is there that 'Alpine' rock plants are found.

(2) Factors depending on the *situation* include (*a*) the effect of the slope, influencing the depth and drainage of the surface soil in which plants can find a foothold and mineral salts; and (*b*) the effect of the direction in which it faces, on which depends the amount of light and warmth that the plants can receive. The degree of exposure to wind also depends on the situation, and partly explains the xeromorphic character of moor and heath plants, which occur, as a rule, on wide, flat, unsheltered expanses.

(3) The *nature of the soil* is of very great importance in determining the kind of plant associations which are found in a district.

On its *physical properties* depend the water supply and the drainage: clay soils hold water but tend to get water-logged in wet weather, sandy and gravelly soils allow the water to run through rapidly so that the plants found upon them are strongly xeromorphic.

The *chemical nature* of the soil may also be important. In limestone districts and on chalk, the soil contains a large proportion of calcium carbonate (calcareous soils) and the vegetation is largely characteristic, many of the meadow plants and weeds of cultivation, as well as wild plants, differing from those found elsewhere. By the sea the presence of salt in the soil, and on moors the acidity of the peat, keep out many plants; acid peat, moreover, is poor in mineral salts. In the fens, on the other hand, the soil is peaty but rich and alkaline, and supports a type of vegetation very different from that of the moors.

The most important factor in nearly all soils is the amount of *available* water (p. 75). A peaty or clay soil may appear damp, yet hold the water so tenaciously that plants find great difficulty in absorbing any from it, even though it be not acid or salt. This helps to explain why plants growing in some apparently moist soils are xeromorphic. It is always necessary to take into account also how the amount of available water varies at different periods. Plants living on a soil which is liable to occasional drought must be adapted to withstand this drought or they will succumb, however abundant the water supply may be at other times.

(4) *Biological factors* include the influence of *animals* and *man* and the influence of *plants* upon each other.

Not only cattle and sheep, but wild herbivorous animals, especially rabbits, and even slugs and snails (as every gardener knows) may interfere with or prevent the growth of some plants. Earthworms have an enormous influence on the nature and composition of the soil when they are abundant[1]. The dispersal of fruits and seeds by birds and other animals, and pollination by insects, sometimes limit the spread of particular plants to the haunts of the animals on which they depend (e.g. Monkshood, p. 213), or determine their relative abundance. In the *Origin of Species* (Ch. III), Darwin has given very striking illustrations which show how complex are the relations which may exist between animals and plants.

The influence of plants upon each other, in determining the composition of a plant community, has already been referred to. Parasitic plants are a special instance: their distribution depends on that of suitable victims or *host plants*, and, on the other hand, they may play havoc with their hosts. Root-parasites, like the Yellow Rattle and the Eyebright, are especially frequent.

Hints for studying the vegetation of a district.

The study of vegetation can be begun in a small district within easy reach of school or home. Much may **The study of a community.** be learned by selecting one small spot for observation throughout the year and finding out as much as possible about the life of all the plants which occur in it—how they are grouped amongst and affect each other, when they begin active growth in spring, whether they flower and fruit at the same or at different times of year, what kind of underground parts and root-systems they have, whether they are perennials or annuals, how severe weather affects them, and what special features their structure shows which may fit them for their particular environment; how their flowers are pollinated, how their seeds or fruits are dispersed and to what distances, how their seedlings fare in the same spot and among neighbouring communities; how wide-spread the species are, and what differences in habit, size, and

[1] The action of worms on the soil is described in Charles Darwin's book on *Earthworms*.

other characters they show in different situations; where species come from which are found only occasionally and how they are brought, what other species are likely to have their seed carried to the spot, whether these ever germinate and if so why the seedlings do not succeed in establishing themselves. Some of these problems may be investigated experimentally by clearing small patches of ground of all vegetation and observing what seedlings spring up. The gradual colonisation of such a piece of ground often takes place in an interesting succession of stages before the vegetation reaches its original composition again.

In a rather larger area the effect of local differences of level, aspect, drainage, etc., on the composition of the vegetation should be recorded.

For the whole neighbourhood a vegetation map is a convenient way of recording the distribution of the main **Distribu-** associations. It is best to use an Ordnance **tion of** Survey map (on a scale of one, or six, inches to the **associa-** mile), distinguishing on it by different tints the **tions.** cultivated land, permanent grassland, woods and plantations, and the areas covered by different natural associations.

Notes should be made of the well-marked societies in each association, and the species composing them should be arranged in order of their relative abundance, distinguishing (a) the dominant species from (b) those which are dependent on them and (c) those which are only found here and there.

In most districts there are species which only occur very locally or are rare. The characters and distribution of these species are worth observing, but special care should be taken not to disturb or injure them.

Collections of common plants, on the other hand, may usefully be made to illustrate the structure, and the condition at different seasons of the year, of plants typical of different habitats and associations, or the variation in form and size of the same species in different situations.

In considering what factors determine the distribution of associ- **The** ations the nature of the soil is of first importance. **deter-** A Geological Survey map of the district will show **mining** the kinds of rock which underlie the surface soil. **factors.** From the colour and texture of the different soils, when moist and when dry, can be roughly inferred the proportions of sand, clay, humus, fibre and even chalk which they contain. Peaty soils and others containing much humus should be tested with litmus paper: the very acid soils of most moors turn blue litmus red; rich fen-peat is distinctly alkaline, turning red litmus blue. Soils containing more than about three per cent. of calcium carbonate (chalk or limestone) can be distinguished by pouring a ten per cent. solution of hydrochloric acid on them: carbon dioxide is liberated rapidly enough to make visible bubbles. When still more lime is present a hissing sound is heard and froth may be formed.

Succession. Where two different communities occupy adjoining areas on the same kind of soil the border is often the scene of a struggle between them for mastery. Where one is encroaching on the other the transition is sometimes marked by one or more intermediate temporary cummunities, which succeed each other in definite order: this transition by stages is called *succession*. It requires experience and careful judgment, however, to distinguish between true succession and the mixed condition of the vegetation, even showing well-marked zones (*zonation*), which is sometimes found between two communities which are both holding their own.

Simple examples of succession are met with in the gradual colonisation of waste ground, the reclaiming of salt marshes, and the fixing of sand-dunes. The pioneer plants bring about changes in the habitat, such as alterations of level or composition of the soil, as a result of which other plants are able to grow there. The vegetation may go through a series of changes before it reaches a final *stable* composition. In some cases, especially in salt marshes, and on sandy shores where sand-dunes are forming, the successive communities are often very sharply marked off from one another.

Further hints and more detailed instructions will be found in *Practical Plant Ecology* by A. G. Tansley (Geo. Allen and Unwin, 1923).

NOTES ON SOME NATURAL PLANT ASSOCIATIONS.

In the following pages are brief notes on some of the chief types of natural vegetation, intended merely to suggest the kind of points which should be looked for. It is not necessary that all these types should be studied; attention should be concentrated on those which are represented in the neighbourhood, though others should be examined when opportunities occur. Only work in the field can be of much value, and these notes are given as helps to such study. A more detailed account will be found in *Types of British Vegetation*, by A. G. Tansley.

Woodland and Grassland Associations.

The character of a woodland community depends usually upon one or a few dominant species **Woods.** of large trees. In England, the most important kinds of natural or semi-natural woods are dominated by the Oak, the Ash, or the Beech. Their distribution is found to be connected with the character of the soil and the rocks underlying it. Oak woods are characteristic of soils containing little calcium carbonate. On limestone, the Ash usually dominates. On the chalk in the south and south-east of England natural Beech woods occur: in other parts of the country the Beech has frequently been planted, but is probably not native. In the Highlands of Scotland, where the climate is colder, there are considerable areas of natural Pine woods.

The chief factor in determining what plants will grow in these different woods is the depth of shade in them. Oak woods, for instance, usually have a dense undergrowth of shrubs and a rich carpet of plants on the ground, contrasting strongly with the almost bare ground of many Beech woods. The ground vegetation of woods contains many plants which flower and come into leaf in the early spring, while plenty of light can still reach them between the bare branches.

It is instructive in studying the vegetation of woods to estimate roughly the intensity of light which reaches the ground in different spots and at various times as compared with full sunlight. This may be done very simply as follows: A narrow strip of photographic P.O.P. is covered with a piece of card and gradually exposed, a portion at a time, so that the different portions are acted upon by the light for successively shorter times. For instance, if the first portion is uncovered at the beginning of a minute, the next after 20 seconds, and the following portions after 30, 40, 50, 55, 57, and 59 seconds, the whole strip being covered again at the end of the minute, the successive portions will have received 60, 40, 30, 20, 10, 5, 3, and 1 seconds exposure respectively. On comparing strips so prepared in different places, including open sunshine, portions of about equal depth of colour can be selected, and the time of exposure which was necessary in each case to produce this tint will be inversely proportional to the intensity of the light. Thus in a place where it took 60 seconds to darken the P.O.P. to a given tint, which was reached in open sunshine in five seconds, the intensity of the light must have been only $\frac{5}{60}$ or $\frac{1}{12}$ of that of full sunshine.

Wood-land, grassland and heath. There can be little doubt that the different kinds of woodland which are now found to be characteristic of calcareous and non-calcareous soils respectively represent more or less closely the kinds of woodland which at the dawn of history covered a large part of the country. On clay and sandy soils where to-day small Oak woods are scattered, large tracts were then occupied by Oak forests; similarly in limestone districts large areas were probably covered with Ash woods, while continuous Beech woods stretched for miles over the chalk of the southern and south-eastern counties of England.

To-day, where the land which has been cleared is least interfered with, permanent grassland is found, which probably represents fairly closely the type of association which occurred in open glades in the primeval forest and replaced the woodland when clearings were

made. Like the woodland, the grassland on calcareous
soils is very different from the grassland of non-calcareous
soils. There are thus distinct types of grassland standing
in close relation to the types of woodland which are
characteristic of the same soils. On dry and poor sandy
soils in the east of England, heath sometimes stands in a
similar relation to woodland.

Non-calcareous soils.

Woodland. Oak woods are characteristic of moist clay soils.

Phot. S. Mangham

Fig. 203. Oak wood with thick undergrowth, on clay soil in the
Weald, Surrey. The trees are *Quercus Robur*: Hazel, Hawthorn,
Brambles, Bracken, and Foxglove are seen below.

As they exist to-day they usually consist of large Pedunculate
Oaks (*Quercus Robur*) at intervals, interspersed with Hazel and
other shrubs and small trees, such as Ash, Maple, and Hawthorn. On the ground are various societies of plants, their
distribution depending mainly on the water-content of the soil
and the amount of light which reaches it: among the dominant
plants are Wild Roses, Brambles, and Bracken in the more open
areas; Dog's Mercury, Wood Anemone, Bluebell, and Primrose,
where less light penetrates. Ivy and Honeysuckle are common.

On shallower, drier soils, the Sessile Oak (*Quercus sessiliflora*) is found: in many woods on such soils, among the Pennines for instance, it is dominant, and the common Pedunculate Oak almost absent. In such woods Birch and Holly are abundant, and Mountain Ash is frequently met with. On the poorer sandy soils, plants from neighbouring heaths, like Ling, Bilberry, and Heather, as well as the Silver Hair-grass (*Deschampsia flexuosa*) and Bracken, dominate the ground vegetation, and transitions are found through pure Birch wood to heath. Sometimes local patches of Bluebells and other typical woodland plants remain as evidence that the heath has been encroaching on the woodland.

Where in Oak woods marshy places occur, Alders and Willows become dominant. Along the banks of streams an Alder-Willow association can often be distinguished. On marshy land, woods occur formed entirely of this association.

Pine woods, of Scots Pine (*Pinus sylvestris*), are more frequent in Scotland. They are probably native only in the Highlands. In England they have all been planted, at least originally, but on heaths, in the neighbourhood of Pine plantations on sandy soils, seedling Pines spring up, and convert the heaths into woods. Owing to the rather deep shade cast by Pines, not only in summer but in spring, as well as to the dense carpet of Pine needles which resist decay, the ground flora is usually scanty; but Bracken and Bilberry often occur, with other heath plants, in the more open spots.

Grassland. In the rides and glades of woods of Pedunculate Oaks and on grazing land once covered by such woodland grow our best known wayside and meadow grasses, such as Foxtail, Cocksfoot, and Sweet Vernal Grass (*Anthoxanthum odoratum*), mixed with other well-known meadow plants like Buttercups, Clovers, Silverweed, Sorrel (*Rumex Acetosa*), and Yarrow. In very moist low lying places, the common Rushes become dominant. On sandy, drier soils, where Pedunculate Oaks are mixed with Sessile Oaks in the woodland, the grassland is dominated by Fine Bent-grass (*Agrostis*), and Bracken, Broom, and Gorse are often abundant. Sandy commons are frequently covered with an association of this composition.

On harder, siliceous rocks, which weather less readily and so give rise to the shallower soils (like those of the Pennines), on which the woods are of Sessile Oaks, the grassland is dominated in dry places by Mat-grass (*Nardus stricta*) which forms in summer a

dry and slippery grey-green turf. Along with it Silver Hair-grass
(*Deschampsia flexuosa*) is abundant, and Bracken, Gorse, or Rushes
are common in parts. When the soil is damp the Purple Moor-
grass (*Molinia*) is dominant, and with it are found Mat-grass,
Ling, Pink Bell-heather and Crowberry (*Empetrum*), or Rushes in
wetter spots. On poor, sandy soils heath is found, with Ling
dominant in most parts.

Calcareous soils.

Woodland. In Ash woods, Wych Elms and Hawthorn re

Phot. S. Maugham

Fig. 204. Interior of a Beech wood in Kent, on chalk, showing almost
bare ground in the deep shade under the trees. In the lighter
space in the foreground is a patch of Dog's Mercury.

especially frequent, but many other trees and shrubs occur. The
ground vegetation varies with the dampness: by streams and in
marshy places Marsh Marigold, Meadow-sweet, and Water Avens,
in less moist places the Lesser Celandine and Garlic, are very
abundant; drier places are carpeted with Dog's Mercury, Ground
Ivy, etc.

Beech woods differ greatly from any of the other kinds of
woodland. The Beeches cast a dense shade, and come into lea.

usually much earlier than either the Oak or the Ash: the ground flora is therefore scanty or quite absent over considerable areas, even spring flowers being often unable to survive. Dog's Mercury covers the ground where the foliage is less dense. The Bird's Nest Orchid (*Neottia*) and Yellow Bird's Nest (*Monotropa*), saprophytic plants without chlorophyll (p. 369), are often found. There is usually no undergrowth, except of young Beeches, which can stand being shaded. The only other tree or shrub commonly occurring is the Yew, which itself casts a dense shade, and like the young Beeches can grow in a comparatively feeble light.

Grassland. The grassland associations found on chalk and limestone have much in common, and are very different from any of the grassland associations on non-calcareous soils. The dominant grass is the tough and wiry Sheep's Fescue (*Festuca ovina*). Numerous other plants are found, some of them specially characteristic of calcareous soils, such as Rock Rose, Salad Burnet (*Poterium Sanguisorba*), the Hairy Violet (*Viola hirta*) and the Hoary Plantain (*Plantago media*). Thyme, Bird's Foot Trefoil, Mouse-ear Hawkweed and others are also common.

Moorland Associations.

Moors and heaths are developed on flat, unsheltered expanses on poor, peaty, acid soil. Especially in the damper western parts of the country, on mountains, wherever there are level expanses from which water does not readily drain, the wet remains of plants accumulate and form peat. *Moors* are found on nearly pure peat, with very little admixture of mineral matter, *heaths* on peaty soils containing more mineral matter.

The characteristic plants are xeromorphic. Most dry heath Grasses have their leaves rolled up, with the stomata inside the groove so formed. Ling, the Heathers, and the Crowberry (*Empetrum nigrum*), plants characteristic of heather moors and heaths, have very small narrow leaves, with the stomata sheltered from wind in a narrow groove on the under side, further protected in some cases (e.g. Ling, p. 110) by hairs in the groove.

There are several kinds of moorland associations.

Sphagnum bogs. In very wet places bogs are found, of spongy Bog Moss (*Sphagnum*) with some Pink Bell-heather (*Erica tetralix*), Cranberry and Cloudberry (*Rubus Chamaemorus*).

Cotton-grass moor. On the Pennines, Cotton-grass moors cover wide areas, consisting almost entirely of Cotton-grass (*Eriophorum*, not really a Grass but a Sedge). The peat is often many feet in depth and where it has been cut through by streams or removed for fuel the history of the vegetation of the moor can be read by examining the plant remains. In some districts the whole of the peat below the Cotton-grass consists mainly of the remains of Cotton-grass; but in others the remains of trees are found abundant at certain levels, showing that forests once covered the area. Often there is evidence of a succession of different associations, showing that a gradual change of conditions has taken place.

Sedge moor. Elsewhere the Cotton-grass is not so abundant, and Sedge moors are found dominated by another Sedge, *Scirpus caespitosus*, mixed with Ling, Cotton-grass, and Pink Bell-heather.

Bilberry moor. On ridges and summits, where the peat is better drained and not as deep, the Bilberry (*Vaccinium Myrtillus*) becomes dominant; at lower levels it is associated in different places with Ling or Crowberry or with Purple Bell-heather (*Erica cinerea*).

Heather moor. On shallower drier peat, at rather lower levels still, Heather moors are found; these are usually strictly preserved for grouse shooting. The dominant plant is Ling (*Calluna vulgaris*) which grows, when left to itself, to a low bush; but on grouse moors it is regularly set fire to every few years to clear the ground for the growth of young plants. Associated with it in different parts, sometimes equally abundant, are Bilberry, Bracken, and Purple Bell-heather, and heath grasses like Mat-grass and Sheep's Fescue. In the peat below these moors, remains of Birches are frequently found. In the wetter places the Common Rush is dominant, while stagnant hollows are occupied by *Sphagnum* and plants like Pink Bell-heather and Cotton-grass characteristic of the wet moor of higher levels.

Lowland moors. Heather moors occur at low levels in some parts of the country where peaty deposits have been formed. There is often evidence that these lowland moors were once fens developed through the gradual drying or silting up of estuaries and lakes; but the accumulating plant remains have lifted the vegetation above the underlying richer soil, the peat has become poor and acid, as well as drier, in consequence, and has been invaded by a succession of moorland associations.

Sandy heaths. On poor sandy soils the humus that collects under trees or elsewhere tends to become acid and favour the growth of Ling and other heath plants. Under these conditions, in our eastern districts especially, heaths occur such as have already been mentioned as invading dry Oak and Birch woods.

Fens.

The Fens occur on low-lying, undrained, marshy soil that is peaty but rich and alkaline. Compared with elevated moors, the conditions are not so severe. A corresponding difference is found in the vegetation, for the strongly xeromorphic moorland plants like the Heathers, Ling, etc., are absent.

Fens were once very extensive in the eastern counties of England, and formed almost impassable barriers around the higher ground of the Isle of Ely, famous in history as the place where Hereward the Wake made his last stand against William the Conqueror. Most of this fenland has now been drained, and forms very rich soil for cultivation. A few very small areas of unreclaimed fen remain, like Wicken Fen, and larger fens around the Norfolk Broads.

Fenland has probably been formed where once were large estuaries and deltas, which have gradually been silted up with rich alluvial soil, brought down by the rivers, and the accumulating remains of aquatic plants. To-day, transitions are found between typical associations of water-plants and the fen associations as the level of the soil rises above the water level.

The most characteristic fen plants are the Reed (*Phragmites*), the Fen Sedge (*Cladium Mariscus*) and a common Rush (*Juncus obtusiflorus*); other Grasses and Sedges are also abundant, together with moisture loving plants like Meadow Sweet, Valerian, Water Forget-me-not, Ragged Robin, and here and there the shrubs Buckthorn and Sweet Gale. Under the shelter of the taller plants, Marsh Pennywort (*Hydrocotyle vulgaris*) is abundant. (For the stratification of the vegetation in fenland, see p. 449 and Fig. 202.)

Aquatic vegetation.

In water, different associations of plants may be distinguished according to the depth and character of the water. On the margins of lakes and streams, where the bottom is gently sloping, a well-marked zonation is often to be seen. Submerged plants grow in the deeper water; plants like Water-lilies with floating leaves are found nearer the margin where the water is not so deep and moves more slowly; then come plants like Bulrushes and Reeds, or other plants with only their bases in water. Nearest the margin, in very shallow water or mud, are marsh plants, like Marsh Marigold, Frog-bit, Water Forget-me-not, and Water Mint.

The vegetation of stagnant ponds differs from that of streams and lakes; plants like Duckweed and Water Starwort (*Callitriche*) often cover the surface, Water Dock, Water Plantain, etc., growing in the margins.

Ponds or lakes which are slowly drying up are of especial interest as showing examples of succession of plant associations, zones nearer the margin encroaching on those farther out, and being in turn superseded by the drier associations of the surrounding land as the water recedes.

In the sea very few flowering plants occur among the sea-weeds; Grass Wrack (*Zostera*, allied to Pondweeds), with narrow grass-like leaves, is very common on mud below the level of the Glasswort association in salt marshes. The seaweeds form distinct zones, the Bladder Wrack and other large brown seaweeds, Grass Wrack, etc., occurring between high and low water marks, where they are left high and dry twice a day. The red seaweeds are found below low-water mark, or in pools where they are always submerged.

Sea-coast vegetation.

By the sea very special conditions exist, and special types of plant association are found.

In many situations the soil water is highly charged with salt, which makes it difficult for the roots of plants to absorb water. The plants characteristic of this kind of habitat are called halophytes (Gr. *halos* = salt). Many of them are succulent.

Another important factor is the salt spray to which sea-side plants are often exposed: this spray may do great harm to other plants in the vicinity, causing the leaves on the seaward side of trees to shrivel, sometimes for miles inland, after a storm.

There are several types of associations depending upon the kind of habitat. Rocks and cliffs, sandy shores and sand hills (sand *dunes*), shingle beaches, and salt marshes are four chief classes of habitat with widely differing conditions and characteristic plant associations. Below the high-water mark grow green, brown, and red seaweeds.

Salt marshes. These are sandy or muddy flats sheltered from the direct wash of the tides but covered periodically, at least by the highest spring tides. Several plant associations are usually found distributed according to the level of the soil and the amount of submergence it receives. In muddy salt marshes at lower levels, just out of reach of the neap tides, Marsh Samphire or Glasswort (*Salicornia*) usually grows, practically alone. It is an annual with jointed succulent stems and very reduced opposite leaves that appear merely like small swellings of the upper part of each joint. It has a very meagre root-system and if left long exposed to the air and sun gradually wilts, for it differs from most succulent plants in having a very thin cuticle; but it recovers again when submerged, being capable of absorbing water through the whole of its surface. The succulent habit enables it to store enough water to last it between the times of submergence.

Below the Glasswort association Grass Wrack is often found abundantly. In sandy salt marshes the Sea Sweet-grass (*Glyceria maritima*) occupies the outermost zone, corresponding to that occupied by the Glasswort in muddy salt marshes.

Phot. A. G. Tansley

Fig. 205. Sand-dunes at Hemsby, Norfolk. In front of the main range of dunes are smaller ones in process of formation, that on the left by Marram-grass (*Ammophila*), those to the right by Lyme-grass (*Elymus.*)

Sand or mud gradually collects around the plants of the pioneer association and their remains, which hold it together and gradually raise the level of the marsh so that it is submerged for shorter periods. As this change occurs other associations gradually encroach upon the Glasswort or Sweet-grass association, forming raised carpets of vegetation. The composition of these associations varies at different parts of the coast; among the plants commonly found are Sea-blite (*Suaeda maritima*) with narrow fleshy leaves, Sea Arrow-grass (*Triglochin maritimum*), Sea Sweet-grass (*Glyceria*

maritima), Michaelmas Daisy (*Aster Tripolium*), Thrift or Sea
Pink, and Sea Lavender. Where the Sea Pink is abundant it
forms beautiful sheets of pink blossom in early summer. Differences of level are associated with differences in the relative
abundance of the various species. At the highest level, reached
only by the highest spring tides, the Sea Rush (*Juncus maritimus*)
is often the dominant plant.

Sand-dunes and sandy shores. On shores where much sand
has been washed up by the sea, the dry sand above high-water
mark is blown about by the wind. This loose blown sand is the
feature which chiefly determines the character of the vegetation.
The wind blows the sand up into more or less unstable ridges and
hills, known as sand-dunes, which are colonised and held together
by characteristic sand-dune Grasses. The Marram-grass (*Ammophila arenaria*) is usually the most important of these : its long
rhizomes and roots penetrate the dune in all directions, and in
places where an unusually violent wind has broken away part of
the dune the oldest parts of such rhizomes can be seen to reach
right to the bottom of the dune, showing that the Marram-grass
has been responsible for the gradual building up of the dune.
Small clumps of Marram-grass can be seen below the dunes, round
which sand has collected to form little mounds, illustrating the
manner in which the formation of a dune begins. In some parts
of the coast the Sea Couch-grass (*Agropyron junceum*) begins to
form dunes nearer the sea, and the Marram-grass colonises these
as the sand accumulates, and sometimes a line of dunes formed
by the Sea Couch-grass is found on the seaward side of the
Ammophila dunes. Along with these the Sand Sedge (*Carex
arenaria*) is abundant, while, among other plants, the Sea Holly
is often common.

In addition to the shifting sand, the water supply on such
soil is scanty and precarious: accordingly dune plants are strongly
xeromorphic; the dune grasses in particular are capable of rolling
up their leaves in dry weather, so retarding transpiration; but
they are not necessarily halophytes, for the fresh rainwater
washes the salt from the sand, so that the water supply contains,
as a rule, but little salt.

Dunes built up by plants like Marram-grass may become stable
—*fixed dunes*—covered and protected from the shifting action of
the wind by a carpet of other Grasses, with Rest Harrow (*Ononis
repens*), Bird's Foot Trefoil, etc.: in some places a dense scrub

of Sea Buckthorn and other shrubs is found; in others the fixed dunes become converted into heath. Behind the dunes, stretches of fixed sand are covered by a dry pasture, forming excellent golf links, or, in places, by heath vegetation.

Between the dunes and the sea, where the water supply is salt, associations of *strand* plants, typical halophytes, many of them with fleshy leaves, are found, such as Sea Rocket (*Cakile maritima*), Sea Purslane (*Arenaria peploides*), with short broad leaves in four ranks, and Saltwort (*Salsola*). Where sand is being washed up by the sea the strand plants may begin to form small dunes, and then the dune plants themselves step in and become dominant.

Shingle beaches. On pebbly shores, the sea heaps up the pebbles in banks above high-water mark. Between the pebbles, fine humus collects, derived largely from the air-dried and powdered remains of seaweeds washed up by the sea. Many of the plants found, especially at higher levels, farther away from the sea, are plants found also inland in stony and waste places. Others, like the Sea Campion and Sea Horned Poppy, are characteristic of pebbly beaches.

Where much sand is mixed with the pebbles, the plants are such as are found on sandy shores.

Sea cliffs and rocks. Wherever these are exposed to spray from the waves, the conditions differ from those of similar situations inland, and plants like Sea Pink, Sea Lavender, Samphire (*Crithmum maritimum*, with leaves divided into narrow thick and fleshy segments: Family *Umbelliferae*), and the fleshy-leaved Sea Plantain (*Plantago maritima*) are found.

SUPPLEMENT

SEEDLESS PLANTS

CHAPTER XXVI

ALGAE

Seed-bearing plants to which we have hitherto confined our study have a complicated construction. At the outset we re-cognised certain organs which perform various functions in the life of the plants and in form and structure are wonderfully fitted for the parts they play.

There exist other kinds of plants which do not bear flowers or seeds. The construction of some of them is extremely simple. Our object in these supplementary chapters will be to study and compare a few examples, in order to obtain a better idea of the diversity of plant life.

Close inspection of almost any lake, stream, pond, or ditch, reveals the presence of other green matter besides the leafy weeds which are the seed-bearing water-plants. When samples are examined under the lens or microscope green threads are seen and other smaller objects of various shapes and sizes. The threads are rows of green cells. The others consist of green cells variously arranged, or of only a single cell, often far smaller than the green cells of an ordinary leaf. These are Green Algae. We have learned to associate green colour with the power to assimilate carbon dioxide by means of radiant energy of sunlight. It is an interesting fact that in some of Priestley's earliest observations on the renovation of air it was a green scum of Algae growing in his gas jars which decomposed the carbon dioxide, and he noticed bubbles of gas forming, entangled in the scum, in the sunlight.

Pleurococcus.

Our first example grows not in water but most commonly on tree trunks, wood palings and damp walls, forming a green covering on the windward or shady side. If a small speck be scraped from a piece of bark on the point of a knife and mounted in a drop of water under the microscope, numerous tiny green cells are to be seen, some single, but most arranged in groups of two, four, eight or more. The free single cells are round.

In cool damp weather, which favours the growth of *Pleurococcus*,

Growth and multiplication. cells can be found with a cross wall dividing them into two, and others showing various stages between this first division and the complete separation of the daughter cells from each other. The larger groups have been formed by repeated divisions without rapid separation.

Under the high power each cell is seen to possess a protective cell-wall enclosing dense contents, part green, part colourless. In favourable specimens the green part is seen to be a large lobed chloroplast. A nucleus can be demonstrated by proper staining in the middle of the cell. There is no vacuole: the whole cell is filled with protoplasm, of which the nucleus and chloroplast are special parts.

Each cell is an independent living being. It grows and

Nutrition. eventually divides to form two beings like itself. To grow it must feed. Like other green plants it assimilates carbon dioxide. It also respires, so utilising the stored energy of the starch and other products of photosynthesis.

Consisting largely of water, it requires water for its growth. There is no vacuole of sap; but protoplasm can itself absorb and hold water like gelatine.

It also requires the other substances which rooted plants absorb from the soil. These it can only obtain from the moisture trickling down the trunk or paling on which it grows; but being so small it only requires them in very minute quantity.

At times, the moisture dries completely from its habitat. If

Survives drying. Dispersal. some of the grey-green dust from dry bark be moistened, almost at once the dry cells appear as fresh as ever and will resume their growth. *Pleurococcus* has thus the remarkable property of retaining its life even when dried up. Seeds have this power,

but not the growing parts of most plants. When quite
dry and powdery *Pleurococcus* is easily caught up by wind
and borne away. This accounts for its very wide distribution
throughout the temperate regions of the globe, where a cool and
relatively moist climate favours its healthy growth.

In this organism structure and life history are alike extremely
simple. The single cell carries out all the processes essential to
its nutrition. There is no special preparation for dispersal, nor
for survival through summer drought or winter cold.

Spirogyra.

Spirogyra is found in more or less stagnant waters as bright
green clouds of delicate threads, slimy to the touch, buoyed to
the surface in the sunshine by the many bubbles rich in oxygen
which they entangle. Under the microscope each thread (or
filament) is seen to be a row of cells, all alike, and marked by green
spiral bands to which the name *Spirogyra* refers[1].

By focussing for the edges of a cell these large spiral chloroplasts can be seen to be ribbon like, and embedded
in a thin layer of protoplasm that lines the cell-wall.
In iodine, starch grains, stained blue, are seen at
intervals in the chloroplasts, closely grouped around
definite lumps called *pyrenoids*. Here we have a differentiation
of structure not found in the small rounded chloroplasts of Seed-
plants. While the whole expanded surface of the band absorbs
light and probably assimilates carbon dioxide, the function of
forming a temporary reserve of starch from the excess of soluble
carbohydrate is performed only by the pyrenoids (p. 40).

**Structure
of cells.**

Each cell has a conspicuous nucleus which in many species
lies in the middle of the cell, suspended by strands of protoplasm
which connect it to the outer protoplasm near pyrenoids.

The bulk of the cell is occupied by sap. When a 5 % solution
of salt is run in the cell is plasmolysed. The protoplasm is an osmotic
membrane, and the healthy filament retains its shape owing to
the osmotic inflation of its walls (p. 84). Only vacuolated cells
reach so large a size as these; cells like those of *Pleurococcus*,
without vacuoles, are generally small.

[1] There are many different species of the genus *Spirogyra*, varying
in size and in the closeness and number of the chloroplasts. Large
species with loosely coiled bands are easiest to examine.

Here and there may be seen pairs of cells, separated by a thin
wall, which have only recently been formed by the
Growth and division of a fully grown cell.   All the cells of the
Nutrition.   filament grow and divide, the division of a cell being
preceded by division of its nucleus.

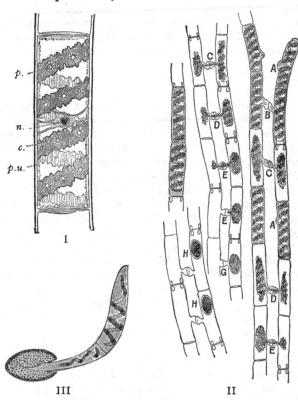

Fig. 206. *Spirogyra.* I, cell showing nucleus (*n.*), chloroplast (*c.*),
protoplasm lining cell-wall (*p.u.*) and pyrenoids (*p.*) surrounded by
small starch grains.   II, conjugating filaments: A–G, stages in
conjugation; H, zygospores.   III, germinating zygospore. (I and II
after Darwin, III after West.)

As in *Pleurococcus*, what each cell needs for growth it absorbs
from the water independently; indeed, under some unfavourable
conditions the filaments break up into separate cells; but usually

they remain attached end to end.   They grow only in length, and
the dividing walls are all across the filament parallel to one another;
whereas the spherical cells of *Pleurococcus* grow equally in all
directions and the new cell-walls are formed at right angles to those
already laid down.

The filaments grow actively in the spring, but die down almost
completely during the heat of summer.   Towards
Conjuga- the latter end of spring filaments may be found lying
tion. side by side in pairs, linked together by connexions
between opposite cells like the rungs of a ladder.
Often the cells of one of the filaments are empty, while those of
the other contain, lying loose within them, oval, dark green, thick-
walled *spores*.

It is not difficult to find filaments showing various stages in
the formation of connexions and spores.   The process is as
follows.   Two filaments somehow come to lie side by side.   Bulges
appear on the walls of opposite cells and grow towards each other
till they meet; then their ends are dissolved leaving an open tube
connecting the two cells.   Meanwhile a great change comes over
the protoplasm, for instead of remaining pressed against the
wall it contracts away as if plasmolysed, although still in water,
and rounds itself off.   When the tube is complete one of the two
protoplasts, as if attracted by the other, squeezes itself through
the tube.   The two protoplasts then fuse completely, their nuclei
fusing also.   The resulting mass rounds itself off and surrounds
itself with a thick wall, forming the spore.

This union of two similar protoplasts is called *conjugation*;
and the resulting spore a *zygospore*.   Conjugation
Conjuga- is a sexual process in the sense that, as in the
tion a fertilisation of an ovule, two cells unite to form a
process of reproductive cell which is the beginning of a new
sexual re- generation.
production. The zygospores, like seeds, last through seasons
adverse to the active growth of the organism.   They
turn reddish or orange, their starch changes to oil, and they sink
to the bottom.   There they remain as a rule till the following
spring.   Then they burst their thick walls, except for a thin inner
layer, and grow out into new filaments.

Vaucheria.

Vaucheria is a genus of Green Algae some species of which grow in water, others on damp earth. *Vaucheria sessilis*, for instance, can often be found on damp paths, forming a dark green covering of interwoven threads. When a little is removed, as much earth as possible gently washed away and some of the alga carefully drawn on to a glass slide in water, it is seen under the lens to consist of branching threads, with occasional smaller finely branched colourless threads which fastened the green threads to the earth.

Under the microscope no cross walls can be seen in healthy filaments. Each branch has a tubular wall, lined with proto-plasm, enclosing cell-sap. The vacuole is continuous through all the branches of a plant. Near the wall are many small chloro-plasts; proper staining reveals also numerous nuclei embedded a little deeper in the protoplasm. At the rounded tip of each branch the protoplasm is thicker and the chloroplasts rather less numerous: this is the apical growing point. Iodine reveals no starch; but numerous small oil-drops can be seen. Oil-drops are in fact formed by the chloroplasts like starch in other plants.

In vegetative structure *Vaucheria* thus presents many points of difference from *Spirogyra*. Instead of a filament of cells it is a branching tube. There are green assimilating branches and colour-less rooting branches. Growth is apical, whereas in *Spirogyra* no distinction exists between apex and base (except immediately after the germination of the zygospore) and all parts of the filament share in its growth.

The tubular structure must obviously be very delicate. It is indeed very difficult to mount specimens undamaged. When a branch suffers injury, however, the protoplasm rounds itself off away from the injury and eventually growth continues once more. When in danger from drought portions can contract in a similar way and protect themselves with thick walls till the unfavourable season passes.

Vaucheria has two special methods of reproduction, one of which is not accompanied by the union of cells and **Zoospores.** is therefore distinguished as *asexual*. Some of the branches become swollen and darker green, and the club-shaped end is cut off by a cross wall. Eventually a small hole appears at the tip and the contents begin to squeeze out through it. The

escaped protoplast is capable of movement: it swims slowly away,
and for perhaps a quarter of an hour wanders about. Then it
settles down and soon begins to germinate, putting out usually
two tubular branches on opposite sides, one of which sooner or
later forms a colourless rooting branch.

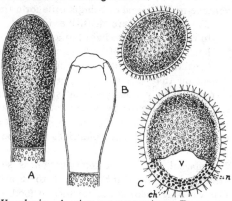

Fig. 207. *Vaucheria*. A, ripe zoosporangium; B, zoospore escaped;
　　C, zoospore further enlarged, the lower part in sectional view,
　　showing a nucleus (*n*) below each pair of cilia, chloroplasts (*ch*), and
　　vacuole (*v*). (After West.)

This reproductive body, capable of locomotion, is curiously
animal-like. It is called a *zoospore*. It moves by means of
numerous very fine threads of protoplasm called *cilia*. They are
so fine and transparent as to be scarcely visible except after
careful fixing and staining[1]; but then they are seen to be arranged
in pairs, each opposite one of the nuclei, which lie nearer the
surface than the chloroplasts.

There are many organisms such as *Volvox* and numerous uni-
cellular forms, containing chlorophyll and having
the power to assimilate carbon dioxide, which remain
motile throughout their life. Thus they combine in
one and the same organism plant-like with animal-
like characteristics. *Vaucheria* retains the power of
locomotion only for a very brief period in its life
history, yet long enough for us to realise that here
we are not far from the border line where no sharp distinction

**Algae
with
power
of loco-
motion.**

---

[1] They may sometimes be distinguished after killing and staining
in iodine, if the width of the beam of light from the mirror of the
microscope is cut down by nearly closing the iris diaphragm.

can be drawn between the animal and vegetable kingdoms. The ideas we form of the nature of a plant and an animal from the study of the more complex organisms do not help us much in considering the relationships and classification of these lowly forms.

The advantage to *Vaucheria* of being motile for a time is clear.

Response to stimuli. The zoospores are sensitive to light, swimming towards a moderate light but away from very bright light. If a dish of water containing *Vaucheria* is kept in a poor light coming from one side the zoospores collect on that side as a green scum. They are also attracted or repelled by various chemical substances in solution in the water. Thus their movements are not always aimless, but are influenced by the environment. The result is that in nature they select situations where the conditions best promote their growth.

The other special method of reproduction is sexual. The conditions that induce the plant to resort to it are not fully understood, but among them are abundance of food, bright light and approaching drought, conditions which are not infrequent in the situations where *Vaucheria sessilis* grows. Two

Sexual reproduction. short lateral branches appear close together, one slender and curved, the other short and globular. In the former a cross wall appears, cutting off the curved part, which loses its chlorophyll. This is called an *antheridium*. It contains many nuclei, each of which separates along with a little protoplasm to form a minute motile male cell, with two cilia, called an *antherozoid*. The antherozoids eventually begin to move about, and the bursting of the antheridium at the tip allows them to escape.

In the globular branch there are also at first many nuclei; all but one of these wander back into the main branch and a cross wall cuts off the *oogonium*, which now contains one large deep green cell, with a single nucleus, numerous chloroplasts, and many large oil drops. This is the *egg-cell*.

As the oogonium ripens a colourless spot appears on one side behind a beak-like protrusion of the oogonium wall. Here the wall is dissolved later and a little colourless protoplasm comes out. From the protoplasm of the oogonium some substance attractive to the antherozoids must diffuse, for they swim toward it in the film of moisture that covers the plant. Only one enter the receptive spot and passes into the egg-cell where its nucleus approaches the nucleus of the egg-cell and finally fuses with it.

The fertilised egg-cell surrounds itself with a thick wall inside the oogonium and becomes an *oospore* which rests for a few months, withstanding drought, heat or cold. When it germinates, a tubular filament, like those of the mature plant, grows out directly from it.

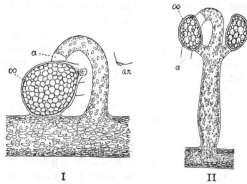

Fig. 208. *Vaucheria.* I, sexual organs of *V. sessilis*: *oo*, oogonium ready for fertilisation; *a*, empty antheridium, partly hidden; *an*, antherozoid. II, fertile branch of *V. geminata*, bearing a terminal antheridium (empty) and two lateral oogonia (fertilised). (After West.)

In contrast with *Spirogyra* there is here a great difference between the cells that unite. In *Spirogyra* the two are very similar, only differing in that one exercises a very limited power of movement. In other Algae closely allied to *Spirogyra* there is not even this difference, for the two protoplasts meet half way, in the connecting tube, and there form the zygospore. In *Vaucheria*, on the other hand, the female cell is large and non-motile, with an abundant supply of food; the male cell very small and actively motile. Their union reminds us in its essentials of the fertilisation of the egg-cell of an ovule by the male nucleus of the pollen grain ; but it is far simpler.

Fucus.

In the sea we find other seedless plants classed as Algae— green, brown and red. Most of them are larger than the fresh water Algae and some attain a considerable size. The red sea-weeds are found chiefly in the deeper waters and rock pools, the brown between high and low tide-marks, the green chiefly in shallow waters. They vary greatly in form, structure and life-history; but we must confine our study to one very common genus of brown seaweeds, *Fucus*. One well-known species is the Bladder-wrack, *F. vesiculosus*, a very much branched ribbon-like plant with bladders of air that buoy it in the water. It grows in great profusion on rocks which the sea leaves high and dry at ebb-tide, but covers again at the flow.

Bladder-wrack.

At the base is a broad holdfast by which it is attached firmly to a rock. From this proceeds a thick but flexible stem-like portion which forks repeatedly. The upper branches are thin and flat, with a slight midrib. The ends of some of them may be warty, or reddish in colour: these are reproductive branches. The smooth tips of vegetative branches end in a slight depression, in the centre of which can be seen with a lens a tiny slit: this is where the growing point is situated. If we examine carefully a number of such tips we find some with two slits close together, others with two farther apart. In other branches the midrib has already forked, and in still others the branch as a whole has begun to fork. We have here stages in the forked branching so characteristic of *Fucus*.

Mode of growth.

Farther back from the tip the branch is coarser and the midrib thicker. Farther back still the thin wings show traces of hard usage and finally disappear, leaving only the much thickened midrib[1]. We can readily convince ourselves that even the stem-like base has gone through these same changes if we examine the young plants, which usually abound on the rocks or attached to the holdfasts of the larger plants. The youngest of them are like single delicate miniature branches.

The Bladder-Wrack has thus only two kinds of vegetative organs, the holdfast and the branches. Such a plant body, which is not differentiated into leaves, stems and roots, is called a *thallus*. This term can also be applied, for instance, to the *Vaucheria* plant.

Thallus.

[1] The great increase in thickness of the midrib is the result of secondary growth, of a peculiar kind very different from that of a dicotyledonous stem. The cells grow out into filaments which run in between the other cells and so fill out the whole mass.

The thallus of *Fucus* varies in colour from olive green to reddish brown. If dipped into boiling water a rearrangement of the pigments takes place and it becomes green. It contains chlorophyll and can assimilate carbon dioxide from the sea-water. The soft gelatinous thallus is readily permeable to water, and through its whole surface can absorb carbon dioxide, oxygen and the salts necessary for its nutrition.

In a section across a young branch, an outer layer of closely packed cells is seen, in which are numerous chloroplasts deeply coloured. In the natural state they are olive-green or brown owing to the presence of a large proportion of yellow and orange pigments along with the chlorophyll. These are the chief assimilating cells, in direct contact with the water whence they derive carbon dioxide, and also fully exposed to the light. Below this outermost layer the tissues are less deeply coloured and more and more loose and irregular. They are not however separated by air-spaces, but have thick and gelatinous walls. The midrib consists of elongated cells which facilitate the conduction of food to the growing tips.

Fertile branches. The reproductive branches when ripe are orange-tinted on some plants, dark olive on others. Under a lens, minute pores can be seen dotted over the surface[1], leading to small cavities in the thallus. If left for a few hours exposed to the air orange or dark coloured mucilage oozes from the pores. Under a high power of the microscope the orange mucilage is seen to contain large numbers of pear-shaped *antheridia* filled with minute orange specks. They soon burst in sea-water and liberate numerous *antherozoids*, each with an orange chloroplast and two long cilia (Fig. 209, I), with which they swim actively about.

In the mucilage from other plants dark dots are plainly visible even to the naked eye. They are large olive spherical egg-cells without any cell-wall, with no power of movement, but with dense contents. Some of the larger dots are packets of eight egg-cells enclosed in a membrane, but in sea-water the membrane soon bursts.

Conceptacles. How the spermatozoids and eggs are produced can be seen in sections of the fertile branches. The pores lead into round chambers, called *conceptacles*, from the

---

[1] Somewhat similar pores with protruding hairs are scattered over the rest of the thallus, but they are barren.

walls of which numerous hairs radiate inwards, some protruding from the opening of the female conceptacle. In the male plant the hairs are much branched, and many of the smaller branches become *antheridia*. In the female the hairs are unbranched. Between them grow dark oval *oogonia*, the contents of which divide into eight eggs. Oogonia in different stages, some un divided, others divided into two, four, and eight, can easily be found. The hairs produce mucilage which fills the conceptacles. The mucilage exudes at low tide and antherozoids and egg-cells are washed off together when the tide returns.

The egg-cells are a little heavier than sea-water, so that they slowly sink. The antherozoids are repelled by light, so that they too move downwards, and the chance of their coming within reach of the attractive influence of an egg-cell is thus increased.

If mucilage from a male plant is added to sea-water in a

Fertilisa-tion. watch glass containing egg-cells from a female plant, the way in which fertilisation occurs can be watched under a low power of the microscope.

The antherozoids are attracted to the eggs. Around each egg

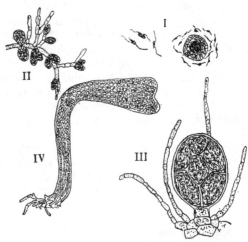

Fig. 209. *Fucus vesiculosus*. I, egg-cell and swarming antherozoids. II, branched hair from male conceptacle, bearing antheridia III, oogonium from female conceptacle with contents already divided to form egg-cells. IV, young plant or sporeling. (I after Thuret, II–IV M. G. T.)

numerous antherozoids swarm, seeking to gain admission. The active lashings of their cilia even move the egg and cause it to rotate. At last one favoured individual begins to enter the egg-cell, and at once the others leave it, being apparently repelled where the moment before they were strongly attracted. The fertilised egg-cells reach the bottom and there germinate, attaching themselves by short filaments, and growing directly into young *Fucus* plants.

CHAPTER XXVII

FUNGI

Moulds and mildews, mushrooms and toadstools, classed together as *Fungi*, differ from Algae in possessing no chlorophyll. Like the few Seed-plants that have no chlorophyll and depend for their nutriment either on other plants (parasites, p. 178) or on organic substances which they find in humus (saprophytes, p. 369), so also Fungi are parasitic or saprophytic: being unable to assimilate carbon dioxide they must absorb organic nutriment.

Fungi, along with other organisms without chlorophyll, still more inconspicuous and minute, called Bacteria, are of immense importance to mankind. Many of them infect men and animals, crops and fruit trees with disease, or spoil gathered produce. Others, on the other hand, are of value to mankind.

Bacteria are so very small that even under a high power of the **Bacteria.** microscope they appear only as tiny specks and lines. Most human and animal diseases are due to bacteria, and some diseases of plants. Other bacteria live normally and even beneficially in the human and animal intestines. Many live in soil and are responsible for the decay of animal and vegetable remains, and for various changes that occur in the carbonaceous and nitrogenous constituents of the soil: these include the nitrifying and nitrogen-fixing bacteria and the bacteria of root nodules (pp. 95–7). The activity of others turns milk sour and butter rancid, gives butter and cheese their characteristic flavours, makes meat and fish go bad, and is responsible for many other important chemical changes.

Fungi are responsible for most diseases of plants and a few of animals. Others help in the processes of decay, or ferment fruit juices. Vast numbers live on the humus of the soil, especially in woods, or on rotting wood and other vegetable and animal remains beginning to decay.

Mucor.

Our first example grows as a thick white mould, for instance on bread. If a slice that has been left on the table for a time is moistened and kept quite damp under a bell-jar, the mould wil usually appear very soon. On old growths black dots are seen, like pin heads, on white stalks.

The white growth, called the *mycelium*, consists of interwoven branching threads, called hyphae, with no cross walls. The plant is a branching tube, containing protoplasm with many vacuoles. In suitably stained preparations numerous very small nuclei are seen.

The hyphae grow at their tips. Some branches penetrate the bread and absorb nourishment from it. As the bread consists chiefly of colloidal substances, starch and gluten, the nutriment can only be absorbed after digestion by enzymes (see pages 39 and 44) which are produced by the hyphae.

Nutrition.

The black heads, borne on stout erect branches of the mycelium, are full of black spores and are called *sporangia*. In young stages, before the spores turn black, it is easily seen that the sporangium is cut off by a cross wall, and that this wall gradually bulges into it. The dense protoplasm that fills it meanwhile breaks up into many portions, each of which becomes a small round spore with its own cell-wall; but a little protoplasm remains and forms mucilage. The sporangium wall is brittle and breaks into many pieces when the sporangium is moistened and the mucilage swells.

Sporangia.

Spores are produced in large numbers. Some may germinate as soon as they are liberated, others are dried up and are dispersed in the air as a constituent of dust. Gradually the dust settles and any exposed surface like a slice of bread may receive stray spores and become *infected*. Once infected, if the required conditions of moisture and temperature are provided, it will become covered with a mycelium.

If, however, the moist bread be first heated to the temperature of boiling water or a little higher for half an hour, being scrupulously kept during the heating and afterwards from all possibility of fresh contamina-tion, no growth appears. The spores have been killed and the bread is *sterile*. In this condition it will keep indefinitely. Antiseptics and poisons, too, like carbolic acid, boracic acid, mercuric chloride, etc., prevent the growth of mould either by killing the spores or by hindering them from germinating.

Sterilisa-tion.

Occasionally, when mycelia derived from two different spores meet, dark thick-walled *zygospores* are found. They are formed by the conjugation of two similar hyphae end to end. These hyphae swell up as

Conjuga-tion.

they approach each other, a small cell is cut off from the end of each, and these cells expand and unite. The ripe zygospore has a thick, dark brown, warty coat and is very resistant. After a long resting period, when conditions favour germination, the zygospore bursts its outer wall. It does not, however, form a mycelium, but begins at once to multiply and spread the mould by producing one or more hyphae, each of which ends in a sporangium.

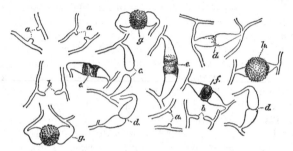

Fig. 210. *Mucor*. Stages in conjugation (*a–g*); *h*, zygospore. (After Darwin.)

Penicillium.

On bread which is kept less moist, though under cover to prevent it from drying up, other moulds appear. They include a green mould, *Penicillium*, which also occurs very frequently on jam, cheese, mouldy boots, cloth, etc.

Under the microscope, when the mould is examined dry and lighted from above, green cylindrical powdery bodies are seen arising from a mycelium. For further examination a little of the mould is picked off with a needle, moistened in alcohol and mounted in water. The hyphae have cross walls, though at rather long intervals, and are more slender than those of *Mucor*, but they attack the material on which they grow in a similar way.

From the vegetative mycelium arise stouter hyphae. These branch repeatedly, and the branches grow erect and close together. Each ultimate branch bears a bead-like string of spores, called *conidia* to distinguish them from spores that are formed inside a sporangium. The end of the branch forms a small swelling, which grows like a bubble and having reached a certain size is cut off by a constriction of the

Conidiophores.

branch immediately under it. Meanwhile below the constriction another conidium has begun to form, and as, under natural conditions, they do not immediately break off, long chains of conidia are formed, the oldest at the top, the youngest next the branch. It is the cluster of parallel chains borne on the branches of one fertile hypha (or *conidiophore*) that are seen as cylindrical bodies in a surface view of the mould.

These conidia are dry, light and very numerous. Like the spores of *Mucor* they are present almost everywhere in the dust of the air and germinate as soon as they settle on suitable damp material.

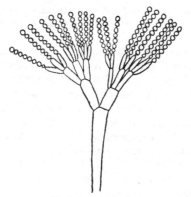

Fig. 211. Conidiophore of *Penicillium* (M. G. T.)

Penicillium has another mode of reproduction, in the course of which it forms spherical "fruits" called *perithecia*, large enough to be seen with a lens; but they are very rarely found. It depends almost entirely upon its conidia for multiplication, dispersal and survival.

Aspergillus (Eurotium).

Another mould very closely allied to *Penicillium* called *Aspergillus* forms perithecia readily. They are found as tiny yellow balls on old mycelia, for instance on mouldy bread that is getting dry[1]. The conidiophores of this mould have swollen heads bearing many projecting points, from each of which

[1] Their formation is promoted and hastened by incubating at 35° C. for several days.

conidia are abstricted, the radiating chains together forming a spherical mass. The colour of the mould changes from white through blue (conidia) to yellow (perithecia).

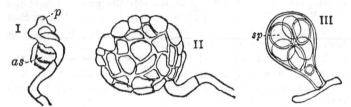

Fig. 212. *Eurotium.* I, coiled female hypha (*as*) with male hypha (*p*). II, perithecium. III, ascus containing ascospores (*sp*). (After De Bary.)

Perithecia. The perithecia are attached to the mycelium by thicker brown hyphae. When the coverslip is tapped with a needle the brittle wall is burst and small oval sporangia escape, each containing eight spores. Similar sporangia with eight spores are characteristic of a large class of Fungi, and have therefore received a special name, *ascus*; the spores are called ascospores, and the class Ascomycetes.

Sometimes stages in the formation of perithecia may be found. A hypha coils up into a close spiral, and another grows up over it. Occasionally they unite, and it seems therefore that the branches are sexual organs; but more usually the second, male branch, or antheridium, withers before any union has occurred. In either case, however, a perithecium develops. Sterile branches grow up around the coil and closely invest it, forming a mass of interwoven nutritive hyphae enclosed in a hardened outer layer. From the female branch, or *ascogonium*, branches grow which push their way into the nutritive tissue, branching repeatedly. Finally at the end of each branch an ascus is formed. The female organ here forms no resistant resting spore, but provides at once in a peculiar way for the multiplication and spread of the mould, fed and protected meanwhile by the investing branches of the mycelium that bore it.

Though sexual organs are formed, fertilisation does not always occur, so that the true sexual process is often omitted from the life-history. In very many Fungi we find traces of sexual processes which have degenerated, but which probably played in the life-history of their ancestors a much more importan part. *Penicillium* seldom resorts even to the degenerate mode of sexual reproduction of which it is still capable.

Yeast.

Brewers' yeast is obtained as a thick creamy froth from fermenting liquor. Shaken up with the liquor it makes the latter cloudy. Under the microscope the cloudiness is seen to be due to numerous very small cells, some of them single, others **Budding.** attached together in branching chains. Under the high power, cells are found here and there with small protuberances. When one of these is kept under observation the "bud" is seen to grow into a cell like the one producing it, much as a conidium is formed by *Penicillium.* The new cell in turn produces another and, by rapidly repeated budding, chains and groups are formed; but sooner or later the cells separate.

Each cell is rich in protoplasm, and has a small round vacuole, on one side of which is a nucleus which stains deep yellow in iodine. In budding cells the nucleus divides, and one daughter-nucleus passes into the bud along with part of the vacuole.

When well nourished Yeast is mounted in iodine, deposits of a carbohydrate called glycogen are seen, stained orange-red. This is the form in which Yeast, and also many other Fungi, store carbonaceous food. They do not form starch, but glycogen and fat. In this they are like animals, which, in addition to forming fat, store glycogen in the liver. For this reason glycogen is also called *animal starch.*

Yeast is remarkable because it is able, if supplied with sugar, **Alcoholic** to live and grow for a long time without oxygen[1]. **fermenta-** Life under such conditions is called *anaërobic.* **tion.** Ordinarily growth is only possible in presence of oxygen (i.e. under *aërobic* conditions) because only by respiration can the energy that is stored in carbonaceous food be liberated and made available for the work of growth (p. 47). In fermentation, however, yeast liberates some of this stored energy by decomposing sugar into carbon dioxide, water and alcohol. As in respiration, some of this energy is lost as heat, which raises the temperature of the fermenting liquor.

---

[1] A suitable culture solution is Pasteur's, made up as follows: 1 g. potassium phosphate, $\frac{1}{10}$ g. each calcium phosphate and magnesium sulphate, 5 g. ammonium tartrate, 75 g. sugar, in 450 c.c. water.

The rise of temperature can be readily observed if two small flasks of Pasteur's solution[1], to one of which yeast has been added, are packed in cotton wool to hinder the escape of heat, a thermometer placed in each (the necks being plugged with cotton wool) and the temperatures recorded at intervals.

That carbon dioxide is formed can be shown by filling a test-tube with fermenting solution and inverting it in a beaker of the solution. The gas thus collected dissolves if potash is added.

In bottled liquors fermentation continues slowly for a time and the carbon dioxide accumulates in solution. It can only escape when the bottle is opened and then forms a frothy "head" on beer or brewed ginger-beer, and makes cider or champagne sparkle.

Yeast can, however, under other conditions live an aërobic life, respiring in the ordinary way. If spread on filter paper and in this condition put into a bottle of air, the oxygen is used up (see Expt 17, p. 49 for tests, etc.).

Fermentation can only take the place of respiration and make anaërobic life possible if the Yeast is supplied with sugar of the right kind. Grape sugar and cane sugar, for example, are fermentable, but not milk sugar. Yeast can on the other hand use milk sugar as a source of carbon and energy in respiration, and in presence of oxygen will flourish upon it if given a little nitrogenous food, such as peptone, in addition. Even peptone alone will serve as both nitrogenous and carbonaceous food under aërobic conditions.

Some species of *Mucor* and the conidia of some moulds form yeast-like budding cells and chains in sugary liquids and ferment them; in fact Mucor in this form is known as Mucor-yeast. For this reason the Yeasts are thought to be degenerate moulds that have taken permanently to life in liquids.

If Yeast is starved and well supplied with oxygen, so that it uses up its reserves, it forms spores within the cell, four in number, each with its own cell-wall[2]. In a nutritive solution they germinate by budding.

---

[1] See note on p. 487.

[2] Spores can usually be seen if active brewers' yeast is spread thinly on a glass slide and kept in moist air at 25° C. for a few days. In Pasteur's solution they germinate in a few hours at room temperature.

Peronospora.

Peronospora parasitica infects Shepherd's Purse, Field Mustard and other Cruciferae with Downy Mildew. The infected parts of stem, leaf or even fruit are very much swollen, smooth and whitish, with a slight mildew on the surface. *Peronospora* is often associated with another fungus, *Cystopus* (or *Albugo*); but this breaks through the epidermis and exposes a powdery white surface of conidia that have formed underneath—hence the name White Rust given to the disease.

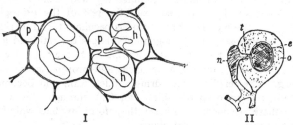

Fig. 213. *Peronospora*. I, hypha with haustorium in cell of host-plant. II, sexual organs after fertilisation showing antheridium (*n*) and oogonium (*o*), fertilising tube (*t*), and fertilised egg (*e*). (II diagrammatic M. G. T.)

In a section through an internode infected with Downy Mildew (with unbroken epidermis) *Peronospora* hyphae are seen filling the spaces between the living cells, and sending into the cells lobed structures, of dense protoplasm. These absorb food from the cells of the host plant and are called *haustoria*. It is remarkable that they do this without killing the cells they enter, or even at first sight greatly disturbing the protoplasm. But, compared with a section through a corresponding healthy internode, the infected cells are much larger, and the mechanical tissues are poorly developed.

At the mildewed surface hyphae project through the stomata and form conidiophores, which fork repeatedly and produce conidia singly at the tips of the branches. These are dispersed readily by the wind. If they fall on other suitable plants they germinate at once, requiring little moisture, and produce hyphae which enter by way of the stomata. In this way the disease rapidly spreads from plant to plant. Conidia must of course fall

Disease spread by conidia.

on many plants other than those few species of Cruciferae which are liable to the disease, but for various reasons cannot infect them—in other words these plants are *immune* to the disease. Parasitic fungi, especially those which like *Peronospora parasitica* show a very delicate adjustment to their hosts and infect without killing, are usually limited to a few host-plants.

When the disease is well established, sexual reproduction occurs within the host-plant. At the tips of certain **Sexual repro- duction.** hyphae oogonia are formed and beside each a slightly swollen club-shaped antheridium grows up. Within the oogonium a rounded egg with a single nucleus is differentiated from the rest of the protoplasm around it. The mode of fertilisation is peculiar: the antheridium applies itself closely to the oogonium, pierces the wall and forms a narrow tube through the outer protoplasm to the egg. By way of this fertilising tube a male nucleus and protoplasm pass into the egg. The fertilised egg forms a thick wall, and becomes a resistant resting oospore which can last through the winter. When these oospores germinate they form hyphae which can enter and infect young host-plants.

Agaricus.

The edible Mushroom grows up very rapidly from the ground among the dark green grass of "fairy rings." The thick stalk bears the umbrella-like *pileus*, on the under side of which are the *gills* that darken from pink to black as the mushroom ripens. If the pileus be cut from a ripe mushroom and laid on a piece of white paper, after a few hours a pattern in black dust is found corresponding to the radiating gills. Under the microscope the dust is seen to consist of dark oval spores.

A section across the gills shows how the spores are formed. **Gills.** The gill is made up of hyphae which pass along the middle and turn outwards forming at the surface a close palisade-like layer called the hymenium. Here are found slightly swollen club-shaped hyphae, surmounted by two curved points, each bearing a spore at the tip. These spores are evidently conidia. Conidiophores of just the same type, though more often with four spores, are characteristic of a large class of Fungi called the Basidiomycetes, including the toadstools, bracket-fungi, and many others. The conidiophore is called a *basidium*, and the conidia *basidiospores*.

The mushroom, then, is a large spore-bearing organ. It requires for its growth quantities of food which must be somehow absorbed and passed up the stalk to the pileus and gills. If we trace the stalk below ground we find it attached to many white strands of varying thickness. Each strand is a bundle of hyphae, and is connected with smaller strands and single hyphae which penetrate the earth in all directions.

Vegetative mycelium.

Humus containing an abundance of this mycelium is what we know as "mushroom spawn." To cultivate mushrooms the spawn is added to a mixture rich in manure. The mycelium absorbs organic substances for its nutrition from the humus: like *Mucor* and *Penicillium* it is a saprophyte. This mycelium is the vegetative body of the fungus, *Agaricus campestris*, of which the edible mushroom is the fructification.

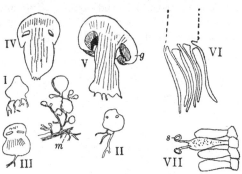

Fig. 214. *Agaricus campestris*. To the left are shown, I, the mycelium (*m*) with young fructifications, and (II–V) longitudinal sections through fructifications of various ages; to the right, sections through the fructification showing the gills (*g*): VI, several gills seen with a lens; VII, a portion of the surface of a gill seen under a high magnifying power, bearing two spores (*s*). (I–V modified from De Bary; VI and VII, M. G. T.)

The fructification appears first as a little white button which gradually enlarges. The upper part swells and within it the gills are developed. The thin layer of tissue that covers them in below becomes stretched as the pileus expands, forming a thin veil which is left as a frill round the stalk when the pileus finally tears away from it.

Fructification: mode of growth.

How regularly the gills and the hyphae that compose them are arranged we have already seen. The remarkable feature of *Agaricus*, and indeed of all the numerous fungi with conspicuous fructifications, is the way in which hyphae, which are apparently quite independent of one another, cooperate to form so highly complex a structure. The association of the vegetative hyphae into strands is a step in this direction. But what stimulus or impulse starts the growth in unison of numerous hyphae and how their combination is regulated are mysteries—though, indeed, we are as far from understanding the orderly development of any organism.

Agaricus campestris is propagated chiefly vegetatively. In fact, until comparatively recently no one had succeeded in getting the basidiospores to germinate. It has been found, however, that they will germinate if pieces of the fructification are placed near them. If this is the essential condition, *Agaricus* must depend for dispersal on the fructification as a whole—the dispersal of the basidiospores alone will not distribute the species.

CHAPTER XXVIII

MOSSES AND LIVERWORTS (BRYOPHYTA)

Pellia.

The Mosses and Liverworts are small leafy plants or flat green thalli. *Pellia* is found in abundance on damp shady banks and by ditches, or growing in woods on damp soil, where the atmosphere is moist. The thallus is of a dark translucent green, with a midrib which bears on the under side numerous rooting hairs called *rhizoids*.

The apical growing point is situated between two lobes which extend beyond it, and it is protected by short glandular hairs. Like *Fucus* the thallus forks repeatedly, but except when it grows under water the branches are short; in drier situations they are stunted, more numerous, and overlapping.

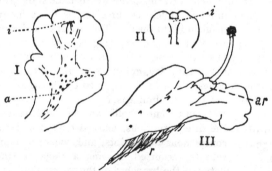

Fig. 215. *Pellia.* I, thallus about July showing many antheridia (*a*) and an involucre (*i*) (compare Fig. 217, I). II, a branch in spring showing young involucre between two growing points. III, plant in spring bearing a sporogonium with stalk elongated and capsule dehisced; *ar*, broken venter of archegonium; *r*, rhizoids.

In a section across the thallus the wings are seen to be a few cells in thickness, the midrib much thicker. There are no air-spaces through which gases can circulate to and from the inner cells. Round chloroplasts of the ordinary form are numerous in the two or three uppermost layers, less numerous in the lowest layer, and few in the interior where the supply of carbon dioxide is poor. The greenest, assimilating cells have small starch grains in their chloroplasts. The inner cells contain large grains and act primarily as storage cells: the chloroplast that has formed each large grain can be seen on one side of it, stretched over its surface.

"Fruits." In winter and early spring very dark green knobs are found partly covered by a green flap that forms a pocket between two branches. When the flap, or *involucre*, is removed a white stalk is disclosed with its lower end buried in the thallus. The stalk is rather easily broken; but by carefully dissecting the thallus away the lower end can be discovered to be not actually a part of the thallus though embedded in it. It is a conical moist absorbing organ, called the *foot*.

The contents of the head, or *capsule*, when dissected out under the microscope are seen to consist of large, dark green spores, each already divided into several cells, intermingled with curious hair-like *elaters*, with spiral markings.

The whole structure, including capsule, stalk and foot, is called a *sporogonium*. It is already fully developed in the autumn. At the first approach of warm spring weather the stalk grows very rapidly, sometimes as much as three inches in as many days, carrying the capsule up into the air. There it dries and soon opens by splitting into four valves, exposing the spores in a fluff of elaters attached to the base of the capsule. The elaters execute jerky movements as they dry, gradually liberating the dark green spores from amongst them to be dispersed by the wind.

How does this sporogonium arise?

Origin of sporogo-nium. About June, thalli can be found, especially in the drier situations, marked on the upper side by conspicuous raised dots; and, where branching has recently occurred, bearing a ridge or small flap between the branches of the midrib.

Sexual organs. Vertical sections along the midrib of such thalli show that each raised dot is a narrow entrance to a hollow cavity, from the bottom of which arises a shortly stalked oval *antheridium* (Fig. 216, *a*). This is more complex in structure than the antheridium of *Vaucheria* or *Fucus* for its wall is a layer of cells, and the interior also is divided into a large number of small cells. In each of these cells is formed an antherozoid like a spirally coiled thread with two cilia. The cell walls dissolve into mucilage, and when the thallus is wetted by rain or dew the antheridium opens at the tip and the swelling mucilage exudes, carrying with it the antherozoids.

Where a section passes through the fork, the midrib appears to end abruptly with a steep face, from the top of which the flap, the young involucre (Fig. 216, *i*), projects over a thin continuation of the thallus below. On the steep face are thick hair-like structures, bulbous at the base, called *archegonia* (Fig. 216, *ar*). The long *neck* has a central canal with granular contents; in old archegonia it is open at the top and the contents are brown, but in younger ones it is still closed. The neck-canal gives access to an enlarged cavity in the base (the *venter*) in which a lump of dense granular protoplasm can easily be distinguished if the section be mounted in dilute potash: this is an egg-cell. Like the antheridium, the archegonium is a more complex structure than the oogonium of *Fucus* or *Vaucheria*.

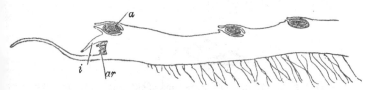

Fig. 216. *Pellia*. Longitudinal section through thallus; *a*, antheridia; *ar*, archegonia; *i*, involucre.

The contents of the neck-canal consist of several cells that disorganise and form mucilage, which swells and exudes when the plant is wetted. Water is readily held within the involucre and between the overlapping branches, and in it antherozoids, liberated at the same time, make their way towards the open neck, guided by the attraction of some chemical substance[1] diffusing from it. On reaching it an antherozoid swims down the neck-canal and enters the egg-cell.

The sexual fusion is followed by the formation of a cell-wall round the fertilised egg-cell; but it is not liberated from the archegonium to live an independent life. It remains within the venter and there grows. Very soon the conical foot is formed, and as the embryo is confined by the venter its growth in length forces the foot through the thick stalk of the archegonium into the cushion of tissue below. Its progress is not however merely mechanical: its outer layer of

Mode of growth and nutrition of sporogonium.

[1] Probably sugar, by which they are strongly attracted.

cells becomes a digestive epithelium full of dense protoplasm, with large nuclei, and around this the digested and compressed remains of cells are clearly seen in longitudinal sections.

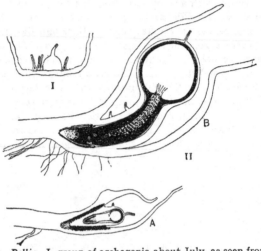

Fig. 217. *Pellia.* I, group of archegonia about July, as seen from above under a lens after cutting away the involucre: one has been fertilised and the venter is expanding round the growing embryo. II, vertical longitudinal sections through branches bearing sporogonia of different ages, *both drawn to the same scale*: A, late summer; B, late winter, ready for elongation of stalk. The regions where starch is abundant are marked by stippling. In A, except for the nutritive tissue surrounding the foot, and also in B, the unshaded parts of the thallus contained scattered starch grains. In B the capsule contained mature spores and elaters.

The young sporogonium is a welcome parasite, for before any egg-cell had been fertilised a special nutritive tissue was formed, small-celled and thin-walled, with prominent nuclei. Around it the ordinary cells of the thallus store an abundance of very large starch-grains. As the sporogonium grows the foot enlarges, until eventually it has absorbed and usurps the place of the nutritive tissue. Meanwhile starch disappears from the neighbouring store and appears abundantly in the foot, stalk and capsule wall.

Notwithstanding the increasing size of the sporogonium the venter of the archegonium is not burst: it grows, and the neighbouring tissue of the thallus grows up also, helping to enlarge the cavity. In this way a thick protective envelope is formed; at the top of it the archegonium neck still remains and the other arche-gonia are carried up on its sides. This envelope is only broken through when the stalk begins its rapid elongation in the spring, using up in the process its abundant store of starch.

Formation of spores. Meanwhile within the capsule the cells have separated from one another. Some elongate to form the elaters; others richer in protoplasm become *spore-mother-cells*, each of which bulges out in four places to form four spores. The spores finally separate, but before they are released from the capsule they begin to grow, dividing into several cells; in this, *Pellia* is exceptional, the spores of most Liverworts being single cells.

When a spore reaches a suitable moist spot it grows into a new thallus.

Alternating generations contrasted. This life history illustrates the remarkable phenomenon of one plant producing as its offspring another bearing no resemblance to it—contrary to the law of heredity, as ordinarily expressed, that the offspring is like its parent.

The difference is extreme. In one generation we have the thallus which lives an independent life, grows flat on the ground and bears sexual organs. In the next appears the sporogonium which is erect, showing radial symmetry; is throughout its life largely dependent, feeding like a parasite on the tissues of its parent (though it contains some chloroplasts and the spores are deep green); and gives rise to asexual spores which are dispersed in the air. These two distinct generations alternate regularly, the sexual thallus producing the asexual sporogonium and the latter in its turn producing the thallus. The reproduction of either requires two generations for its completion.

Marchantia.

The life history of *Marchantia* corresponds very closely with that of *Pellia*.

The thallus is larger and coarser, with longer branches. Structures like rings or cups[1] are often found on the upper surface in which are little flat green bodies, called *gemmae*. They grow

---

[1] *Marchantia* is readily distinguished by the form of its gemma-cups from another very similar common Liverwort, *Lunularia*, where the gemmae are only enclosed by a crescent-shaped ridge on one side.

up from the bottom of the gemma-cup on short stalks. They are
easily detached, and if laid on damp earth grow into new thalli.
They may be compared to the bulbils of some seed plants, serving
as a means of vegetative propagation.

Underneath, the midrib bears rhizoids; and in addition there
are whitish scales running across the thallus, which when young
closely overlap the growing point.

The upper surface is marked out into roughly rhomboidal
areas with a large pore in the centre of each. A
Assimi- section shows that these areas are chambers, the
lating pore in the roof communicating with the outer air.
chambers. From the floor of each chamber rise many short
rows of cells with numerous chloroplasts. These
assimilating cells thus have direct access to the carbon dioxide of
the atmosphere. The bulk of the thallus below has very little
chlorophyll, and like *Pellia* no air-spaces.

The construction of its assimilating chambers gives *Marchantia*
distinct advantages over *Pellia*. Plant membranes
Assimila- highly permeable to carbon dioxide are also per-
tion and meable to water, and a thallus like *Pellia* with its
transpira- absorbing surface outside and fully exposed cannot
tion. protect itself against evaporation of water without
permanently hindering the entry of carbon dioxide
and retarding assimilation. In *Marchantia* the assimilating surface
is *internal* and is many times greater than the external surface
which protects it. Such a division of labour between an internal
assimilating surface and an external protective surface, with pores
for the entry of carbon dioxide into the internal atmosphere, is
also characteristic of the higher land-plants. But there a system
of intercellular spaces provides the internal absorbing surface,
protected by the cuticularised epidermis with its adjustable
stomata. These, in combination with the very efficient water-
absorbing and conducting systems possessed by Seed-Plants,
make it possible for them to grow up into dry air.

It is interesting to find that, although our common British
Marchantia polymorpha frequents moist habitats, many species of
Marchantia and other genera with similar structure grow in dry
situations and have the power to close their pores in times of
great drought.

Even in *M. polymorpha* the possibilities of this type of structure
are realised to some extent in the extraordinary reproductive

branches, called archegoniophores and antheridiophores, on which the sexual organs are borne aloft. The archegonia and antheridia themselves are very like those of *Pellia*, but they are borne on different plants, the thalli being either male or female.

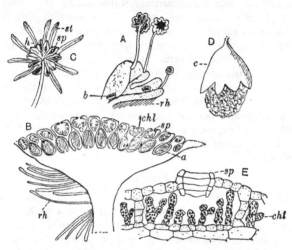

Fig. 218. *Marchantia polymorpha*. A, thallus with gemma-cups (*b*), rhizoids (*rh*) and antheridiophores. B, section through young antheridiophore, showing antheridia (*a*) alternating with assimilating chambers (*chl*). C, head of archegoniophore seen from below, showing rays (*st*), flaps (*h*) and young sporogonia (*sp*). D, sporogonium dehisced, *c*, ruptured venter of archegonium. E, Section through wing of thallus, highly magnified, showing one chamber with pore (*sp*) in roof, *chl*, assimilating tissue in roofed chamber. (M. G. T.)

The *antheridiophore* is a lobed disc on a long thin stalk. The **Sexual organs.** cavities in which the antheridia are formed are on the upper side of the disc, surrounded by assimilating chambers. The *archegoniophore* is more complicated. From the top of a similar stalk project a number of narrow rays, between which are flaps of tissue enclosing radial rows of archegonia, below, in different stages of development.

The stalks of both antheridiophore and archegoniophore are seen in section to be narrow much elongated portions of the thallus. On one side are assimilating chambers, on the other are two strands of rhizoids enclosed in grooves. The fertile expansions above are repeatedly branched and much modified parts of the thallus, the under surface bearing rhizoids that run down the grooves in the stalk to the soil. These enclosed strands of rhizoids act like a rudimentary water conducting system for the supply of the structures above. When moisture is abundant it is drawn up among the rhizoids like oil up a lamp wick.

Sporo-gonium. The sporogonia develop much like those of *Pellia*, enclosed in the enlarging venter of the archegonium, with the absorptive foot buried in nutritive tissue, but head downwards. When they are ripe their stalks grow just long enough to burst the envelopes and push the capsules beyond the screening flaps, where they open irregularly and expose a golden yellow mass of spores and elaters.

As in the case of *Pellia* the spores give rise on germinating to new thalli.

Funaria.

The Mosses are very small leafy plants that grow crowded together, often in much drier places than *Pellia* or *Marchantia polymorpha*. In dry weather they may even dry up completely, showing that the loss of moisture from their leaves is more than their absorbing and conducting systems can cope with. Many of them, however, are capable of surviving for some time in the dry condition and recover when wetted again. Their leaves readily absorb rain or dew, and depend very largely on such direct absorption, the supply of water from below being very inadequate.

They vary in size and habit, some growing erect and tufted, others creeping; but the main features of structure and life history are very similar in all.

Moss-plant. *Funaria hygrometrica* has fairly broad pointed leaves borne spirally on a thin brown erect stem, the youngest at the top forming an apical bud round the growing point. They have a midrib, but otherwise are only one cell in thickness.

The plant has no roots. Rhizoids grow from the lower part of the stem and attach the plant to the earth or the crumbled mortar of a wall.

The plants grow in tufts. Some of the shoots in a tuft are branches arising low down on others, not in the axil of a leaf but immediately below it.

The leafy plants, like the thalli of *Pellia* and *Marchantia*, are
sexual. At the top of some shoots are found rosettes
Sexual   of expanded leaves. If a rosette is teased apart
organs.   with needles under the microscope, among the
leaves are seen long oval antheridia along with
curious green hairs with swollen heads. Except for their length
the antheridia are similar in structure to those of *Pellia*, with a
thin wall of large tabular, or flat, cells enclosing numerous small
cells, from each of which comes an antherozoid. The walls of these
antherozoid mother-cells become mucilaginous, and when moisture
is abundant the mucilage swells, bursts the antheridium at the
apex and exudes.

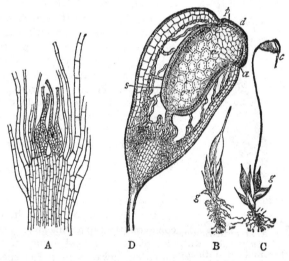

Fig. 219. A, section of stem of moss bearing archegonia. B, moss
plant bearing young sporocarp formed from one archegonium. C,
moss plant with mature sporocarp. D, section of sporocarp. *s*, layer
of cells which form the spores. *c*, cap. *p*, teeth; *d*, lid; *a*, rim.
(After Sachs.)

Lateral branches growing from these male shoots end in a
green bud, of large overlapping leaves. If this bud is dissected,
archegonia like those of *Pellia* are found seated on the tip of the
branch surrounded by the leaves.

Fertilisation occurs when the plants are wet. The fertilised
egg-cell grows rapidly in length, and the lower end,

**Sporo-
gonium.**
or foot, becomes sunk in the tip of the shoot, from
which it absorbs nourishment. For some time the
venter of the archegonium grows and keeps pace
with the sporogonium, but long before the latter is fully grown it
tears the sheathing venter across and carries the upper part on
its apex like a fool's cap (*c*, Fig. 219 C).

Unlike the sporogonium of *Pellia* that of *Funaria* grows
apically and postpones the development of the capsule till rela-

Fig. 220. *Funaria hygrometrica*. A, an antheridium
bursting; *a*, mucilage with
antherozoid. B, anthero-
zoids enlarged, *b*, still enclosed in mother-cell; *c*,
free. (After Sachs.)

tively late. The mature capsule has
a complicated structure. The lower
third consists of assimilating tissue,
with large intercellular air-spaces communicating with the atmosphere through
stomata (easily seen in a surface section)
very like those of Seed Plants. In the
middle of the capsule is a column of
parenchymatous tissue. Outside this is
a thin cylinder of spore-forming cells,
closely sheathed by other cells on both
sides, and separated by very large air-
spaces from the outermost tissues. Each
spore-mother-cell divides to form four
spores.

By the time the ripe capsule dries,
all the tissues inside have shrivelled and
the spores have fallen loose in the cavity.
At the top of the capsule is a lid which
comes off (the cap having already fallen)
and discloses a double row of teeth,
radiating inwards from the specially
strengthened rim. These teeth are hy-
groscopic: in dry weather they bend
upwards, allowing the spores to escape,
but in moist weather they close the capsule again.

When the spores of *Funaria* germinate they do not immediately
grow into moss plants but form branching *filaments*

**Proto-
nema.**
that might easily be mistaken for an alga. They
are composed of cells, end to end, with oblique
dividing walls and chloroplasts like those of the

leafy plant. Each branch grows at the tip, the apical cell alone dividing. Some branches penetrate the soil like rhizoids and lose their chlorophyll.

This filamentous *protonema* is found as a green tangled covering on moist earth, rather brighter in colour and more delicate than *Vaucheria*. In damp, shady places it may continue for a long time growing and branching. But in more open situations short stout branches presently arise here and there, very different from the ordinary filamentous branches. At their growing point a *three-sided* apical cell is cut off by walls inclined to one another. This growing point builds up a cylindrical stem clothed with leaves—the ordinary moss plant.

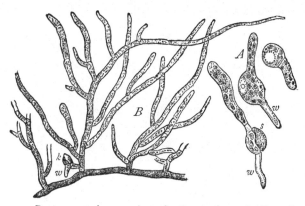

Fig. 221. Protonema of moss. A, early stages of germination of spore. B, developed protonema on which young bud, *k*, *w*, is formed, which will give rise to the moss plant. *s*, spore; *w*, rhizoid.

The rhizoids that grow from the stem are just like the rooting branches of the protonema; and, indeed, if they find an exposed patch of soil may turn green, branch and form a protonema like that which develops from the spore. Any living part of the plant if laid on a damp surface will produce similar branching filaments, and on them moss plants may develop. This is a ready and effectual method of vegetative propagation.

In *Funaria* we thus have a life history with an additional complication. One generation begins as the alga-like protonema; and this gives rise vegetatively to the sexual leafy moss-plants.

The spore-bearing generation is dependent on the sexual plant;
but in later life is only a partial parasite. Both in
The two its mode of growth and in its structure it is more
genera- highly organised than the sporogonium of *Pellia*
tions. or *Marchantia*.

The capsule possesses a system of intercellular
air-spaces, stomata and assimilating tissue which are very similar
indeed to those of an ordinary leaf. Unlike the parent moss-plant
it draws its water-supply chiefly from below through its foot. It
can close its stomata and, being provided with a protective epidermis, is able to withstand temporary drought, even after the
archegonial cap has fallen, and can maintain its turgidity, although
exposed above the general level of the moss from which it arises,
until the spores are ripe for dispersal.

CHAPTER XXIX

PTERIDOPHYTA

The Bracken Fern.

Ferns grow most abundantly in the western parts of the country where the rainfall is high. They frequent woods and shady banks, or nooks and crannies where humous soil accumulates. Their leaves, the fern fronds, grow from underground rhizomes or root-stocks, usually in radical tufts; for the size of the stem they are very large, and few are produced in one season.

The Fern chosen for our study, which is in some respects exceptional, is the most abundant, vigorous and widespread among British Ferns. It is found in the drier eastern parts of the country almost as abundantly as in the west. The rhizome penetrates to some distance below the soil, sending into it long branching adventitious roots, and its rapid growth spreads the plant till it covers wide expanses. It not only occupies open spaces in woods but invades heath and moorland, where the dense shade cast by its fronds in summer and the covering of dead brown fronds on the ground combine to smother most other plants and change completely the character of the vegetation[1].

The growing point of the rhizome is covered thickly with brown hairs. Each growing point produces annually only a single frond; but the fronds are exceptionally large and spreading, even reaching in places the height of a man.

The frond emerges in spring from the ground with the tip turned over and tightly coiled. Gradually it uncurls and expands; but a comparison of young with older fronds makes it clear that in addition to the uncurling and expansion of parts already formed, new parts are continually being added at the margins. For a long time the upper parts of the frond remain coiled up and continue to grow.

The ultimate branches of the frond, called *pinnae*, have a midrib, along either side of which the blade is very deeply divided into segments, called *pinnules*. The small lateral veins of the pinnules fork in a manner which is characteristic of the Ferns.

[1] On the distribution of Bracken see pp. 458—460, and Fig. 203, p. 458.

Branches arise in the Bracken Fern not in the axils of the leaves but as outgrowths from the base of the leaf-stalks.

Ferns not only have true roots but are also provided with very efficient vascular conducting tissues. In a transverse section of the rhizome of the Bracken Fern is seen an outer zone of dark brown protective tissue with fairly thick walls, enclosing a parenchymatous ground tissue, of starch-filled cells, in which are arranged, as shown in Fig. 222, groups of conducting tissue and dark brown bands of fibres.

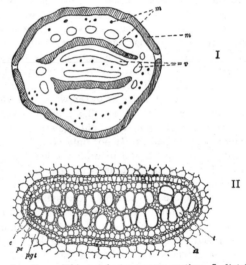

Fig. 222. Bracken Fern. Rhizome in transverse section. I, distribution of tissues: *v*, groups of vascular tissue; *m*, mechanical tissue. II, single group of vascular tissue: *c*, endodermis; *t*, tracheid of xylem; *s.t.* sieve-tube; *p.c.* pericycle; *p.g.t.*, ground tissue.

Each group of vascular tissues is surrounded by its own endodermis. It has a central mass or band of xylem. Sieve-tubes surround the xylem, separated from it by starch-containing cells. In longitudinal sections the xylem is seen to be composed of tracheids with ladder-like markings, thickening bands being laid across between elongated pits. The sieve-tubes are like narrow tracheids in shape: they have no prominent transverse sieve plates, but the greater part of the walls between adjacent sieve tubes are covered with perforations.

The leaf is very similar in general structure to the leaves of herbaceous seed plants, having an epidermis, assimilating tissue, and vascular bundles in the veins. The assimilating tissue has a palisade-like arrangement above and is more spongy below. Stomata are present in the lower epidermis; the ordinary epidermal cells as well as the guard-cells contain chloroplasts.

Sporangia. The edges of most pinnules of mature fronds appear as if folded under. As the summer advances, the folds become more prominent and eventually are lifted, disclosing numerous pale green and brown, stalked *sporangia*. The ripe sporangia are shaped like a thick lens and have a conspicuous rib, called the *annulus*, running up one edge and over the top. Half way down the other edge is a weak place in the wall of the sporangium where it breaks open to liberate the spores. In sporangia that have dehisced, the annulus is often found bent right back and carrying the upper half of the torn sporangium wall.

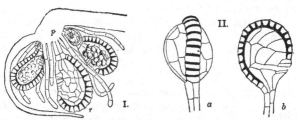

Fig. 223. I, section through edge of fertile pinnule of Bracken Fern, showing placenta (*p*), bearing sporangia and hairs under folded edge of pinnule. II, *a*, sporangium seen from back of annulus; *b*, sporangium seen from one side after dehiscence and the return of the annulus to its original position.

It is the annulus which brings about dehiscence. Under the microscope its cells are seen to have thin outer walls but thick walls separating them from one another next the interior. As the water dries from the cells and diminishes in volume it pulls on the cell-walls, to which it adheres. It thus tends to draw the thick cell-walls together and bend the annulus upwards; but, so long as the sporangium is whole, it is only the thin, extensible cell-walls that yield, and they are pulled inwards. Eventually the strain is too great and the sporangium gives way at the weak spot already mentioned. If the dehiscence is sudden enough the jerk may break the water in the cells of the annulus away from the cell-walls and then the annulus flies back again into its original position. In this double movement the spores are flung out very effectively.

In sections across the pinnule, sporangia in various stages of development are seen, attached to a thickened placenta in the angle of the flap. During the development of the young sporangium a central mass of spore-mother-cells and a layer of nutritive cells, which are eventually absorbed, are differentiated within the wall. As in the spore-capsules of Mosses and Liverworts, the spore-mother-cells separate from one another and each forms a group of four spores.

The formation of sporangia in large numbers on the backs of the fronds is characteristic of Ferns. In most Ferns they are found in small groups, called *sori*, on the under surface of the pinnules, each over a vein, and covered in usually by a special membrane called the indusium. The classification of Ferns depends very largely on the shape of the sorus and indusium, and of the sporangium and annulus.

Fern spores, like those of Mosses and Liverworts, are very small and light, and float like dust in the air, only gradually settling when the air is still. Hence it is that ferns are among the earliest plants to arrive on newly-formed volcanic islands at a long distance from land.

Prothallus. When fern spores are sown on moist earth or on pieces of tile or flower pot standing in water, and kept in moist air, they germinate and form small thin green thalli, very like diminutive branches of *Pellia*. They attach themselves by rhizoids underneath. The apical growing point is in the angle between two projecting wings. The thalli—or *prothalli*, as they are called—are at first only one cell in thickness; but well grown prothalli develop, behind the growing point, a thicker region, like a very short midrib, called the cushion.

On the *under* side of the prothallus, sexual organs, antheridia and archegonia, are borne.

The antheridia may be found anywhere on the thinner parts, especially towards the base among the rhizoids. They are small outgrowths, with a dome shaped wall composed of few cells enclosing a central group of small cells that form antherozoids. The antherozoids are set free by the collapse of the middle cell of the dome, when water bathes the surface and the mucilaginous contents swell. They are spirally coiled and have many cilia.

Antheridia are often formed while the prothallus is still small, especially if it is poorly nourished. Archegonia are only found on large prothalli, on the cushion. They have very short necks bent over away from the growing point. Only the neck projects from the surface of the prothallus. The neck canal leads down below the surface to a cavity, the venter, in which lies the egg-cell.

Apart from these details of size and form, the general construction of the archegonium, as well as the mode of fertilisation, are remarkably similar in the fern prothallus and in *Pellia* or *Funaria*.

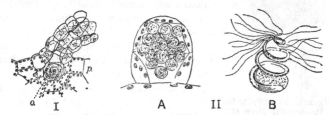

Fig. 224. Sexual organs of a Fern (*Polypodium vulgare*). I, ripe archegonium in section; *p*, prothallus; *o*, egg-cell. II, A, antheridium; B, single antherozoid, much enlarged.

Fern antherozoids afford us an easy example of the attraction which particular chemical substances exert on antherozoids. Fine capillary tubes are made (by drawing out ordinary glass tubing in a flame and cutting into short lengths), closed at one end and filled with about 1 per cent. solution of a sodium salt of malic acid by boiling and subsequently cooling them in the solution. When ripe antheridia and free antherozoids have been found, a tube is taken with forceps and after carefully rinsing it in water the open end is introduced below the coverslip and watched under the microscope. Antherozoids can be seen collecting near the mouth and entering the tube.

The fertilised egg develops within the cushion of the prothallus which grows with it; and for a time it takes all its **Fern embryo.** nourishment from its parent. The embryo, however, differs in important respects from the embryos of Mosses and Liverworts. The embryo sporogonium of *Funaria*, for instance, forms an absorptive foot at one end and a growing point at the other. The fern embryo, on the other hand, forms three growing points and grows out in three different directions. The side next the prothallus swells and becomes the haustorium, or foot. Near the foot grows out a *root* which pushes its way out of the prothallus down into the soil. Meanwhile opposite to the foot another outgrowth appears, turns up into the air and expands as the first leaf. It is very small and has a very simple fan-like form: all its veins branch by forking. Between the leaf and the foot, opposite to the root, the growing point of the stem appears and forms eventually the rhizome bearing other leaves. Seedling ferns can often be found in greenhouses on the soil of fern-pots, still attached by the foot to the under side of the old prothallus, but already rooted in the soil.

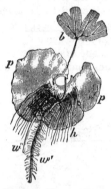

Fig. 225. Young plant of Maidenhair Fern still attached to prothallus (*p*); *h*, rhizoids; *w'*, *w''*, first and second roots; *b*, first leaf. After Sachs.

The two generations.

Thus arises the ordinary fern plant as the offspring of the prothallus. The alternating generations are even more strikingly different than those of *Pellia*. The sexual plant is a small thallus of simple construction adapted like the thallus of *Pellia* to a moist habitat. The spore-bearing plant is a large long-lived land plant, equipped with leaf, stem and root. Because of its root-system, its well-developed vascular tissue and its internal assimilating surface protected by an epidermis it is capable of living with its large fronds expanded in dry air.

Like a sporogonium it is for a time parasitic upon the sexual plant; but by means of its first root it is soon able to make itself completely independent, whereas a moss sporogonium which forms no root even though it has a very efficient assimilating system with epidermis and stomata can only become partially self-supporting. The spore-bearing generation of the Fern is indeed the predominant generation, and the sexual generation is inconspicuous and ephemeral; while in Mosses and Liverworts on the other hand it is the sexual plants which have a long and vigorous life in the course of which they may bear sporogonia time after time.

Selaginella.

The best known species of *Selaginella*, *S. Kraussiana*, which is grown in greenhouses, is rather like a large branching moss creeping over the ground on stilts. The stem forks at intervals, more or less unequally—the branching is not axillary. Below each fork grows down to the ground a white organ, like a root but without a root-cap, called a *rhizophore*. When it reaches the soil, true roots spring from it. The stem is covered with numerous small leaves arranged in four ranks, and of two sizes. There are two ranks of very small leaves above, lying over the stem, and two ranks of larger leaves below, spreading horizontally on either side. At the base of each leaf facing the stem is a white fan-shaped scale. This is best seen on young leaves where it is moist and mucilaginous, doubtless helping to keep the apical bud moist. It dries up very quickly after the leaf expands.

Selaginella, like a fern, is a vascular plant. In a cross section of the stem the conducting tissues are arranged in groups rather like those in the Bracken rhizome, but they are few in number. In *S. Kraussiana* there are two, each with a central core of xylem composed of tracheids (mainly scalariform) surrounded by phloem. In our only British species, *S. spinosa*, which is a small erect plant with leaves all alike, there is only one group. Each group is surrounded by a large air-space bridged by single rows of cells. Chlorophyll is abundant in the cells adjoining these air-spaces.

The leaves are simple in structure. They are thin, though more than one cell thick, and have a single vein. A small vascular bundle in this midrib is continuous with a leaf-trace that runs through the cortex, crosses the air-space and joins the margin of one of the groups of vascular tissue in the stem. The assimilating cells have few chloroplasts but these are unusually large and flat.

Cones. Here and there short branches grow erect, bearing leaves all alike, arranged evenly round the axis and overlapping one another. These are reproductive branches called *cones*. If one of the upper leaves of a cone is bent back a large round sporangium is seen. It is attached by a short stalk in the axil of the leaf between the stem and the ligule. When ripe it splits across the top and, in most cases, liberates large numbers of small spores, much like moss or fern spores, of a light brown colour.

Some of the sporangia, however, usually found towards the
base of the cone, are larger and, instead of being
Mega- nearly spherical, are bulged on four sides by four
spores very large dark brown spores. The *Selaginella*
and plant is therefore a spore-bearing plant, but it
micro- bears *two kinds of spores*. They are called *micro-*
spores. *spores* and *megaspores*.

The large megaspores have a very thick wall,
marked by three radiating ridges on the sides where they adjoined
each other in the megasporangium. The microspores have

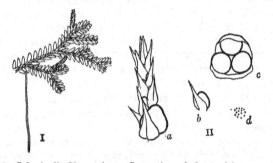

Fig. 226. *Selaginella Kraussiana.* I, portion of plant with two cones and
a rhizophore. II, *a*, a small cone, with a megasporangium and
several microsporangia; *b*, a microsporangium with its subtending
leaf; *c*, dehiscing megasporangium (the fourth megaspore is below
the three shown); *d*, microspores drawn to about the same scale as *c*.

similar marks, for they were grouped in fours in like manner,
tetrahedrally. Each group, called a tetrad of spores, was derived
from one spore-mother-cell. At an early age a megasporangium is
indistinguishable from a microsporangium, for it also contains
numerous spore-mother-cells. Only one of these, however, produces spores, growing at the expense of the others, which disorganise and yield their substance to the one survivor. Each
megaspore is richly supplied with food, including much oil, stored
for its future use.

Although so large the megaspore is at first a single cell; but
before it is shed it begins to germinate, forming a
Sexual tissue of small cells under the radiating ridges. When
genera- it falls on a suitable moist spot the thick protective
tion. outer wall bursts in this region, allowing the tissue
to protrude and grow; but it always remains small

This tissue bears archegonia; the archegonium neck hardly projects, and the neck-canal leads down below the surface. The four uppermost neck-cells can be distinguished, however, in surface view, especially when the archegonium is ripe and they separate, opening the canal to the venter.

The microspores on germinating form only a small oval body rather like a fern antheridium. It consists even when fully grown of little more than an outer layer of cells enclosing a few small antherozoid-cells. The antherozoids are slightly curled, with two long cilia.

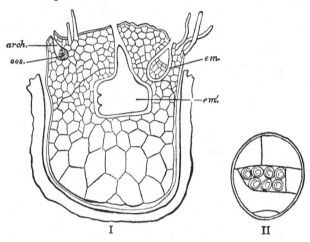

Fig. 227. *Selaginella*. I, germinating megaspore of *S. selaginoides* with tufts of hairs and archegonia borne on exposed prothallus; *em*, embryo; *arch*, archegonium; *oos*, oospore. After Pfeffer. II, germinating microspore.

The sexual plants of Selaginella are thus unisexual, the male plants being produced from the microspores and remaining extremely small; the female, produced from the megaspores, growing to a larger size, but remaining small and insignificant in comparison even with the prothallus of a fern. They are called *male and female prothalli*.

The fertilisation of the egg-cell by an antherozoid is followed
Embryo. by its development for a short time within the female prothallus. First it divides into two cells; the upper cell grows in length, forming a *suspensor*, and pushes

the lower cell down into the store of food in the megaspore where it becomes the embryo proper. Meanwhile the whole megaspore may become filled with cells.

The embryo forms opposite the suspensor a stem apex and two first leaves (*cotyledons*) on either side of it, each with its ligule. A foot is formed on one side as a swelling which by its rapid growth pushes the shoot aside. Finally between the foot and the suspensor the first root appears. The root eventually grows out through the prothallus and enters the soil; and the part of the embryo below the cotyledons grows, forming a hypocotyl, which carries them and the shoot apex upwards into the light and air. Thus the young *Selaginella* plant starts its independent life; but for some time the megaspore remains attached to the foot.

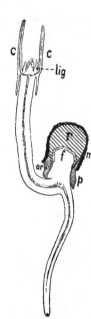

Fig. 228. Young plant of *Selaginella* still attached to megaspore: *c*, first leaves (cotyledons), with ligules (*lig*); *m*, thick coat of megaspore; *p*, protruding prothallus, with an abortive archegonium (*ar*); *f*, foot with epithelium (stippled); *r*, densely packed reserves of food.

We saw how in the Bracken Fern the spore-bearing generation is the conspicuous one, large and long-lived, while the sexual plant is very small and is dependent upon an abundance of moisture at the very surface of the soil. Yet although the prothallus has a short and precarious life it is self-supporting, very little nutriment being stored for its use in the spore, and the nourishment absorbed by the young fern embryo is provided by the prothallus itself.

Comparison with Fern.

In *Selaginella* the sexual generation is still shorter-lived. Moreover, the nourishment on which the embryo depends is supplied by the previous spore-bearing generation. In this sense, indeed, the prothallus of *Selaginella* is only partly independent, although it is not parasitic; it forms archegonia at the earliest possible moment, even before the megaspore is shed, and can hardly be said to have any vegetative life of its own, apart from the reproductive process in which its development culminates. This is even more obviously true of the male pro-

thallus, which is little more than an antheridium, of which all
but one of the cells disorganise by the time the antherozoids are
mature.

General Considerations.

It is very remarkable that the life histories of plants so different
as *Pellia, Marchantia, Funaria,* the Bracken Fern and *Selaginella,*
should show so many parallel features. Alike in all of them we
find an alternation of sexual and spore-bearing generations. The
female sexual organ is an archegonium. The young spore-
bearing plant develops, at least at first, as a parasite upon the
sexual plant. The spores are formed in tetrads from spore-
mother-cells.

Can it be that corresponding features are to be discovered
even among Seed-plants?

It is very interesting from this point of view that pollen-
grains are formed in tetrads from pollen-mother-
cells in the pollen-sacs of all Seed-plants. In size
and form, as well as in mode of formation, a pollen-
grain is indeed very like a fern spore or a microspore
of *Selaginella.*

**Spores
and
pollen-
grains.**

When a pollen-grain of *Pinus* germinates it
forms a few cells around a central cell, the nucleus of which divides
to form two male nuclei. This is not unlike the behaviour of
the microspore of *Selaginella*; only in the latter the male cells are
motile; and in *Pinus* one of the outer cells is large and grows out
to form a pollen-tube.

The pollen-grains of Angiosperms form only a tube nucleus and
two male nuclei not separated by cell-walls; but *Pinus* helps us
to bridge the gap even to this mode of germination and suggests
that here also we have what is really a microspore, with all that
is not essential to the new method of fertilisation by a pollen-tube
left out of its life history.

Can we find anything corresponding to the megaspore and
female prothallus of *Selaginella* in the ovule of Seed-plants?

In the ovule of *Pinus* the embryo-sac becomes filled with a
tissue, which we know later as the endosperm,
before fertilisation takes place. Two cells near the
micropylar end of this mass of tissue become very
large egg-cells. Above each egg-cell, at the surface
of the tissue, a tiny neck can be distinguished like
the neck of the archegonium of *Selaginella.* In sections through

**Embryo-
sac
of Pine.**

these points it is easy to see that the tissue filling the embryo-sac is not organically connected with the nucellus, but lies separately within it.

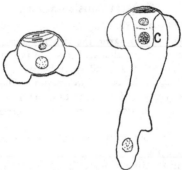

Fig. 229. Diagrammatic drawings of germinating pollen-grains of Pine; *c*, central cell from which come the two male nuclei.

This tissue, bearing archegonia, must be a female prothallus. But the megaspore is here not merely germinating before it is shed—it is not shed, forms no thick wall, but grows as a parasite within the nucellus; the prothallus is not merely dependent for its nourishment on food bequeathed to it by the parent sporophyte —it remains a total parasite throughout its life.

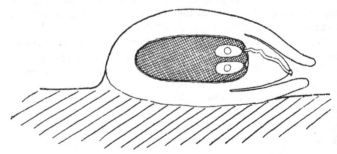

Fig. 230. Diagram of ovule of Pine on ovuliferous scale in longitudinal section, showing prothallus (shaded) filling embryo-sac, with two archegonia, and a pollen-tube reaching from a pollen-grain at the tip of the nucellus to the neck of one of the archegonia.

The fertilised egg in its turn begins life as a parasite within the prothallus. It only becomes independent at a late stage, after a period of rest as the embryo in the dry seed. By this time the well fed female prothallus, now called the endosperm, has absorbed the nucellus, and fills the seed coat, the special covering provided by its spore-bearing parent, the Pine tree.

In Flowering Plants the parallel is hard to trace. There is nothing resembling a prothallus or archegonia in **Embryo-sac of Angio-sperms.** the embryo-sac when the egg-cell is ready for fertilisation; though the endosperm formed *after* fertilisation is very like the prothallus of *Pinus* in mode of growth, serves the same nutritive function and like it grows at the expense of the nucellus. On the female side the sexual generation seems indeed to have been ousted from its proper place in the life history.

Connected with the retention of the megaspore by the spore-bearing plant is the wonderful diversity of pollination mechanisms, which bring the microspore into its near neighbourhood, as well as the formation of a pollen-tube. This organ performs a double function. It penetrates the tissues—of the nucellus in Pinus, of the style in Angiosperms—absorbing food for its growth, like a haustorium, as it proceeds; and it is the channel by which the male nuclei reach the egg-cell.

We cannot here follow further this fascinating subject. It **Evolution.** can only be properly understood in the light of many facts, about which information must be sought in more advanced books and from the plants themselves.

The remarkable correspondences revealed, however, by even so brief a survey in the life histories of plants so widely separated from one another in the Plant Kingdom must be expressions of very fundamental laws of plant development and evolution. Archegonia and a parasitic sporophyte we find from the Bryophyta right up to gymnospermous Seed-plants; the same mode of spore formation in stamen as in sporogonium.

In the formation of true roots, the differentiation of megaspores from microspores, and finally in the retention and parasitism of the megaspore, we see what have probably been three great changes, which, occurring in the life history of their ancestors, have led to the evolution of the Seed-plants we have with us today. As to when these changes came, fossil plants tell us something; but how they came we cannot tell. Having come, their consequences can hardly have been less than revolutionary.

INDEX

For particular plants reference should also be made to the classified notes in the following pages: vegetative organs, 152–79; pollination, 202–18; fruits and seeds, 228–38; trees, 407–20; climbing plants, 421–32; water-plants, 441–5; or under the proper Family in Section IV or Supplement.

For subjects see also the Table of Contents.

Absciss layer, 401
absorption of food by cotyledons, 243–4, 247, 251; of mineral salts, 88–95; of water, 69–77; by seeds, 262
Aconite, Winter, 283
Acorn, 246
actinomorphic, 198
adaptation, 105–11, 382–5
adventitious roots, 9
Agaricus, 490
Agrimony, 233
air in the soil, 76
air-spaces, 34, 113, 116, 118, 498; in water-plants, 434
Albugo, 489
albuminous seeds, 247
Alder, 413
Algae, 469–81
Almond, flower, 189; fruit, 381
alternation of generations, 496, 510
Amaryllidaceae, 360
anaerobic life, 487
anatomy, 112
animal starch, 487
androecium, 181, 186
Anemone, 285
Angiosperms, 375
annual rings, 140, 393
annuals, 181, 382
anther, 182
antheridiophore, 499
antheridium, 494, 501, 509
antherozoid, 476, 480, 495, 502, 509
apex of root and shoot, 147–50
apical growth, 4, 146, 265
apocarpous, 378
Apple, fruit of, 380
archegoniophore, 499
archegonium, 495, 500, 509, 513

Arrowhead, 442
Artichoke, Jerusalem, 168
Arum, 216, 235
ascospores, 486
Ash, 418; leaf-fall, 402; woods, 460
ash of plants, 15, 90
Aspergillus, 485
assimilating tissue, 115–22
assimilation of carbon, 30
associations, plant-, 447
atmospheric pressure, 73
autumn wood, 393
available water, 75
axil of leaf, 4, 7
axile placentation, 224

bacteria, 482; in soil, 95; nitrogen fixing, 96
Banana, 235
Barberry, 162; flower, 216
bark, 141, 394
basidiospore, 490
bast, *see* phloëm
Bean, Broad, 97, 154, 324; Kidney, 239, 324; Runner, 154, 160, 246, 324, 423
Beech, 394–6, 398, 414; woods, 460
bee-flowers, 213
Beet-root, secondary thickening of, 178
berry, 226, 235, 381
biennials, 172, 176
Birch, 232, 412
Birds, dispersal of seeds by, 220, 234, 238
Bird's Nest Orchid, 369, 461
Blackberry, fruit, 225
Bladderwrack, 478
bleeding, 72
bogs, 462
Bracken Fern, 505